# Medicine and Biomedical Sciences in Modern History

The aim of this series is to illuminate the development and impact of medicine and the biomedical sciences in the modern era. The series was founded by the late Professor John Pickstone, and its ambitions reflect his commitment to the integrated study of medicine, science and technology in their contexts. He repeatedly commented that it was a pity that the foundation discipline of the field, for which he popularized the acronym 'HSTM' (History of Science, Technology and Medicine) had been the history of science rather than the history of medicine. His point was that historians of science had too often focused just on scientific ideas and institutions, while historians of medicine always had to consider the understanding, management and meanings of diseases in their socio-economic, cultural, technological and political contexts. In the event, most of the books in the series dealt with medicine and the biomedical sciences, and the changed series title reflects this. However, as the new editors we share Professor Pickstone's enthusiasm for the integrated study of medicine, science and technology, encouraging studies on biomedical science, translational medicine, clinical practice, disease histories, medical technologies, medical specialisms and health policies.

The books in this series will present medicine and biomedical science as crucial features of modern culture, analysing their economic, social and political aspects, while not neglecting their expert content and context. Our authors investigate the uses and consequences of technical knowledge, and how it shaped, and was shaped by, particular economic, social and political structures. In re-launching the Series, we hope to build on its strengths but extend its geographical range beyond Western Europe and North America.

***Medicine and Biomedical Sciences in Modern History*** is intended to supply analysis and stimulate debate. All books are based on searching historical study of topics which are important, not least because they cut across conventional academic boundaries. They should appeal not just to historians, nor just to medical practitioners, scientists and engineers, but to all who are interested in the place of medicine and biomedical sciences in modern history.

More information about this series at
http://www.palgrave.com/gp/series/15183

Christos Lynteris
Editor

# Plague Image and Imagination from Medieval to Modern Times

palgrave
macmillan

*Editor*
Christos Lynteris
Department of Social Anthropology
University of St Andrews
St Andrews, UK

Medicine and Biomedical Sciences in Modern History
ISBN 978-3-030-72303-3 ISBN 978-3-030-72304-0 (eBook)
https://doi.org/10.1007/978-3-030-72304-0



Cover illustration: Wellcome Collection / Engraving by M. Raimondi after Raphael after Virgil.

This Palgrave Macmillan imprint is published by the registered company Springer Nature Switzerland AG.
The registered company address is: Gewerbestrasse 11, 6330 Cham, Switzerland

*This book is dedicated to the memory of John Berger (1926–2017) for having taught us how to see differently.*

# Acknowledgements

The chapters in this volume first appeared as papers in the final conference of the project *Visual Representations of the Third Plague Pandemic* at the Department of Social Anthropology of the University of St Andrews funded by a European Research Council (ERC) Starting Grant under the European Union's Seventh Framework Programme/ERC grant agreement no. 336564 (PI Christos Lynteris). I would like to thank Teresa Abaurrea, Maurits Meerwijk and Abhijit Sarkar for their help in organising the conference, the Department of Social Anthropology at St Andrews for its support of the event, as well as all speakers, discussants and participants for their generous contributions to the discussion of 'plague image and imaginary'.

# Contents

# Notes on Contributors

**David Arnold** is Professor Emeritus in History at the University of Warwick and a fellow of the British Academy. He has written extensively on science, medicine and environment in British India. His published works include *Colonizing the Body: State Medicine and Epidemic Disease in Nineteenth-Century India* (1993), *Everyday Technology: Machines and the Making of India's Modernity* (2013), *Toxic Histories: Poison and Pollution in Modern India* (2016) and *Burning the Dead: Hindu Nationhood and the Global Construction of Indian Tradition* (2021).

**Sheila Barker** is Director of the Jane Fortune Research Program at the Medici Archive Project. An art historian specialized in the study of plague art, early modern women artists and the cultural history of Renaissance Italy, her study of plague art began with her 2002 dissertation, 'Art in a Time of Danger: Urban VIII's Rome and the Plague of 1629–1634'. Her most recent contribution to this literature is her chapter 'Miraculous Images and the Plagues of Italy, c. 590–1656' in *Saints, Miracles and the Image* (2017).

**Ann G. Carmichael** emerita faculty of the History Department, Indiana University, Bloomington, Indiana (USA), holds MD and PhD degrees from Duke University. Her primary publications relate to 'Second Pandemic' plagues, and to state registration of causes of death. Recent publications focus on fifteenth- and sixteenth-century Milanese civic mortality registers and the ecology parameters of Western Europe's recurring plague outbreaks. Thus, in the former category: 'Registering Deaths and Causes of Death in Late Medieval Milan', in J. Rollo-Koster, ed., *Death in Medieval*

*Europe: Death Scripted and Death Choreographed'* (2017), pp. 209–36; in the latter category, 'Plague Persistence in Western Europe: A Hypothesis', in M. H. Green, ed., *Pandemic Disease in the Medieval World: Rethinking the Black Death* (2014), pp. 157–92.

**Samuel Cohn Jr.** is Professor of Medieval History at the University of Glasgow, fellow of the Royal Society of Edinburgh and honorary fellow of the Institute for Advanced Studies in the Humanities, Edinburgh. He has taught at the universities of Harvard, Wesleyan, Brandeis, Brown and Berkeley, and was the first Federico Chabod Visiting Professor at L'Università degli Studi, Milano (Statale) in 2017. From the late 1990s, he has published and taught on two broad themes: popular insurrection in medieval and early modern Europe, and the history of plague and other diseases from antiquity to the present. His most recent books include *The Black Death Transformed: Disease and Culture in Early Renaissance Europe* (2002); *Popular protest in late medieval Europe: Italy, France, and Flanders*, Medieval Sources Series (2004); *Lust for Liberty: The Politics of Social Revolt in Medieval Europe, 1200–1425* (2006); *Cultures of Plague: Medical Thinking at the End of the Renaissance* (2010); *Popular Protest in Late Medieval English Towns* (2013); and *Epidemics: Hate & Compassion from the Plague of Athens to AIDS* (2018). He has just submitted to press *Popular Protest and Ideals of Democracy in Late Renaissance Italy.*

**Lukas Engelmann** is a historian of medicine and epidemiology. His research covers histories of epidemics such as HIV/AIDS and the third plague pandemic (1894–1952), the history of epidemiological reasoning as well as the digital transformation of public health in the present. He is a Chancellor's Fellow and Senior Lecturer at the University of Edinburgh, based in Science, Technology and Innovation Studies as well as the Centre for Biomedicine, Self and Society.

**Christos Lynteris** is Professor of Medical Anthropology at the University of St Andrews (UK). A medical anthropologist investigating epistemological, biopolitical and aesthetic aspects of infectious disease epidemics, he is the author of *The Spirit of Selflessness in Maoist China* (2012), *Ethnographic Plague* (2016) and *Human Extinction and the Pandemic Imaginary* (2019) and co-author with Lukas Engelmann of *Sulphuric Utopias: The History of Maritime Sanitation* (2020). He edited and co-edited the volumes *Histories of Post-Mortem Contagion* ( 2018), *Plague and the City* (2019), *The Anthropology of Epidemics* (2019) and *Framing Animals as Epidemic Villains*

(2019). He was the principal investigator of the European Research Council funded project *Visual Representations of the Third Plague Pandemic* (Cambridge 2012–2018; St Andrews, 2017–2018) and is the principal investigator of the Wellcome Investigator Award-funded project *The Global War Against the Rat and the Epistemic Emergence of Zoonosis* (St Andrews 2019–2024).

**Maurits Bastiaan Meerwijk** is a historian of medicine working on Southeast Asia with a broad interest in vector-borne disease, science and technology, and the environment. Upon completing his doctoral thesis on the history of dengue fever in Asia at the University of Hong Kong, Maurits joined the University of St Andrews as a research associate on the ERC-funded project *Visual Representations of the Third Plague Pandemic*. He is an affiliate scholar with the Centre for the Humanities and Medicine at the University of Hong Kong and is developing a new project on public health messaging in Southeast Asia.

**Z. R. Dieudonné Rasolonomenjanahary** is a sociologist based in Madagascar. He has served as the President of the Red Cross of Moramanga District and the President of the Child Protection Network of Moramanga, and is the former manager (now retired) of the Moramanga branch of the Pasteur Institute of Madagascar. He is training as a consultant in community first-aid and risk and disaster management.

**Abhijit Sarkar** is a full-time British Academy fellow at the University of Oxford. Previously, he held full-time positions at the University of Exeter and University of St Andrews. He obtained his doctorate from Oxford University as a Clarendon Scholar, for a thesis on famine and famine relief in India during the Second World War, which won the Best Doctoral Thesis Award from the British International History Group, among all the theses in international history submitted at all the UK universities.

**Genese Marie Sodikoff** is Associate Professor of Anthropology at Rutgers University, Newark. Since 1994, her research has focused on Madagascar's political ecology, including the labour politics of biodiversity conservation and cultural and biotic extinction events. Her recent work concerns the impact of zoonotic diseases on Malagasy funerary practices and beliefs in the afterlife. She is the author of *Forest and Labor in Madagascar: From Colonial Concession to Global Biosphere* and the editor of *The Anthropology of Extinction: Essays on Culture and Species Death*.

**Nükhet Varlık** is Associate Professor of History at Rutgers University—Newark and the University of South Carolina. She is a historian of the Ottoman Empire interested in disease, medicine and public health. She is the author of *Plague and Empire in the Early Modern Mediterranean World: The Ottoman Experience, 1347–1600* (2015) and editor of *Plague and Contagion in the Islamic Mediterranean* (2017). Her new book project, 'Empire, Ecology, and Plague: Rethinking the Second Pandemic (ca.1340s–ca.1940s)', examines the 600-year-old Ottoman plague experience in a global ecological context. In conjunction with this research, she is involved in developing the Black Death Digital Archive and contributing to multi-disciplinary research projects that incorporate perspectives from palaeogenetics (ancient DNA research in particular), bioarchaeology, disease ecology and climate science into historical inquiry. She is the editor of the *Journal of the Ottoman and Turkish Studies Association* (*JOTSA*).

# List of Figures

# List of Tables

CHAPTER 1

# Introduction: Imaging and Imagining Plague

*Christos Lynteris*

Plague, the cause of some of the most devastating pandemics in human history, is a disease whose image and imagination have become intricately interlinked for centuries. Since medieval times, religious and secular images have taken plague as their subject. Similarly, in creating the first systematic visual record of a global pandemic, the photography of the third plague pandemic (1894–1959) has had an unparalleled impact on the way epidemics are understood and imagined in the contemporary world. Visual images of plague have left deep traces in the societies receiving, using and interpreting them. At the same time, from a global-historical perspective, this shifting visual register has contributed to the development of a pervasive 'plague concept' whereupon the disease functions as an icon of existential threat to humanity.[1] This volume *Plague Image and*

Research leading to this chapter was funded by a European Research Council (ERC) Starting Grant under the European Union's Seventh Framework Programme/ERC grant agreement no. 336564 for the project *Visual Representations of the Third Plague Pandemic* (PI: Christos Lynteris)

C. Lynteris (✉)
Department of Social Anthropology, University of St Andrews, St Andrews, UK
e-mail: cl12@st-andrews.ac.uk



C. Lynteris (ed.), *Plague Image and Imagination from Medieval to Modern Times*, Medicine and Biomedical Sciences in Modern History, https://doi.org/10.1007/978-3-030-72304-0_1

*Imagination from Medieval to Modern Times* brings together new works by world-leading historians and anthropologists to examine the interaction between the image and the imagination of plague in different historical and social contexts. The book illuminates the extent to which the 'image' of plague (both visual and not) has reflected and impacted the experience of the disease in different publics. It asks what is the contribution of the entanglement between epidemic image and imagination to the persistence and transformation of plague as an epistemic, aesthetic and social category.

*Plague Image and Imagination* is not a collection of visual analyses of plague; its aim is not to study plague's visual culture. Instead the goal of this book is to bring together historians and anthropologists to unsettle our understanding of the relation between its image and imagination. Some of the volume's chapters are more visually focused than others; some give more weight to a broader conceptualisation of the 'image' of plague, and others on plague's imagination. All, individually as well as together, take the study of plague's image and imagination beyond our analytical comfort zones. Rather than studying 'visual sources as sources to analyse visual culture', or examining the symbolic efficacy of imagining plague, the volume interrogates the entanglement between plague's image and imagination as constitutive of epistemic, political, affective and aesthetic affordances, accommodations and unsettlements that lay at the heart of societies affected by and interacting with the disease.[2]

Much more than just images of illness, plague images may be heuristically said to form a field of entanglement and tension between cosmological, anthropological, socio-political and medical modes of configuration. In terms of its cosmological grounding, imag(in)ing plague has come to rely upon, foster and unsettle relations between humans and the divine and/or the supernatural. Anthropologically speaking, plague's image and imagination fostered ideas, questions and debates regarding what it is to be human, while, from a socio-political perspective they have in turn fuelled questions about governance and social organisation, including liberty, tyranny, civic duty, stasis, sovereignty and lawlessness. Finally, medically speaking, plague's image and imagination have become significantly entangled with understandings and problematisations of the human body, health and illness, including ideas about contagion, infection, prevention, and cure.

These registers are heuristic, for they were not distinct or conceived, articulated or acted out in exclusion to one another. Indeed they may be best appreciated through analyses that focus on their entanglement in situated images and imaginations of plague. In Chap. 3, Sheila Barker examines a major visual transition in the imaging and imagination of plague in early modern Europe when a new category of painting arose around the depiction of epidemics as historical phenomena. Examining key visual works from an angle that brings together their epistemological, theological and affective efficacies of plague images, Barker argues that this new art form involved the representation of multiple victims in a particularised setting. It thus visually negotiated questions around the miasmatic nature and the contagiousness of the disease, integrating these with religious forms of aetiology. And, at the same time it problematised and acted upon epidemic-related fear by means of 'arousing a charged emotional response in the viewer'. In turn, in Chap. 5, David Arnold shows how photographs of epidemic control in colonial settings, such as those of depicting the cremation of plague victims in British India, impacted colonial configurations of Hindu otherness. Arnold, argues that these photographs unsettled the Orientalist gaze by showcasing the transformation of the much-exoticised practice into a modern one, equipped with corrugated walls, iron stakes, gas light and so on. This not only provided Western audiences with a spectacle of epidemic control efficiency, but also challenged official views that wanted to ban or at least curtail religious rituals on account of the supposed infectiousness of the human corpse. Examining a similar range of sources as Arnold, but focusing on photographs of plague hospitals, in Chap. 6, Abhijit Sarkar examines how photography captured social realities at play in clinical settings. Sarkar shows how hospital photography and its complex medico-political meanings flourished in the context of a shared epistemic anxiety that framed hospitals not solely as sites of cure, but also as sites and media of infection, thus fuelling broader colonial visions of the epidemic as an arena of modernisation, guardianship and mastery.

Yet plague's image is not simply a *visual representation*. Even if we limit ourselves to visual images of plague, these have been far more than just objects for contemplation. Whether functioning as intermediaries with the divine (votive images) or as regulators of humoral imbalances (Barker, this volume), plague images did much more than simply represent the dreaded disease.[3] As Franco Mormando has argued, 'during the time of crisis, the role of plague-related art [...] was above all, to be an instrument of healing and encouragement, a mirror and a channel of society's search for solace and cure from the heavens'.[4] Images of the Virgin and of Christian saints were used so as to amplify their role as intercessors between an afflicted city and a wrathful God, or to thank the patron-saints for the end of an epidemic.[5] Visual images moreover played a key role in the development of plague-related saint cults (on Saints Sebastian and Roch see: Barker and Carmichael, this volume).[6] At the same time, secular images of plague, such as Rouhier's etchings of anti-plague measures in 1657 Rome, provided intricate and 'detailed visual narrative[s] through time and space of the development of the epidemic within its physical context', while at the same time, 'rework[ing] the visual vocabularies of the modern city'.[7] Rather than these being simply visual representations, they constituted a broader imaginary field of plague as an urban disease.[8]

In modern times, the third plague pandemic ushered in photography in recording and configuring epidemics: never before the Hong Kong plague of 1894 had the camera been used to capture an epidemic outbreak of an infectious disease.[9] Plague photography differed considerably from what is more generally understood as medical photography. Its focus was not the human body or symptoms, but rather the causes, mode of transmission and persistence of the disease. Plague photography as well as plague maps and diagrams thus came to play an important role as investigative tools in the quest for understanding plague's complex epidemiology and for clarifying the 'epistemic uncertainties that surrounded the origins and transmission of plague' (Meerwijk this volume). In Chap. 8, Maurits Meerwijk examines plague photography in Dutch Java, and interrogates the way in which colonial doctors and public health officers used photography to problematise native houses as conduits of plague transmission. Meerwijk asks how photography was employed not only to record the demolition of 'infected houses', but also to provide scientific evidence of the implication of bamboo and of native ways of home-making and habitation in the spread and maintenance of plague. Rather than stabilising the disease, the discovery of plague's causative bacillus by Alexandre Yersin (1894) was followed by an explosion of uncertainty: was plague contagious? Was it transmitted by rats and their fleas, or from the soil? Could plague assume a stealthy, benign form? Could it become naturally attenuated in organic or inorganic matter? What led to the seasonality of outbreaks? In Chap. 7, Samuel Cohn Jr. critically approaches the historical debate over contagionism and anti-contagionism by taking seriously the discussion over the transmissibility of plague between British bureaucrats, colonial doctors and indigenous intellectuals and activists during the first years of the third pandemic in British India. Urging for a social history of plague-related contagion, Cohn effectively challenges the pervasive 'image' of resistance to anti-plague measures and of flight in the face of outbreaks as essentially premodern responses to be improved by education and cooperation.

A range of visual devices contributed to posing, legitimising, revising and answering questions regarding the transmission and maintenance of plague in the course of the third pandemic. Maps were used not only as apodictic tools, establishing cause or correlation, but also as colonial tools for establishing and defending epistemic and political authority in the context of aetiological uncertainty.[10] In Chap. 9, Lukas Engelmann examines how paper technologies contributed to epidemiological reasoning and imagination around plague 'as a global object of research'. Asking how epidemics became objects of scientific knowledge in their own right, Engelmann focuses on a plague report from Queensland so as to unravel how lists, tables, graphs and maps were mobilised and entangled in a casuistic 'portrait of plague' that fixed the social structure of affected populations and developed a generalisable 'model' of plague.[11]

At the same time, in the context of the third pandemic, photography was used to foster a completely new notion and experience of human interconnectedness and vulnerability: the pandemic. A term hitherto rarely used, the 'pandemic' rose to prominence at the end of the nineteenth century to describe the spread of plague across the globe.[12] Photography played a significant role in the creation of this experience of global interconnectedness as, through its employment in medical publications and the daily press, it generated a spectacle that portrayed every outbreak, even of the lowest intensity, as part of a global, catastrophic march of the disease.

If Renaissance paintings in Italian palazzos and modern photos of plague in illustrated weeklies seem aesthetically and epistemologically incommensurable, we need to remember that in the global plague imaginary emerging out of the third pandemic the two would be habitually and intricately juxtaposed and combined in both medical and lay publications.[13] This mix-media method achieved a visual narrative that consolidated plague as a world-historical agent, marking a culmination of what Faye Marie Getz has coined 'gothic epidemiology'.[14] As Nükhet Varlık shows in Chap. 2, this distinctly modern and teleological image of plague relied on the Orientalisation of the disease, which rendered it, in European imaginations, an extraneous threat, emerging and re-emerging in Asia. Varlık examines a crucial question when it comes to the image and imagination of plague: why is the Black Death *black*? Arguing that this is not a colour designation but an affective one, Varlık draws a genealogy of the term up until the homonymous book by Justus Hecker (1832), thereby establishing that, while the term was not new per se, Hecker's Orientalising rendition had a transformative impact on epidemiological and historical thinking, rendering the Black Death 'a separate historical and nosological category'.[15]

Plague remains a problem is many regions of the world into the twenty-first century. Whereas news media continue to use images that 'medievalise' the disease and invoke the spectre of the Black Death, in Chap. 10, Genese Marie Sodikoff and Dieudonné Rasolonomenjanahary demonstrate how plague survivors in the Malagasy highlands use photography to produce their own representations and foster their own understandings of the disease.[16] Sodikoff and Rasolonomenjanahary examine the use of ethnographic photography in negotiating the crisis surrounding a seven-year ban on traditional reburials of plague victims in familial tombs, with a particular focus on plague pits as 'haunts, sites of unresolved emotion and immanent risk to the living'. Engaging with the recent turn in anthropology in examining photographs not simply as representations but also as tools for epidemic control, the authors explore how ethnographic photography can provide 'a means by which outsiders can gain access […] to visual representations of the plague for survivors'. At the same time, the twenty-first century has provided novel ways for visualising past plague epidemics. In Chap. 4, Ann Carmichael examines pre-modern lazarettos or pesthouses as 'plaguescapes', arguing that recent, artistic and archaeological aerial photography of pesthouses in Italy, facilitated by the proliferation of camera drones and the use of digital mapping, can help us unsettle 'the static visual and analytical approaches of traditional museum displays and historical analysis'. Advantageous in terms of public outreach due to the ongoing fascination with digital technologies, these hybrid visualisations, Carmichael argues, can help us reimagine the medicalised space of the built pesthouse.

Although not a subject covered in this volume, it needs to be noted here, in light of the current COVID-19 pandemic, that the image and imagination of plague continues to inform what I have elsewhere termed our 'pandemic imaginary'.[17] Hundreds of articles during the pandemic have juxtaposed our living experience with the historical one from the three plague pandemics, while images from the latter have been liberally used in the press and the social media so as to illustrate opinions or make arguments for or against epidemic control measures. Rather than simply being of historical interest, the image and imagination of plague continues to form a powerful ground for reflecting upon the relation between humans and the world and for unsettling given ways of conceiving this relation. It is by bringing together historians and anthropologist with an expertise in different eras and regions that this volume aims to show that the image and imagination of plague are not simply representations of illness, but catalysts of social relations, forms of governance, relations with the non-human realm, and understandings of what it is to be human.

## Notes

1. Merle Eisenberg and Lee Mordechai, "The Justinianic plague and global pandemics: The making of the plague concept," *Historical American Review* 125, no. 5 (2020): 1632–1667.
2. Sander L. Gilman, *Picturing Health and Illness. Images of Identity and Difference* (Baltimore: Johns Hopkins University Press, 1995), p. 10. This volume recognises as its limitation the fact that it does not engage with the study area of the impact of plague on art, something explored in works including: Samuel K. Cohn Jr, *The Cult of Remembrance and the Black Death: Six Renaissance Cities* (Baltimore and London: Johns Hopkins University Press, 1997), chapter 7; Millard Meiss, *Painting in Florence and Siena after the Black Death* (Princeton: Princeton University Press, 1951); Judith B. Steinhoff, *Sienese Painting After the Black Death. Artistic Pluralism, Politics and the New Art Market* (Cambridge: Cambridge University Press, 2006).
3. Martha R. Baldwin, 'Toads and Plague: Amulet Therapy in Seventeenth-Century Medicine', *Bulletin of the History of Medicine* 67, no. 2 (1993): 227–247; Sheila Barker, 'Poussin, Plague, and Early Modern Medicine', *Art Bulletin* 86, no. 4 (December 2004): 659–689; Christine M. Boeckl, *Images of Plague and Pestilence: Iconography and Iconology* (University Park, PN: Pennsylvania State University Press, 2001); Andrew Hopkins, 'Combating the Plague: Devotional Paintings, Architectural Programs, and Votive Processions in Early Modern Venice'. In Gauvin Alexander Bailey, Palema M. Jones, Franco Mormando and Thomas W. Worcester (eds.) *Hope and Healing. Painting in Italy in a Time of Plague 1500–1800*, pp. 137–152 (Chicago: University of Chicago Press); Louise Marshall, 'Manipulating the Sacred: Image and Plague in Renaissance Italy', *Renaissance Quarterly* 47, no. 3 (1994): 485–53; Alexandra Woolley, 'Nicolas Poussin's Allegories of Charity in The Plague at Ashdod and The Gathering of the Manna and Their Influence on Late Seventeenth-Century French Art'. In Jutta Gisela Sperling (ed.) *Medieval and Renaissance Lactations. Images, Rhetorics, Practices* (London: Routledge, 2013).

4. Franco Mormando, 'Introduction: Response to the Plague in Early Modern Italy: What the Primary Sources, Printed and Painted, Reveal'. In Bailey et al. (eds.) *Hope and Healing*, pp. 1–44.
5. Boeckl, *Images of Plague and Pestilence*; H. Avery, 'Plague Churches, Monuments and Memorials', *Proceedings of the Royal Society of Medicine* 59, no. 2 (1966): 110–116; John Henderson, *Florence Under Siege. Surviving Plague in an Early Modern City* (New Haven: Yale University Press, 2019).
6. Boeckl, *Images of Plague and Pestilence*; Palema M. Jones, 'San Carlo Borromeo and Plague Imagery in Milan and Rome'. In Bailey et al. (eds.) *Hope and Healing*, pp. 65–96; Louise Marshall, 'A Plague Saint for Venice: Tintoretto at the Chiesa di San Rocco', *Artibus et Historiae* 33, no. 66 (2012): 153–187; Thomas Worcester, 'Saint Roch vs. Plague, Famine, and Fear'. In Bailey et al. (eds.) *Hope and Healing*, pp. 153–176.
7. Lukas Engelmann, John Henderson and Christos Lynteris, 'Introduction: The Plague and the City in History'. In Lukas Engelmann, John Henderson and Christos Lynteris (eds.) *Plague and the City*, pp. 1–16 (London: Routledge, 2018), p. 9; Rose Marie San Juan, *Rome: A City Out of Print* (Minneapolis: The University of Minnesota Press, 2001), p. 221.
8. For discussion see chapters in Engelmann et al. (eds.) *Plague and the City*.
9. On photography in the Hong Kong outbreak see: Robert Peckham, 'Plague Views. Epidemic, Photography and the Ruined City'. In Engelmann et al. *Plague and the City*, pp. 92–115.
10. Lukas Engelmann, '"A Source of Sickness". Photographic Mapping of Plague in Honolulu in 1900'. In Engelmann et al. *Plague and the City*, pp. 139–158. Nicholas H. A. Evans, 'The Disease Map and the City: Desire and Imitation in the Bombay Plague, 1896–1914'. In Engelmann et al. *Plague and the City*, pp. 116–138; Christos Lynteris, 'Vagabond Microbes, Leaky Laboratories and Epidemic Mapping: Alexandre Yersin and the 1898 Plague Epidemic in Nha Trang' *Social History of Medicine* (2019). https://doi.org/10.1093/shm/hkz053
11. For relevant literature on plague diagrams see: Lukas Engelmann, 'The Configuration of Plague: Spatial Diagrams in Early Epidemiology', *Social Analysis* 63, no. 4 (2019): 89–109; Christos Lynteris, 'Zoonotic Diagrams: Mastering and Unsettling Human-Animal Relations', *Journal of the Royal Anthropological Institute* 23, no. 3 (2017): 463–485. By contrast, caricatures provided powerful counter-narratives to plague aetiologies or to scientific proclamations of specific outbreaks as ones of true plague; Maria Antónia Pires de Almeida, 'Epidemics in the News: Health and Hygiene in the Press in Periods of Crisis', *Public Understanding of Science* 22, no. 7 (2013): 886–902; Lukas Englelmann, 'A Plague of Kinyounism: The

Caricatures of Bacteriology in 1900 San Francisco', *Social History of Medicine* (2018). https://doi.org/10.1093/shm/hky039

12. Myron J. Echenberg, *Plague Ports: The Global Urban Impact of Bubonic Plague, 1894–1901* (New York: New York University Press, 2007); Mark Harrison, 'Pandemic'. In M. Jackson (ed.) *The Routledge History of Disease*, pp. 129–146 (London: Routledge, 2017).
13. See, for example: Christos Lynteris, 'Plague Masks: The Visual Emergence of Anti-Epidemic Personal Protection Equipment', *Medical Anthropology* 37, no. 6 (2018): 442–457.
14. Faye Marie Getz, 'Black Death and the Silver Lining: Meaning, Continuity, and Revolutionary Change in Histories of Medieval Plague', *Journal of the History of Biology* 24, no. 2 (1991): 265–289.
15. J. F. C. Hecker, *Der schwarze Tod im vierzehnten Jahrhundert* (Berlin: Herbig, 1832).
16. For a critique of the use of historical images unrelated to plague in plague-related media pieces see: Monica H. Green, Kathleen Walker-Meikle, and Wolfgang P. Müller 'Diagnosis of a "Plague" Image: A Digital Cautionary Tale', *The Medieval Globe* 1, no. 1 (2014). https://scholarworks.wmich.edu/tmg/vol1/iss1/13; Lori Jones and Richard Nevell, 'Plagued by Doubt and Viral Misinformation: The Need for Evidence-Based Use of Historical Disease Images', *Lancet Infectious Diseases* 16 (2016): e235–40. https://scholarworks.wmich.edu/tmg/vol1/iss1/13
17. Christos Lynteris, *Human Extinction and the Pandemic Imaginary* (London: Routledge, 2019).

CHAPTER 2

# Why Is Black Death *Black*? European Gothic Imaginaries of 'Oriental' Plague

*Nükhet Varlık*

*But should not it be worthwhile learning about a disease that has acquired such a dreadful name?*
—Kurt Polycarp Joachim Sprengel, *Beiträge zur Geschichte der Medicin* (Halle in der Rengerschen Buchhandlung, 1794), p. 36.

*This disastrous pestilence, known everywhere under the name of the* Black Death, *as one of the great events in the world's history, has fixed he attention of writers in a high degree, and has been thought worthy to be painted in minutest details and in the most vivid colours.*
—August Hirsch, *Handbook of Geographical and Historical Pathology*, translated by Charles Creighton (London: The New Sydenham Society, 1883), p. 497.

An earlier version of this chapter was presented at the conference 'Visual Plague: Image, Imagination and Imaginary', held at the University of St Andrews, on

N. Varlık (✉)
Rutgers University, Newark, NJ, USA
e-mail: varlik@rutgers.edu



C. Lynteris (ed.), *Plague Image and Imagination from Medieval to Modern Times*, Medicine and Biomedical Sciences in Modern History, https://doi.org/10.1007/978-3-030-72304-0_2

## Introduction

The term Black Death has become virtually synonymous with plague (the infectious disease caused by the bacterium *Yersinia pestis*). As a modern historical term used by specialists, its current meaning is rather narrow: the pandemic of plague that affected Afro-Eurasia in the mid-fourteenth century (1346–1353) and killed 40–60 percent of its population. The Black Death was the initial outbreak of the Second Plague Pandemic, which continued in a series of repeated outbursts for several centuries.[1] A historical event of enormous significance, the Black Death is a topic of great importance, and one that attracts commensurate interest. It is not only taught at the college level but also included in high school curricula, and frequently referenced in popular media; as a consequence, students and the general public are quite familiar with the term—if not necessarily its specific historical context.[2] It is therefore not surprising that non-specialists use the term much more liberally and often ambiguously. To the non-specialist, the Black Death can mean anything from medieval plague to any plague outbreak in any time period, and may even be used in reference to epidemic outbreaks caused by other diseases, that is, a generic disease. Of even greater concern is the frequent use of the misnomer 'black plague' in common parlance.

What everyone agrees on, however, specialists and non-specialists alike, is that the Black Death is *black*. But why the colour designation? When did the term arise and become accepted, then ubiquitous? And what does it imply? A now-common folk etymology has it that the Black Death takes its name from the black spots and necrosis that appear on some plague victims' bodies. It is not clear where this commonly held belief originates, but it seems that it has been around for at least half a century. Writing in 1969, historian Philip Ziegler criticised this 'traditional belief' and aptly observed that none of the medieval sources used this term in reference to the epidemic.[3] Notwithstanding occasional references to plague as black in late medieval and early modern Latin and European vernacular texts, as it will be discussed below, the Black Death is primarily a historical term. It refers to a historical phenomenon: a pandemic, an epoch, and not a clinically defined nosological category. As I will argue below, the 'black' in the

12–14 July 2018. I thank the audience there for stimulating discussions and great feedback. Special thanks to Christos Lynteris, Ann Carmichael, Monica Green, Lori Jones, and Matthew Melvin-Koushki for their comments on this chapter.

Black Death does not refer to the 'blackness' (much less to any specific colour, for that matter) of plague's clinical manifestations, but rather enshrines an exclusively nineteenth-century European historical imagination of past epidemics; the latter became in turn a cultural marker fully integrated into the lineage of plague scholarship that stemmed from that imagination, which has hence remained dominant to the present. In what follows, I will first offer an overview of how 'black death' was used in reference to plague before the modern era; trace the genealogy of both the term and its specific historical definition in the historical scholarship, beginning with the nineteenth-century German medical historian Justus Friedrich Carl Hecker and his followers; discuss why this particular historical imagination and definition of the Black Death succeeded while other competing ones failed; and explore the epistemological problems this particular imagination of the Black Death—termed appropriately 'Gothic epidemiology' by medical historian Marie Faye Getz in 1991—entails.

The upshot of this study is thus the living legacy of this nineteenth-century Gothic epidemiology as it relates to contemporary plague historians and practicing scientists alike. It asks: How did Gothic epidemiology establish itself as historical-scientific orthodoxy? And how does it restrain our historical imagination of past plagues in certain directions, making some ideas more difficult to accept, even in the face of overwhelming evidence?

To clarify, my goal here is not to present an exhaustive survey (historical or contemporary) of the use of terms 'black death' or 'black plague'.[4] Neither is it to propose a replacement for the term Black Death, as it is already a well-defined, established historical and pedagogical category; as a scholar and teacher of plague, I recognise the value of using consensus terms both for scholarship and in teaching. Yet it is imperative to acknowledge that the term Black Death does come with rather weighty ideological baggage—it is the product of a specific moment in the nineteenth-century Eurocentric understanding of plague's history developed by a few European historians, and has since been universalised. To understand the extent to which other avenues of inquiry have thereby been precluded, it is critical that we map that particular historical imagination of past plagues, make explicit its inherent biases, and trace its implications for modern historical and scientific plague studies.

## Overview: 'Black' Plague Before the Modern Era

The Black Death is not a term contemporary with the event it refers to; it does not appear in medieval chronicles or medical literature. During the fourteenth century and the centuries that followed, the epidemic was instead known as the 'great death' or 'great pestilence' in English, *mortalitas magna* or *pestilencia mortalitatis* in Latin texts, and similar terms in European vernacular languages (e.g. *mortilega grande* in Italian, *grande pestilence* or *grande mortalité* in French, *großen Sterben* or *großen Pestilenz* in German, etc.). Despite occasional instances of a comparable term being used in the early modern period, it is generally accepted that the term Black Death (or rather its German original, *der schwarze Tod*) was coined in the nineteenth century by the German physician and medical historian Justus Friedrich Carl Hecker (more on whom below), and was later translated to other languages. Today, in many languages of the world, the translation or adaptation of the term is being used (e.g. *peste nera* in Italian; *peste noire* in French; *peste negra* in Spanish, *al-mawt al-aswad* in Arabic, *marg-i siyāh* in Persian, *kara ölüm* in Turkish, etc.; see Table 2.1).[5] Despite its seemingly universal acceptance today, there is no consensus about when the term became a convention in English-language scholarship, much less among non-academic audiences.[6]

Before the nineteenth century, there are scattered references to plague as black in texts composed in late medieval and early modern Latin and European vernacular languages. For example, a mid-fourteenth-century Latin poem by the physician and astrologer Symon de Covino uses *mors nigra* (black death) in reference to the epidemic.[7] Similarly, *angau du* (black death) and *du bla* (black plague) appear in a fourteenth-century Welsh poem. In an elegy composed in the 1360s, the Welsh poet Llywelyn Fychan laments the loss of his children to plague in the following words:

> A shower of peas giving rise to affliction,
> messenger of swift *black death*;
> parings from the petals of the corn-poppy,
> murderous rabble, evil omen;
> *black plague*, they don't come with any good intent,
> halfpennies, seaweed scales;
> a grim throng, humble speech,
> berries, it is painful that they should be on fair skin.[8]

**Table 2.1** Wikipedia list of 'Black Death' in difference languages

| Language | Title |
|---|---|
| es | Peste negra |
| en | Black Death |
| pt | Peste Negra |
| it | Peste nera |
| fr | Peste noire |
| de | Schwarzer Tod |
| ru | Чёрная смерть |
| pl | Czarna śmierć |
| ar | الموت الأسود |
| id | Maut Hitam |
| zh | 黑死病 |
| tr | Kara Ölüm |
| sv | Digerdöden |
| nl | Zwarte Dood |
| ko | 흑사병 |
| ro | Moartea neagră |
| fi | Musta surma |
| no | Svartedauden |
| he | המוות השחור |
| fa | مرگ سیاه |
| ca | Pesta Negra |
| hu | Fekete halál |
| cs | Černá smrt |
| el | Μαύρη πανώλη |
| vi | Cái Chết Đen |
| sk | Čierna smrť |
| hr | Doba kuge |
| da | Den sorte død |
| uk | Чорна смерть |
| ka | შავი ჭირი |
| bg | Черната смърт |
| hi | काली मौत |
| th | กาฬมรณะ |
| sl | Črna smrt |
| sr | Црна смрт |
| bn | কালো মৃত্যু |
| ms | Maut Hitam |
| lv | Melnā nāve |
| lt | Juodoji mirtis |
| simple | Black Death |
| mk | Црна смрт |
| sh | Crna smrt |
| eu | Izurri Beltza |
| bs | Crna smrt |
| az | Qara ölüm |
| et | Must surm |
| af | Swart Dood |
| gl | Peste negra |
| zh-yue | 黑死病 |
| hy | Սև մահ |
| ur | سیاہ موت |
| kk | Оба індеті |
| ml | ബ്ലാക്ക് ഡെത്ത് |
| tl | Salot na Itim |
| is | Svartidauði |
| arz | الموت الاسود |
| ta | கறுப்புச் சாவு |
| nn | Svartedauden |
| wuu | 黑死病 |
| ast | Peste negro |
| si | කළු මරණය |
| ne | कालो मृत्यु |
| la | Mors atra |
| fo | Sóttin svarta |
| mt | Pesta s-sewda |
| nds | Swart Dood |
| jam | Blak Det |
| als | Schwarzer Tod |
| mn | Хар үхэл |
| my | မဲသေရောဂါ |
| be | Чорная смерць |
| an | Peste negra |
| got | 𐍃𐍅𐌰𐍂𐍄𐍃 𐌳𐌰𐌿𐌸𐌿𐍃 |
| cy | Pla Du |
| pa | ਕਾਲੀ ਮੌਤ |
| ku | Mirina reş |
| new | हाकुगु मृत्यु |
| xmf | ჩა ჭირი |
| kbd | Емынэ фӀ ыцӀ э |
| sco | Black Daith |
| azb | قارا اؤلوم |
| fiu-vro | Must surm |
| sc | Peste niedda |
| zh-min-nan | Hek-sú-pēng |
| eo | Nigra morto |
| fur | Peste Nere |
| gcr | Lapès nwè |
| kab | Tarka taberkant |
| lad | Hazinura negra |
| li | Zjwarte Doeëd |
| mwl | Peste negra |
| scn | Pesti niura |
| so | Cudurka Madow |
| br | Bosenn Zu |
| gd | Am Bàs Dubh |
| os | Сау мæлæт |
| ga | An Phlá Mhór |
| ia | Morte Nigre |
| cdo | Háik-s□-bêng |
| diq | Mergo Siya |
| fy | Swarte Dea |
| mai | ब्ल्याक डेथ |
| nah | 1347 Tlīltic Miquiztli |
| war | Itom nga Kamatayon |
| be-tarask | Чорная сьмерць |
| kbp | Kɩ zɩ kʊkɩ kpɛdʊʊ (Peste noire) |
| wa | Pesse des djins |

It should be noted that these expressions are not clinical designations; rather they reflect the emotions of the grieving poet to express the agony of watching young individuals suffer the pain of the disease and die from it. Still the question remains why 'black' was used to express a certain emotional context.[9]

There are other instances of references to blackness, darkness, bleakness of plague (and other epidemic diseases). For example, it has been noted that the Latin phrase *atra mors* (black death) was used since the Antiquity, not only for plague, but for various epidemic diseases.[10] Similarly, the Danish historian Johannes Pontanus used *atram mortem* (black death) in his 1631 *History of Denmark*, while recounting the events of the year 1348 and the terrible ravages brought about by the plague.[11] In the early modern era, phrases that mean black death appeared in Nordic languages in reference to plague (e.g. *swarta döden* in Swedish, *den sorte Død* in Danish, *svartidauði* in Icelandic).[12] It has been suggested that this was the origin for the German term *der schwarze Tod*, whence its English translation, the Black Death.[13]

What appears to be the first use of *der schwarze Tod* in German was in fact during the late eighteenth century, about sixty years before the publication of Hecker's work. In 1773, the German historian August Ludwig von Schlözer (d. 1809) published a book titled *Islandische Literatur und Geschichte* (Icelandic Literature and History) in which he used this term.[14] Interestingly enough, after using the term once, Schlözer referred to it as *die große Pest*, which suggests that he was translating the term from Icelandic sources for a German audience who was not familiar with its meaning.[15] Two decades after Schlözer's work, the German botanist, physician, and medical author Kurt Polycarp Joachim Sprengel (d. 1833) published his *Beiträge zur Geschichte der Medicin* (Contributions to the History of Medicine), in which he included a chapter titled 'Der schwarze Tod der Jahre 1348–1350' (The Black Death of the years 1348–1350).[16] In that long chapter (pp. 36–117), Sprengel talks about the origins of the Black Death, its spread in Europe, causes of the epidemic, protection and remedies against it, and effects on society. He asks, 'but should not it be worthwhile learning about a disease that has acquired such a dreadful name?' which may indicate that he was (or thought that his readers could be) still somewhat uneasy about using the term and may have tried to normalise its use to justify its study.[17]

A few decades later, it seems the appellation 'black death' was already on its way to becoming accepted in German, and also in English.[18] Hecker published his *Der schwarze Tod im vierzehnten Jahrhundert* (The Black Death in the Fourteenth Century) in 1832. Its translation to English by B. G. Babington appeared the following year, and hence carried the term to English-language medical literature. Yet the first well-known appearance of 'black death' in the English language seems to date rather from 1823—a decade before the translation of Hecker's work. The British historian Mrs. Markham (pseudonym of Elizabeth Penrose, d. 1837) used this term in her *A History of England*: 'a most terrible pestilence—so terrible as to be called the black death—which raged throughout Europe, and proved a greater scourge to the people than even the calamities of war.'[19] Here again, we can feel the author's uncertainty about using the term, given its negative implications, especially because her history was written for a younger audience. A popular work that went through multiple editions, it eventually became one of the most popular school histories of the mid-nineteenth century, and hence likely contributed to the spread of the term in English.

Unlike Sprengel and Markham, however, Hecker was entirely unapologetic about using the term: he did not shy away from pushing it to its fullest emotional extent and painting his narrative in morbid colours. As I will discuss in detail below, his particular treatment of the subject set the emotional tone of almost all later historical scholarship on the Black Death. Hence there are many reasons to situate Hecker's work as an important turning point in Black Death scholarship. But before delving into a more detailed examination of Hecker's work and the intellectual context that informed his writing, it may be helpful to glance at the tremendous success of the Black Death as a subject of intellectual pursuit, along the lines identified by Hecker. The multiple editions and translations of Hecker's works indeed transformed the Black Death and historical epidemiology into a captivating field of study in historical scholarship for the rest of the nineteenth century. But it was with the breakout of the Plague of Hong Kong in 1894 (the beginning of the Third Plague Pandemic) that both scholarly and public interest in the Black Death skyrocketed worldwide. The general public came to imagine the Black Death, as a pandemic now resurgent, along the lines set by Hecker.

An example of how widely Black Death was accepted among readers can be seen in book titles from the period. For example, when the English Benedictine Cardinal and historian Francis Gasquet published his in 1893 book on the plague of 1348 and 1349, the book appeared with the title *The Great Pestilence*. At that time 'Black Death' was known as a term, but the author clearly decided not to choose that as the title. Fifteen years later, however, the second edition of the book appeared as *The Black Death of 1348 and 1349*. His preface to the second edition (1908) explains how research conducted during the Third Plague Pandemic contributed towards a better understanding of the medieval Black Death. Gasquet commented on the term as follows: 'The great epidemic of the fourteenth century, now commonly known as "The Black Death"—a turn of phrase which may suggest that the term was already commonplace among non-academic audiences by the early twentieth century. He wrote: '*Black Death*, by which the great pestilence is now generally known, not only in England, but elsewhere.'[20] In the meantime, further proof for the term being in circulation at the close of the century in English comes from James Murray's 1897 edition of *A New English Dictionary On Historical Principles*, predecessor of the Oxford English Dictionary, which includes an entry on the Black Death (under 'death').[21]

## Making the Black Death Black

As mentioned, the German physician and medical historian Hecker published his pioneering *Der schwarze Tod* in 1832.[22] A physician at the Friedrich Wilhelm University in Berlin, Hecker was interested in all aspects of plague, its history, along with its origins, causes, spread, and treatment, and wanted to establish disease as a force in human history. The book was a spectacular success. It was immediately translated into English, and went through multiple editions; translations into Italian, Dutch, and French soon followed, leading to Hecker being recognised in Europe as the foremost authority on historical epidemiology.[23]

At this point, it might be helpful to think about the intellectual and historical context of the early nineteenth century that informed Hecker's work. The broader intellectual context that informed Hecker's work is to be sought in nineteenth-century European Romanticism. This was a time when academic history started to develop, especially pioneered by German universities, and many of the examples produced during this period adopted a romantic approach to history. In particular, these works share in common a tendency to glorify the past, especially the distant past of the medieval period rather than more recent past, to employ emotions and a fair degree of individualism, or even heroism in historical narratives. This was also a time when rigorous analytical methodologies were adopted and used by historians to explore new subjects. Hecker was not a historian by training. He was a physician in the German university system, and was presumably informed by the new intellectual trends in that context, including what historians were producing.

Beyond this intellectual context, the most immediate historical factor that informed Hecker's work was the global cholera pandemic that first hit Europe in the 1830s. In the decades that followed, outbreaks of cholera have repeatedly affected Europe, much like the rest of the world, and caused a great number of deaths. It was not long before cholera's spread was understood to be associated with water, which made sanitation the main focus of discussions about epidemic diseases. This was a time when epidemiological and sanitary anxieties intensified everywhere in Europe, with ramifications in medicine and policymaking pertaining to public health, border control, and the like. These concerns were also prominent themes in the historical scholarship. Hecker's book is generally regarded as the beginning of modern historical scholarship on plague. A work of history that recognised epidemic disease as a legitimate subject, it was indeed pioneering. In many ways, his work, and the works of those who followed him, are responsible for paving the way for developing plague historiography that would prevail to the present; modern historical plague scholarship still carries that legacy. Moreover, Hecker is not only credited with coining, if not also universalising, the name of the Black Death, but also with setting the emotional tone of historical epidemiology, which is to say, 'Gothic epidemiology', as proposed by Getz, characterised by its emphasis on death, disease, and destruction.[24]

The effects of Hecker's work in European historiography were substantial, but did not singlehandedly forge the historical narrative of the Black Death. Other medical authors of the nineteenth and early twentieth centuries developed the basic tenets of historical epidemiology and helped lay the foundations of the burgeoning field of modern scientific epidemiology. One of most important representatives of the German epidemiological tradition was August Hirsch (d. 1894) who was a professor of medicine at the University of Berlin. Hirsch was part of a medical commission sent by the German government to study the plague epidemic in Vetlianka in 1878–1879, as well as member of the German Cholera Commission sent to examine the conditions in Posen and West Prussia. These experiences allowed him the opportunity to integrate his own observations to his works. His two-volume *Handbuch der historisch-geographischen Pathologie* (Handbook of Historical and Geographical Pathology) discusses a number of infectious diseases including plague, cholera, smallpox, measles, influenza, and yellow fever, with respect to their historical and epidemiological aspects. It was a very successful work that had multiple editions in German and translation into English by Charles Creighton—a British physician and medical author who later published his own two-volume *A History of Epidemics in Britain* in 1891–1894.[25] Needless to say, he followed the main principles laid out by Hecker in using the Black Death as a historical and nosological category.

But what was the historical plague narrative that emerged from the writings of Hecker and his followers? As it will be discussed below, this body of scholarship (1) singled out the Black Death, (2) was Eurocentric and (3) Orientalist, and these were interconnected. To start with the first feature, Hecker and his followers emphasised the Black Death at the expense of other outbreaks of plague. They singled out the fourteenth-century Black Death as unique among all plagues, past, present and future. Hecker writes (here in Babington's 1833 translation):

> The whole period during which the Black Plague raged with destructive violence in Europe, was, with the exception of Russia, from the year 1347 to 1350. The plagues, which in the sequel often returned until the year 1383, we do not consider as belonging to 'the Great Mortality'. They were rather common pestilences, without inflammation of the lungs, such as in former times, and in the following centuries, were excited by the matter of contagion everywhere existing, and which, on every favorable occasion, gained ground anew, as is usually the case with this frightful disease.[26]

In doing so, Hecker positioned the Black Death as a distinct historical event and a separate category of disease, but this characterisation creates the illusion that it was not part of a series of outbreaks that form a pandemic. This lack of continuity renders the Black Death almost a historical anomaly—a timeless event—that is characterised by mysterious, bizarre, and cataclysmic features.

Notwithstanding, Hecker's impulse to single out the Black Death resonated with historical and scientific sensibilities of nineteenth-century Europe. Imbued with a knowledge of modern epidemiology and microbiology, the reconstruction of past plagues now entailed an effort to identify them as distinct historical and biological entities. This moment is perhaps best reflected in a shift in naming practices, ruled by new historical sensibilities for epoch making. For the first time, a past epidemic thus acquired its own proper name, making it a historically identifiable and distinct phenomenon. Until that time, epidemics were merely named after locations they affected, such as the Great Plague of London (1665), the Great Plague of Marseille (1720–1721), or the Great Moscow Plague (1770–1772)—with 'great' added to distinguish them from ordinary outbreaks. In sharp contrast to other past plagues, the Black Death—as a historical term—was now freed from the confines of a specific city or location. At the same time, the eighteenth- and nineteenth-century European scientific practices of identifying, classifying, and naming plants, animals, and diseases (in pre-laboratory context) informed this venture. Much in line with the work of botanists and zoologists towards developing taxonomies of plants and animals, historical epidemiologists sought to develop nosology. In the pre-laboratory context, colour played an important part in the identification, classification, and naming of distinct organisms, including diseases. Black, along with other colours, such as white, red, and brown, were typically used to denote a certain species in association with that colour. Hence a plant or an animal—and for that matter a disease—could be named black, if that was useful for distinguishing it from a lighter variety. Numerous species thus came to be named in this manner, such as 'black pine', 'black oak', 'black bear', and so on. In fact, the European practice of naming diseases by reference to colours was already commonplace before the modern era; examples include yellow fever, jaundice, scarlet fever, green sickness, and others.[27] In this context, it may be helpful to remember that Sprengel, who offered the first extensive historical treatment of the Black Death, was also a botanist actively involved in improving the Linnaean system of classification. Nevertheless, the Black Death was soon to lose its nosological character: once Alexandre Yersin (d. 1943) isolated the plague bacillus (*Y. pestis*) in 1894, the Black Death could only serve as a convenient historical designation, and no longer a scientific one.

Hecker and his followers' efforts to identify the Black Death as both a historically and a nosologically distinct category, different from all earlier and later plagues, was not the only opinion in medical historical literature. There are other nineteenth-century medical historical works that developed different historical approaches to past plagues, and those that flirted with the idea of establishing connections between the Black Death and other nineteenth-century occurrences of plague. For example, in a much lesser known yet colossal two-volume work, *Neuf années à Constantinople*, written in 1836, the French physician A. Brayer discusses the Black Death without singling it out, discussing it as one of the major epidemics of the medieval period.[28] A medical doctor trained in the Medical Faculty of Paris and member of several learned societies, Brayer wrote his monumental work (nearly a thousand pages) based on his observations in Istanbul where he was a physician between the years 1815 and 1824. He wrote close to the time as Hecker, and published his work in Paris in 1836, and yet his writing has none of the characteristics of Hecker's: no usage of the term Black Death, no singling out of the Black Death as a historical and nosological entity, no Gothic epidemiology flavour, and no epidemiological orientalism (as will be discussed below). Brayer's work, as it were, stands as the anti-thesis of Hecker's. Why?

Brayer positioned himself in the opposite camp that maintained that plague was not contagious. He notes that Europe was protected from plague not because of its quarantine policies or heavy regulations, but on account of its regulated agriculture, greater progress in public health and in the medical sciences.[29] 'The distance of places considered as its reservoir, the difference of climates, that of religions, manners, and traditions, leads us to believe that the Oriental plague was a disease quite different from those known hitherto and of an extremely dangerous character.'[30] While he too presupposes a form of European exceptionalism, he nevertheless points out the hypocrisy of European epidemiology on this issue: 'As long as plague does not appear, we attribute its absence to sanitary laws, but when it does manifest, we attribute it to negligence and violations of quarantine regulations.'[31]

This takes us to the question of why Brayer's book never had the same impact as Hecker's did. To be sure, it is a comprehensive medical work that includes discussion of all aspect of plague, including its history, symptoms, treatment, as well as Brayer's own anthropological observations about the beliefs of people in Istanbul about plague, their responses to it, precautions, and so on. Despite its volume and rich content, this work remained almost invisible, which makes it necessary to look at how it was received by his own contemporaries. Perhaps, his book was not deemed worthy of the attention of nineteenth-century western European medical circles, because it took as its subject the Ottoman experience of plague, and thus could be ignored. Moreover, it is not unlikely that Brayer's critique of European practices and his anti-contagionist position impaired the reception of his work. He openly criticised the works of the contagionists he disagreed with and clearly expected his position, very much against the grain, to be challenged by his readership in Europe. Nor was he mistaken: European readers did not receive his ideas well. According to a review of his book that appeared in the *Edinburgh Review* in 1837, his account did not satisfy the curiosity of the western European reader because '[w]ith the exception of a few good stories, there is little of that knowledge of the interior of Oriental life which we are led to anticipate from the opportunities of the writer … His general view of the Turkish character, both in its virtues and its defects, is picturesque, but evidently exaggerated.'[32] The British author of the review thought that Brayer idealised the Ottomans, yet believed that was typical for French travellers. He was entirely dismissive of Brayer's long stay in Istanbul as well as his interactions with the locals as a physician working there, as a cause for suspicion rather than trust. He writes: 'Whether a French man has resided nine days or nine years among a foreign people, the result of his observations is sure to assume the shape of a system.'[33] Brayer was quickly forgotten as a consequence. He is not mentioned in plague scholarship at all, and hence could not be properly positioned as an anti-Hecker voice. That critique would have had to wait until the Annales movement, and late twentieth-century historiography.

## The Black Death and Gothic Epidemiology

In 1991, as noted, historian Faye Marie Getz published a remarkably insightful article titled 'Black Death and the Silver Lining: Meaning, Continuity, and Revolutionary Change in Histories of Medieval Plague', in which she identified the evolution of two divergent traditions of historical epidemiology. One was defined by Hecker's Gothic tenor, stressing death, disease, and destruction; the other, revisionist in character, emphasised resilience in human societies and preferred to study *longue durée* continuities.[34] Simply put, these two different approaches disagree as to where they assign greater agency. The former tends to attribute more agency to disease, imagining it as an alien force on human society, even one that has the power to determine their fate, always for the worse. The latter attributes more agency to human actors, and situates disease as one of the challenges human societies have had to deal with throughout history, more or less successfully. Needless to say, following one tradition at the expense of the other could only result in a rather thin, one-sided understanding of the dialectical relationship between disease and societies: diseases change societies—and are changed by them.

According to Getz, Hecker and his followers constitute the former, reductionist camp: the advocates of what she evocatively calls Gothic epidemiology. But how should we define this rather peculiar and decidedly

Eurocentric form of epidemiology? What makes the work of Hecker and his followers' approach peculiarly *Gothic*? And how does the naming of the Black Death epitomise this Gothic epidemiology? In Getz's words, the 'gothic interpretation of the Black Death comprises themes of teleology, individual heroism, abrupt change, death, and, most notably, a dialectic between opposing forces'.[35] Thus Hecker's emphasis on morbid and bizarre aspects of the Black Death, exclusive to western Europe, such as the Flagellants movement and Jewish pogroms, are also presented as products of the same Gothic imagination.[36] She further qualifies Gothic epidemiology as being fixated on

> distant and exotic places and times, especially in the Middle Ages and the Orient; the celebration of the power of nature and the ineffability of nature's essence; the unity of disparate elements—of good and evil, the hideous and the beautiful, the dead and the living; the seduction of the primitive and wild in nature, of the bizarre; the insignificance of human beings against nature; the existence of geniuses; the importance of individual experience; and finally the emphasis on suffering, death, and redemption.[37]

Taking Getz's argument one step further, one might even argue that the Gothic epidemiology itself was an artefact of the ideas and practices generated by the Black Death and its recurrent outbreaks. Arguably, the mass mortality witnessed during the Black Death was an important factor in shaping attitudes towards death for centuries after. In the case of Europe in particular, the Black Death and the myriad outbreaks of plague that followed contributed to the making of a death culture, of which art offered one form of expression. For example, in images of the *danse macabre* (dance of death) dead individuals from all walks of life are depicted as coming together to dance. Similarly, allegorical references to death, the closeness of the hour of death, and elaborate descriptions of death can be found equally commonly in literary and historical works, poetry, hagiographies, as well as in apocalyptic literature of the early modern era. They functioned as reminders that death was imminent, and one should prepare for it. A certain fascination with death, as illustrated in Hecker's Gothic epidemiology, seems to be reminiscent of the robust European tradition of writing about death since the time of the Black Death.

This particular historical imagination, with heavy Gothic undertones, takes us back to the question of the Black Death's naming and whether the black in it refers to a colour on the visible spectrum, rather than to a cultural construct. As mentioned before, the argument that tied the blackness of the Black Death to the black marks or necrosis on the body seems weak, though admittedly Hecker invokes the idea that the black spots that appeared on the body of the plague victims may have been why it was called the Black Death in Germany, while he also adds that it was named as the Great Mortality in Italy. Granted that plague would manifest in the same manner in Germany as in Italy, this brief allusion Hecker makes here cannot be accepted as an explanation for his adoption of the term that would serve him as a universal historical term.[38] Be that as it may, this usage left behind a historical legacy, even though its usage as a diagnostic or nosological category did not survive. The long-term legacy of Gothic epidemiology, set by Hecker, with its dark, gloomy emotional tone still prevails in historical narratives.

Hence, it may be argued that the blackness of the Black Death is not so much of a colour marker as an emotional one. In European and other languages, black often stands for more than a colour; it represents the absence of colour and the absence of light, and hence stands as metaphor for darkness and non-visibility. By extension, it is used for all things dark, scary, foreign, mysterious, unknowable. Yet black eventually did become the universal colour of plague—a perfectly knowable, indeed familiar, disease. This product of a particular historical moment in nineteenth-century western European modernity was later internationalised through colonialism and attendant academic imperialism. Black came to acquire in western European modernity a symbolic meaning, to a much greater extent than other colours, as to matters of death and dying. This meaning, now embedded in modern scientific and popular knowledge (as illustrated in the very name of the Black Death), was then transmitted globally to other cultures through colonialist processes. The list of world languages that use a translation of the Black Death suffices to prove this point (see Table 2.1). Yet it was not only the name of the Black Death that was translated and adopted internationally. The imagery of dark and gloomy Gothic epidemiology is likewise accepted everywhere in the world today; it evokes shared emotions of terror and despair internationally.

## The Black Death and Epidemiological Orientalism

This Gothic epidemiology was not merely Eurocentric in tenor; it was also heavily Orientalist. In the spirit of *epidemiological Orientalism*—the totality of discursive practices whereby western Europeans viewed, experienced, imagined, reproduced, and represented the 'Oriental' health-scape as the perennially plague-stricken other—of the long nineteenth century, European epidemiologists continued to imagine plague as an 'oriental disease' and the Orient as the source of the Black Death, and by extension that of all past plagues.[39] Writing in the late eighteenth century, Sprengel included comments about the origins of the plague, which unsurprisingly points to nowhere other than the 'Orient'. In his 1795 *Handbuch der Pathologie*, the German epidemiologist writes that 'plague is so common in the Orient because the Turks dispose of all the food and let domestic animals to rot on the streets'.[40] In *Beiträge zur Geschichte der Medicin*, moreover, he notes that most outbreaks of plague come to Europe from the Orient, from places such as Constantinople, Smyrna, Cairo, and Isfahan.[41] Such assertions are not merely commonplace in nineteenth-century European historical epidemiology, but have also stood as consensus facts in modern historiography until recently.[42] Nor did Hecker and his followers refrain from making remarks about the Oriental origins of the Black Death. For example, Hecker wrote: 'Doubtless it is the Nature which has done the most to banish the Oriental plague from western Europe, where the increasing cultivation of the earth, and the advancing order in civilized society, prevented it from remaining domesticated; which it most probably had been in the more ancient times.'[43] It was this particular vision that served as an ideologically necessary *telos*, validating a narrative of progress that eventually takes us to the rise of Western Europe. In this teleology, the Black Death becomes a twist of fate, a challenge to be overcome by European society, and hence celebrated as integral to the history of European supremacy. In the words of David Herlihy, plague presented a multifarious challenge to European society, '[b]ut it also prepared the road to renewal… Europe proved to be a strong patient, and emerged from its long bout with pestilence healthier, more energetic, and more creative than before'.[44]

Hecker's usage of the term Oriental plague was unique, in that it made possible, for the first time, its broad association with the Black Death. Hecker proposed the term *morgenländische Pest* (Oriental plague) in reference not only to the outbreaks of his time, but also to past plagues, opening up its use to new, much more sweeping possibilities. His historical analysis made it possible to imagine pre-nineteenth-century plagues, in Europe or elsewhere, in taxonomically similar terms to nineteenth-century Oriental plague. Under such circumstances, the Black Death could be labelled an Oriental plague without further proof. And yet Hecker seems to have felt the need to give the reader a little extra information to distinguish it from other diseases. He thus noted that the 'great pestilence of the fourteenth century ... was an oriental plague, marked by inflammatory boils and tumors of the glands, such as break out in no other febrile disease' and that the 'buboes ... are the infallible signs of the oriental plague'.[45] In so doing, Hecker irreversibly changed the meaning of the Oriental plague for subsequent medical writers and historians. By the late nineteenth century, the Oriental plague's new temporal scope was entrenched. In an essay written for the Statistical Society of London, British surgeon and medical writer Henry Percy Potter surveyed the great plagues of the Antiquity and the Middle Ages, including the recurrent waves of the Black Death in Europe. He stated: 'Epidemics of Oriental plague were noted in 1407, 1427, and 1478, and during the fifteenth century the plague broke out seventeen times in different parts of Europe'[46] Read in this light, the Great Plague of London, Baltic Plagues, and the Plague of Marseille were all appearances of Oriental plague—and so by definition an alien incursion.

This extension of Oriental plague's temporal scope (by retrospectively attributing the term to the Black Death and its subsequent outbreaks) was nothing if not ironic: for the plague had started to recede from the 'Orient' at precisely the same historical moment, in the second half of the nineteenth century. In 1880, Potter wrote: 'The malady known to us as Oriental plague ... has been described under different appellations of pestilential fever, septic or glandular fever, the black death. ... It is of such rare occurrence nowadays that few modern physicians have had the opportunity of examining or reporting upon a case.'[47] Nevertheless, as the actual Oriental plague became a rare occurrence, its invocation by Europeans became ever more frequent and comprehensive.

With the onset of the Third Plague Pandemic (ca. 1855–1959), however, many long-held assumptions about past plagues gradually came under revision. The geographical, climatic, and ecological differences of the Third Plague Pandemic posed incommensurable challenges that required radical rethinking of received epidemiological wisdom. Discussions about the nosology of the disease (e.g. was it the same old 'true' plague or something different?),[48] about its spatial extent (e.g. would it penetrate the tropical zone?), or its temporality (e.g. is the current plague related to past plagues?) came back in vogue. But more importantly, the same shift paralleled the beginning of modern bacteriology, which culminated in the discovery of the plague pathogen itself (*Y. pestis*), as well as its hosts and vectors, which demanded a radical departure from older terminology, which thenceforth lost their nosological uses—and yet have retained their historical ones to the present.

## The Antihistorical Legacy of Gothic Plague

The particular historical imagination of the Black Death constructed by Hecker and his followers was, in hindsight, rather astonishing in its sweeping historiographical impact. To recap, it is characterised by its singling out of the Black Death both historically and nosologically; by its Gothic epidemiology; and as a definitive statement of epidemiological Orientalism. And despite the abundant historical and epidemiological evidence to the contrary, this particular lineage of Black Death scholarship continues to single out the Black Death as somehow distinct from the centuries-long plague persistence during the Second Pandemic, which has led to a gross neglect and severe circumscription of the latter. Adopting a catastrophe or even apocalypse approach, it treats the Black Death as a foreign, cataclysmic entity, an exogenous, and never endogenous, factor in European history. The Black Death thus serves as a template to explain all subsequent plagues in European history—as well as world history generally. Thus the way Black Death spread continues to be reflexively applied to later outbreaks of the Second Pandemic, even though recent scholarship has demonstrated the illegitimacy of this approach.[49] This facile positing of the (European) Black Death as universal template, without questioning how later waves in various regions were different from it, has serious implications for both current historiography and epidemiology. Because this particular vision, blatantly ahistorical, maintains that all later outbreaks of the Second Pandemic were re-introductions of the plague to Europe from the 'Orient', the possibility of plague persistence spilling over from local, European reservoirs is simply ruled out, and hence unworthy of investigation.[50]

Gothic epidemiology, though seemingly less prominent today than it was in nineteenth- and early twentieth-century historical scholarship, still defines both scholarly and popular understandings of plague. Historical scholarship, textbooks, and works for the general public on the Black Death are still represented to a large extent by iconic imagery of Gothic epidemiology (e.g. *danse macabre*, images of the Grim Reaper or iconic image of the plague doctor—ironically not a medieval phenomenon, but a seventeenth-century introduction). These have become staples of Black

Death imaginaries. Yet Gothic epidemiology imposes serious limitations on the study of plagues past and present. Among the most lasting legacies of this historical imagination is the belief that quarantine was the vanguard of European public health system against the plague. In this narrative, *lazarettos* are portrayed as iconic public health institutions, and proof of European medical superiority to the 'Orient'. Tackling this important question, historian Ann Carmichael tactfully demonstrates that this was a historiographical artefact of thinking about plague as contagion.[51]

As such, the substantial body of scholarship on plague still carries heavy Orientalist tones, not only in its use of terminology, but also in generating sweeping claims as to the origins and spread of past plagues. Given the epidemic of *epidemiological Orientalism*, it is time to rethink this deeply entrenched legacy and its enduring role in the modern historical imagination of past plagues. While nineteenth-century conceptions of this epidemic disease are still rampant, we do not operate in a world where plague is an immediate threat to public health (at least for the majority of the developed world), and we now have ample evidence about its history, origins, transmission, persistence, and extinction, as supplied by scientific and humanistic research.

## Notes

1. Currently, there is no consensus about when the Second Pandemic ended. For a critical discussion of periodisation of plague pandemics and its inherent Eurocentrism, see: Nükhet Varlık, 'New Science and Old Sources: Why the Ottoman Experience of Plague Matters', *The Medieval Globe* 1 Pandemic Disease in the Medieval World: Rethinking the Black Death (2014): 193–227. Newest research suggests that the periodisation of the Second Plague Pandemic will have to be revised radically, with the Black Death starting in mid-thirteenth century, instead of the mid-fourteenth; Monica H. Green, 'The Four Black Deaths', *American Historical Review* 125, no. 5 (Dec. 2020): 1601–1631.
2. For a thoughtful pedagogical treatment of how to teach the Black Death, see: Monica H. Green, 'On Learning How to Teach the Black Death', *HPS&ST Note* (March 2018): 7–33, https://www.hpsst.com/uploads/6/2/9/3/62931075/2018march.pdf (last accessed 11 June 2020). For a comprehensive bibliography on the state of the field of Black Death research, see: Joris Roosen and Monica H. Green, 'The Mother of All Pandemics: The State of Black Death Research in the Era of COVID-19—Bibliography', https://drive.google.com/file/

d/1x0D_dwyAwp9xi9sMCW5UvpGfEVH5J2ZA/view?usp=sharing (last accessed 12 May 2021).

3. Philip Ziegler, *The Black Death* (New York: Harper & Row, 1969), p. 17.
4. Here it may be noted that the term Black Death, along with others used in medievalist terminology, has recently come under closer scrutiny and criticism, especially in the light of the COVID-19 pandemic and the Black Lives Matters protests. See, for example: Mary Rambaran-Olm, '"Black Death" Matters: A Modern Take on a Medieval Pandemic', https://medium.com/@mrambaranolm/black-death-matters-a-modern-take-on-a-medieval-pandemic-8b1cf4062d9e (last accessed 11 June 2020).
5. Needless to say, the Arabic, Persian and Turkish terms resulted from a process of autocolonialism, the acceptance of what I refer to elsewhere as 'epidemiological orientalism'; Nükhet Varlık, '"Oriental Plague" or Epidemiological Orientalism? Revisiting the Plague Episteme of the Early Modern Mediterranean'. In Nükhet Varlık (ed.) *Plague and Contagion in the Islamic Mediterranean*, pp. 57–87 (Kalamazoo MI: Arc Humanities Press, 2017).
6. A search using Google Books Ngram Viewer reveals that the term 'Black Death' was used a few times in the early nineteenth century in reference to the mid-fourteenth-century epidemic. See, for example: John Richard Green, *History of the English People* (Chicago IL: Donohue, Henneberry & Co., 1800), vol. 2, p. 22. There are also slightly earlier uses. See, for example: James Pettit Andrews, *The History of Great Britain* (London: T. Burton and Co., 1794), vol. 1, pp. 371–373. Note that Andrews uses the Black Death in quotation marks and with additional clarification for his readers, which suggests that it was not an established term yet. The two opening quotes above frame that process quite nicely. While the former quote from Sprengel in 1794 marks a moment when the term needed to be explained because it was not yet known, the latter by Hirsch in 1883 makes it clear that it was already in the process of becoming an established term.
7. See: Emile Littré, 'Opuscule relatif à la peste de 1348, composé par un contemporain', *Bibliothèque de l'école des chartes* 2 (1841): 201–243, poem on p. 228:

   Cum rex finisset oracula judiciorum,
   *Mors nigra* surrexit, et gentes reddidit illi;
   Vitaque victa dolens confusa recessit ab aula.
   Postea venerunt Saturnus, Jupiter et Mars
   Coram rege, quibus commisit ut exequerentur
   Haec sua judicia divinis edita verbis. (Italics mine)

8. *Galar y Beirdd: Marwnadau Plant: Poets' Grief: Medieval Welsh Elegies for Children*, edited and translated by Dafydd Johnston (Cardiff: Tafol, 1993),

pp. 56–58; David Hale, 'Death and Commemoration in Late Medieval Wales', PhD diss., University of South Wales, 2018, for the Welsh original see pp. 370–373, for the English translation see pp. 373–375. I thank Sam Cohn for bringing this early poem to my attention. (Italics mine)

9. For a critical examination of cultural, racial, and rhetorical meanings embedded in black metaphors in medieval, early modern, and modern European society, see: Cord J. Whitaker, *Black Metaphors: How Modern Racism Emerged from Medieval Race-Thinking* (Philadelphia: University of Pennsylvania Press, 2019).
10. Stephen D'Irsay, 'Notes to the Origin of the Expression: "Atra Mors", *Isis* 8, no. 2 (1926): 328–332.
11. Johannes Pontanus, *Rerum danicarum historia, libris X* (Amsterdam, 1631), 476: 'Vulgo & ab effectu *atram mortem* vocitabant'. Also cited in Adrien Phillippe, *Histoire de la peste noire*, p. 137. In 1893, Gasquet noted, with a fair degree of skepticism, that the term may have had its origins in Denmark or Sweden, without suggesting a direct connection; Francis Aidan Gasquet, *The Black Death of 1348 and 1349*, 2nd edition (London: George Bell & Sons, 1908), p. 7. He suggests instead that the term Black Death was probably not used before the Great Plague of London (1665), probably to differentiate them; ibid., p. 8.
12. I. Reichborn-Kjennerud, 'Notes and Queries: Black Death', *Journal of the History of Medicine and Allied Sciences* 3, no. 2 (1948): 359–360, p. 359; Philip Ziegler, *The Black Death* (New York: Harper & Row, 1969), p. 17. The term *Svartidauði* (black death) was used in Icelandic sources in reference to the medieval plague, but ironically enough in reference to 1402, the year plague arrived at the island, and not to the Black Death of 1346–1353.
13. Reichborn-Kjennerud argued in 1948 that the name *schwarze Tod* was brought to Germany from Icelandic through the translation of an Icelandic book of travel from 1771 titled *Svarti dauði* (black death); Reichborn-Kjennerud, 'Notes and Queries', pp. 359–360. This 1948 article has remained fairly obscure; the particular lineage proposed in it does not seem to have been taken up by other modern scholarship.
14. August Ludwig von Schlözer, *Isländische Literatur und Geschichte* (Göttingen: Dieterich, 1773), vol. 1, p. 3: 'der schwarze Tod, oder sie große Pest' (the black death or the great pestilence).
15. Schlözer, *Isländische Literatur*, vol. 1, p. 4. Schlözer based his periodisation on the year 1350, as the year of the great Pest in Icelandic history and literature.
16. Sprengel, *Beiträge*, p. 36. His *Handbuch der Pathologie*, published in 1795 does talk about plague but he does not use the term *der schwarze Tod* in it; instead he uses *Pest* or *Epidemie*.

17. Sprengel, *Beiträge*, p. 36.
18. For earliest-known examples of the use of "Black Death" in English language publications, see note 8 above. Similarly, the term enters the French language around that time. *Peste noire* starts to appear in that specific meaning in the 1830s. See, for example: Adrien Phillippe, *Histoire de la peste noire: 1346–1350 d'après des documents inédits* (Paris: Direction de publicité médicale, 1853), which includes documents from Hecker's work. Philippe writes: 'Ils [peste et choléra] … sont sortis de ce sol impur du Levant, sur lequel ont germé, dans le cours des âges, les contagions les plus désastreuses' (p. viii).
19. Mrs. Markham, *A History of England from the First Invasion by the Romans to the 14th Year of the Reign of Queen Victoria* (London: John Murray, 1853), p. 152.
20. Gasquet, *The Black Death*, 7. Here Gasquet cites Karl Lechner's 1884 *Das grosse Sterben in Deutschland*, but Lechner actually does not give an exact date about when the term started being used; he simply notes that he was not able to find when it was used for the first time. He only notes that the term *schwarzer Tod* was not used in the fourteenth century in Germany or elsewhere in Europe, and that it was not in circulation for at least another century after the Black Death. See: Karl Lechner, *Das große Sterben in Deutschland in den Jahren 1348–1351 und die folgenden Pestepidemie bis zum Schlusse des 14. Jahrhunderts* (Innsbruck, 1884), pp. 8–9.
21. James Murray, *A New English Dictionary On Historical Principles*, vol. 3 (Oxford: Clarendon Press, 1897), p. 73. Here Murray cites earlier uses of the term both in English-language publications and in other languages to confirm that the epithet 'black' is 'of uncertain origin' and certainly not contemporary to the pandemic itself.
22. J. F. C. Hecker, *Der schwarze Tod im vierzehnten Jahrhundert* (Berlin: Herbig, 1832).
23. Faye Marie Getz, 'Black Death and the Silver Lining: Meaning, Continuity, and Revolutionary Change in Histories of Medieval Plague', *Journal of the History of Biology* 24, no. 2 (1991): 265–289, p. 275; J. F. C. Hecker, *The Epidemics of the Middle Ages*, trans. B. G. Babington (London: Trübner, 1859).
24. For further discussion on Hecker's 'gothic epidemiology', see: Getz, 'Black Death and the Silver Lining'; for an observation about the orientalist character of this epidemiology, see 276n50. I use the term Gothic here, only in reference to Gothic epidemiology, as proposed by Getz and discussed in greater detail below, to characterise the emotional tenor of the historical narrative, with an emphasis of death, disease, and destruction. The reference to Gothic here should not be confused with Gothic art or architecture, which have different intellectual genealogies.

25. August Hirsch, *Handbook of Geographical and Historical Pathology*, translated by Charles Creighton (London: The New Sydenham Society, 1883); Charles Creighton, *A History of Epidemics in Britain*, 2 vols. (Cambridge: Cambridge University Press, 1891–1894).
26. Hecker, *The Black Death in the Fourteenth Century*, translated by B. G. Babington (London: A. Schloss, 1833), pp. 72–73.
27. For green sickness, see: Helen King, *The Disease of Virgins: Green Sickness, Chlorosis, and the Problems of Puberty* (London: Routledge, 2009).
28. A. Brayer, *Neuf années à Constantinople* (Paris: Bellizard, Barthès, Dufour et Lowell, 1836).
29. Ibid., iii.
30. Ibid., *Neuf années*, vi.
31. Ibid., *Neuf années*, vi.
32. *The Edinburgh Review* 64 (1837): 125–155, quote on pp. 126–127.
33. Ibid., p. 127.
34. These two competing traditions still remain in effect today, with ongoing discussions of 'collapse' versus 'resilience' especially in the field of environmental history. In the context of plague, a recent controversy calls into question the demographic and economic impact of the Justinianic Plague and its overall effect on late Antique society. See, for example: Lee Mordechai, Merle Eisenberg, 'Rejecting Catastrophe: The Case of the Justinianic Plague', *Past & Present* 244, no. 1 (2019): 3–50.
35. Getz, 'Black Death and the Silver Lining', pp. 266–267.
36. It may be important to note here that both the Flagellants movement and the Jewish pogroms predate the Black Death. See, for example: R. I. Moore, *The Formation of a Persecuting Society: Power and Deviance in Western Europe, 950–1250* (Oxford: Blackwell, 1987); Samuel K. Cohn, Jr., 'The Black Death and the Burning of Jews', *Past & Present* 196 (2007): 3–36; David Nirenberg, *Communities of Violence: Persecution of Minorities in the Middle Ages* (Princeton: Princeton University Press, 2015). Hecker may or may not have known that, but his emphasis on these phenomena set the tone for the narrative, making these elements staples of future narratives of the Black Death.
37. Getz, 'Black Death and the Silver Lining', p. 279.
38. See: Hecker, *The Epidemics of the Middle Ages*, p. 4.
39. Varlık, '"Oriental Plague"'. The Plague of Marseille was a turning point for the articulation of a European scholarly consensus that situated the Ottoman Empire as the origin of all historical plague outbreaks that affected Europe. See: Nükhet Varlık, 'Rethinking the History of Plague in the Time of COVID-19', *Centaurus* 62, no. 2 (2020): 285–293.
40. Sprengel, *Handbuch der Pathologie* (1795), pp. 523–524.
41. Sprengel, *Beiträge*, pp. 37–38.

42. Varlık, '"Oriental Plague"'.
43. Hecker, *The Epidemics of the Middle Ages*, p. 60.
44. David Herlihy, *The Black Death and the Transformation of the West* (Cambridge: Harvard University Press, 1997), p. 81.
45. Hecker, *The Epidemics of the Middle Ages*, pp. 2–3.
46. Henry Percy Potter, 'The Oriental Plague in Its Social, Economical, Political, and International Relations, Special Reference Being Made to the Labours of John Howard on the Subject', *Journal of the Statistical Society of London* 43, no. 4 (1880): 606–613, p. 608. In his 'Historical Sketch', Potter relies on William A. Guy's *Public Health* published a decade earlier; William A. Guy, *Public Health: A Popular Introduction to Sanitary Science* (London: Renshaw, 1870).
47. Potter, 'The Oriental Plague', pp. 605–606.
48. See, for example: Adrien Proust, *La défense de l'Europe contre la peste et la Conférence de Venise de 1897* (Paris: Masson, 1897), p. 104: 'mais nous croyons que cette peste offre les mêmes caractères que la peste d'Orient.' About two decades before, Potter had pointed out that plague's nosological classification was controversial among the European medical establishment. He opined that plague was very similar to typhus; Potter, 'Oriental Plague', p. 611. In fact, plague's nosology was being debated for quite some time. An example from the turn of the nineteenth century comes from the writings of the Scottish medical writer James Tytler. Following the 1793 yellow fever epidemic in Philadelphia, Tytler compared plague ('Asiatic or True Plague') to yellow fever and concluded that 'true plague' is distinguished by its high mortality; James Tytler, *A Treatise on the Plague and Yellow Fever* (Salem: Printed by Joshua Cushing, for B.B. Macanulty 1799). Even though Tytler's treatise does not use the term 'Oriental plague', a very critical review of his book published the next year refers to plague as 'Oriental plague'. See: 'Review of *A Treatise on the Plague and Yellow Fever*, by James Tytler', *Medical Repository* 3, no. 4 (1800), pp. 373–379. In a similar vein, Hecker used the term Oriental plague mostly for nosological purposes in the first half of the nineteenth century; Hecker, *The Epidemics of the Middle Ages*.
49. Lori Jones, *Time, Space, and the Plague: Rereading English and French Plague Tracts, 1348–1750* (Montreal and Kingston: McGill-Queen's University Press, forthcoming).
50. See, for example: Schmid et al., 'Climate-driven Introduction of the Black Death and Successive Plague Reintroductions into Europe', *Proceedings of the National Academy of Sciences* 112, no. 10 (2015): 3020–3025; Dean et al., 'Human Ectoparasites and Spread of Plague in Europe', *Proceedings of the National Academy of Sciences* 115, no. 6 (2018): 1304–1309; Namouchi et al., 'Integrative Approach Using *Yersinia pestis* Genomes to

Revisit the Historical Landscape of Plague during the Medieval Period', *Proceedings of the National Academy of Sciences* 115, no. 50 (2018): E11790–7. It is important to note that these publications came even after a fully rationalised, quite brilliantly argued case was made for plague persistence within Western Europe by Ann Carmichael in 2014; Ann G. Carmichael, 'Plague Persistence in Western Europe: A Hypothesis', *The Medieval Globe* 1 (2014): 157–191. Likewise, several paleogenetics studies have convincingly demonstrated such persistence in the meantime. See, for example: Kirsten I. Bos et al., 'Eighteenth-Century *Yersinia pestis* Genomes Reveal the Long-Term Persistence of an Historical Plague Focus', *eLife* 5 (2016): e12994; Lisa Seifert et al., 'Genotyping *Yersinia pestis* in Historical Plague: Evidence for Long-Term Persistence of *Y. Pestis* in Europe from the 14th to the 17th Century', *PLoS ONE* 11, no. 1 (2016): e0145194; Maria A. Spyrou et al., 'Historical *Y. pestis* Genomes Reveal the European Black Death as the Source of Ancient and Modern Plague Pandemics', *Cell Host and Microbe* 19, no. 6 (2016): 874–881.

51. Ann G. Carmichael, 'Pesthouse Imaginaries', in this volume.

CHAPTER 3

# Painting the Plague, 1250–1630

*Sheila Barker*

## Introduction

Visual representations of plague offer rich insights into pre-modern conceptions of epidemic disease, including its presumed causes, consequences, and remedies, in the collective imaginary.[1] This chapter examines a few key artworks made before and during the long extent of Europe's second pandemic (roughly 1347–1772), pinpointing profound changes that occurred in those conceptions, as well as shifts in the kinds of responses that societies believed to be appropriate to an outbreak.[2] As we shall see, artists experimented with visual depictions of plague as a means of both reshaping how epidemic disease was imagined and conditioning the response to it, on a collective scale and within an individual psyche. Although there are manifold points of contrast and comparison among the images under discussion, this chapter prioritizes three pivot points that determined nearly all the changes in the representation of plague across four centuries: the ontological conception of the plague, the attitudes towards its victims, and the role of art during epidemics.

S. Barker (✉)
Medici Archive Project, Florence, Italy



C. Lynteris (ed.), *Plague Image and Imagination from Medieval to Modern Times*, Medicine and Biomedical Sciences in Modern History, https://doi.org/10.1007/978-3-030-72304-0_3

## The Augsburg Pamplona Bible: Plague as an Unknowable Scourge

Several surviving Western European images of plague date from well before the second pandemic and are found in Bibles where they illustrate the plagues sent to Egypt (Exodus 9:1–12) and to Philistia (1 Samuel 5–6). These plagues are presented in the Bible as signs given by God to awaken idolaters to God's power, to demonstrate God's control over the world, and to show Yahweh's superiority over all other gods.[3] In the depictions of the Biblical plagues made in this period, the nature of the disease from a medical perspective is extremely protean and vague so as to confound all attempts to diagnose it. What is instead described with specificity is both the sin that caused the Hebrew God to punish the idolaters and the means by which the victims expiated their sins. In effect, the emphasis is on the spiritual aetiology of the disease, and disease marks the difference between sinful populations who are displeasing to God, and righteous populations who are favoured by God.

One such early depiction of plague can be found in the Augsburg Pamplona Bible, produced in Spain under King Sancho VII in the late twelfth century and probably copied from a much older Jewish illustrated Bible (Fig. 3.1).[4] Two vertically abutting illustrations combine on one page to narrate in sparing, notational imagery the story of the Plague of the Philistines. In the upper image of this two-act drama, three diviners ponder what deity might be responsible for the mysterious epidemic ravaging their nation, unaware yet that the disease was sent by the God of the Israelites in response to their sinful possession of the Ark. In the bottom scene they have found the solution pleasing to God in the form of an expiatory offering accompanying the return of the Ark.

The upper scene concerns us most because this is the only place the epidemic is indicated. It shows three seated Philistines of different ages, all slumping in the same direction, each cradling his head in his left hand while looking forlorn and woeful. Two of the three men rest their right arms listlessly in their laps, signalling their helpless prostration but also indicating the location of their painful disease, which is described in Jerome's Bible and Symmachus as tumours afflicting their 'secret parts'. [5] The third man is the Philistine diviner who has realised why both a plague of rats (as mentioned in the Septuagint Bible) and a plague of tumours have been visited upon his people. He raises his right arm to point to a rat, for he has recognised that the two plagues have been sent by the God of

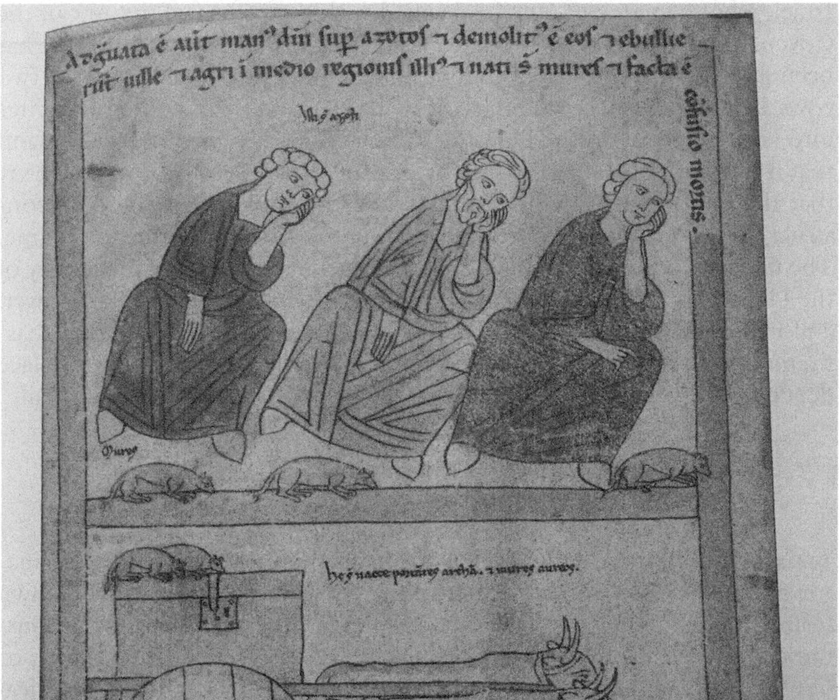


**Fig. 3.1** 'The Plague of Ashdod' and 'The Trespass Offering', Pamplona Bible, Cod.I.2.4.15, fol. 95v, Pamplona, Spain, ca. 1200, watercolour on parchment, 240 × 165 mm, UB Augsburg—Oettingen-Wallersteinsche Bibliothek

Israel, whose anger can only be pacified with the return of the Ark of the Covenant to the Israelites accompanied by a guilt offering of gold that has been formed into the shapes of tumours and rats. In the scene below, two cows by themselves miraculously lead the Ark and the golden offerings into Israel, proving to the Philistines that the plague was sent by God, and that it did not befall them merely by chance (1 Samuel 6.9). We might say that the episode provided later readers of the Bible with a rational construct for understanding large-scale calamitous disasters of their own time. The twelfth-century reader can gather from the imagery that the misery of the Philistines was God's means of prodding them to recognise His power, and to seek His forgiveness for their wrongdoing. Plague is a moral chastisement more than a disease, and the salvation of the afflicted populace depends entirely upon their ability to discover what God wants of them.

## The Morgan Library's Crusader Bible: A Divine Disease with Material Causes

In representing the Plague of the Philistines, the Pamplona Bible illuminator adhered closely to the text without providing further details drawn from either experience or imagination. That is not the case with the illustration of the Plague of the Philistines in the so-called Morgan Bible, or Crusader Bible (Fig. 3.2). This luxurious Bible with abundant gold leaf ornamentation was created at the French court of Louis IX in the middle decade of the thirteenth century, a time when public sanitation laws were being mandated in many European cities for the sake of protecting people from disease. These modern attitudes towards civic public health legislation are clearly reflected in the Crusader Bible's depiction of plague.

To explain the story of the plague unleashed upon the Philistines, the illustrator of the Crusader Bible extrapolated the narrative beyond the storyline provided by the Biblical text. There is, for instance, a portrayal of the city itself—either Ashdod, Gath, or Ecron—in the form of tightly packed gabled buildings towering some five stories above ground and indicating a very densely populated urban centre. The eerily darkened windows of these vast constructions show them to be empty, and their former inhabitants are now among the dead piled high in the streets outside where they are eaten by the rats that correspond to the plague of rats mentioned in the Septuagint.[6]

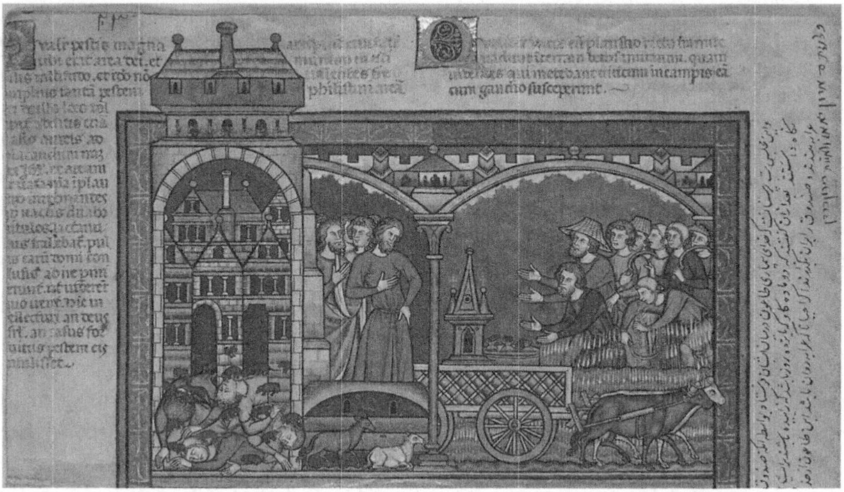

**Fig. 3.2** 'The Plague of Ashdod' 'The Trespass Offering', The Crusader Bible, MS M.638, fol. 21v, Paris, France, ca. 1244–1254, ink, gold leaf, and body colour on parchment, 390 × 300 mm, The Morgan Library and Museum

Even though the Septuagint Bible alludes to panic and confusion in the plague-stricken cities (1 Samuel 5:6), neither it nor any of the other Biblical sources mentions unburied bodies in those same cities. This element thus constitutes the artist's free interpretation of the Biblical text. Moreover, it was a highly significant addition to the story, since the miasmic fumes produced by suppurating corpses were themselves a material cause of deadly illness according to the ancient medical theories that were authoritatively expanded by Avicenna in the early eleventh century. The elite viewership of this expensive Bible would have thus recognised these cadavers as a source of a rotting stench capable of infecting surrounding air with a dangerous pestilence.[7] By this time, several European towns already had municipal ordinances regarding the removal of corpses, garbage, and excrement from public spaces.[8] The artist's inclusion of cadavers in the street shows that the virulence of the Philistines' epidemic was compounded by their lack of civic sanitation. In other words, while the Biblical text correlates the Philistines' plague to their spiritual pollution, the image connects that plague with a miasmic material cause of disease that was both a quotidian concern in any densely populated city and a cause which bore no correlation to religious purity.

In this illuminated page of the Crusader Bible, we find further evidence that educated elites in thirteenth-century France were beginning to regard the Biblical plagues as analogous to the pestilential epidemics with natural causes (and, it follows, with natural remedies) that French physicians of the time would have read about in the medical texts of Rhazes (865–925) and Arnaldus de Villanova (ca. 1240–ca. 1310). Indeed, the illustrator has taken pains to emphasise details that were common both to the medical study of epidemics and to the account of divine plague in the Vulgate. For instance, the artist seized upon the Biblical testimony that Philistines' plague struck both the young and the old alike, and accordingly showed the dead victims as both bearded and beardless.

Not traceable to Biblical sources is the inclusion of a peasant's corpse (distinguished by his rustic cap) near the bottom of a heaping pile of dead townsmen.[9] This curious interloper, among the first to fall victim in the city, raises many questions. His early death might mean that the disease had originated far away in the countryside, and that it had subsequently been introduced to Philistia by sick peasants whose festering bodies released deadly miasmas once they fell dead in the city. Or perhaps we are to understand that the peasant arrived in the city in good health, only to be sickened by the miasmas coming from the unburied dead of the urban populace. In either case, the Crusader Bible disrupts the previously neat equation between sin and plague, and its subtle visual enhancements complicate the purely theological aetiology conveyed by the Pamplona Bible's imagery.

It is important to note that, in keeping with the visual narrative of the earlier Pamplona Bible, the Crusader Bible's sequential scene shows the Ark being returned to the Israelites to underscore that the disease's cause and cure are in the hands of God. Yet despite this continuity, an important conceptual shift has occurred: in the European medieval imaginary, the understanding of a plague sent by God now incorporates the secular and material explanation of epidemic disease as a corruption of the air in a particular place. A century before the onslaught of the Black Death, medical professionals and civic leaders shared a growing concern about the insalubriousness of various aspects of urban life. Physicians, councillors, and priests would have all found themselves in agreement with the ancient Biblical prophecy that 'those in the city will be killed by plague and famine, but those who flee the city will be saved' (Ezekiel 7).

## A Public Image of Plague in the Wake of the Black Death

Both of the Bibles discussed above were commissioned for private use. Only after the Black Death of 1348 did representations of plagues begin to occur in publicly displayed artworks commissioned for use in collective religious rituals. The earliest public representations of plagues tend to emphasise the religious recourses that brought about the end of a calamity. Their purpose was to preserve for public memory a testimony of the power of the divine over the course of human events, while instructing viewers in the pious practices that might prevent future recurrences.[10] They are instruction manuals for survival.

The oldest extant example of a representation of plague for a public space, as well as one of the earliest surviving paintings of a plague set in Christian times, is Giovanni del Biondo's *St. Sebastian Triptych* (Fig. 3.3). This altarpiece was made around 1375 to ornament the Florentine Cathedral's altar of St. Sebastian, recently supplied with a relic of one of

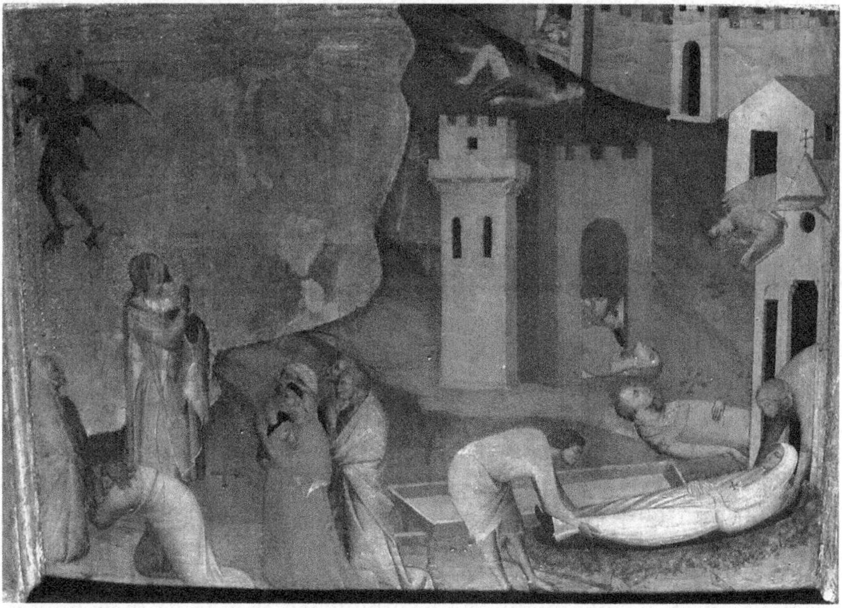

**Fig. 3.3** Giovanni del Biondo, St. Sebastian Triptych, 1375

Sebastian's arrows.[11] Commissioned in the immediate aftermath of the plague outbreak of 1374, the altarpiece has a large central panel showing the saint's first martyrdom. The four flanking scenes show Sebastian's *passio* based on the text of Jacopo de Voragine's *Golden Legend*, a hagiographic compendium composed around 1260. Of these four scenes, the one that interests us is in the lower left panel, illustrating the story of a plague that ravaged Pavia in the year 680, but which ceased following the transfer of Sebastian's arm relic from Rome to a local Pavian church. The miracle confirmed not only Sebastian's supernatural power against plague, but also the efficacy of prayers pronounced in the presence of the saint's holy relics.

Giovanni del Biondo's *St. Sebastian Triptych* makes liberal use of poetic licence to tell its story. The cause of the plague is represented as a demonic angel of divine punishment hurling deadly darts at the city from upper gilded area that represents heaven. His terrible progress is arrested by angel of mercy, almost invisible now except for the lines incised into the gold leaf. The merciful angel seemingly responds to the prayers of two intercessors below on earth, namely the Virgin Mary and St Sebastian dressed in a green robe, both of whom stand before the kneeling citizens of a plague-ridden city of Pavia.

While this spiritual drama unfolds above, the plague's material presence in the world below is known through the super-mortality it unleashes, for the painter has seemingly depicted more dead than living in this scene.[12] In the right foreground, a cadaver has been carried in a litter to the sacred ground of a country church; however, for want of a priest and mourners, the funeral has been omitted and the gravediggers hastily place the shrouded body into the ground without so much as a coffin.[13] As the viewer's eye follows a painted road winding towards a city, it encounters many human corpses bodies piled together in the doorway of a red towered castle. Here we are reminded of the contemporary observation from Mariano di ser Jacopo that plague sometimes killed everyone in a household, regardless of their humoral constitution or age.[14] In this sense, plague was thought of by physicians as a kind of airborne poison with no known antidote rather than a humoral imbalance that their medicines could correct. Further along the road is a dead man in a pink tunic lying in the middle of the road: he must have succumbed as he travelled. From this small detail vignette, the viewer understands the rapidity of the plague's onset: people might feel healthy when they leave their houses, but the disease can stop them dead in their tracks before they have time to find shelter.

The figure of the traveller is a significant sign of changing conceptions of plague due to his association with movement, and his potential role in spreading the disease from one place to another through contagion, a process that pertained to materialistic aetiologies of epidemics rather than religious ones. The defenestration of plague-infected bodies to prevent further contamination in a household is another detail that indicates plague to be a contagious disease. Particularly shocking in this regard is the detail near the top of the panel in which a woman pushes her female relative out of a window onto the street, unconcerned about the dishonour done to the woman. If the victims of plague in this painting are reviled, it is not because of an association with sin or spiritual pollution; rather it is because they are seen as capable of carrying the plague to any place and infecting any person, quite accidentally and circumstantially, and without regard to moral status.

This notion of human-to-human contagion was circulating in Italy at the time this painting was created. In 1360, the Milanese physician Cardone de Spanzotis wrote *De preservatione a pestilentia* for Gian Galeazzo Visconti, citing the transmission of plague through human-to-human contact, and the utility of fumigation and incineration to stop the spread of the contagion through circulated objects.[15] The theory of human-to-human contagion was also the basis of the plague-time regulations designed to seal up the intramural populations of both Milan and Mantua in 1374.[16] As Ann G. Carmichael has demonstrated, two of the earliest cities to adopt plague-time travel restrictions were under the rule of autocrats: Bernarbò Visconti in Milan, and Ludovico Gonzaga in Mantua.[17] Florence, by contrast, had a communal government in which power was shared by the various guilds. Probably because of the keen desire to foster trade on the part of the merchants and manufacturers who steered Florence's public policies, plague-time legislation limiting access to the city was not imposed until much later, in 1450. Giovanni del Biondo's image confirms that already in the late fourteenth century, most Florentines—except special interest groups economically invested in keeping cities open—perceived plague as a contagious disease and strongly feared being in proximity to those stricken with a disease for which physicians had no cure. Those who had the means, such as a self-sufficient farmhouse in the countryside, followed the universal medical advice to move one's family out of the city for as long as possible, as encapsulated in the dictum 'mox longe tarde, cede recede redi' (flee fast, flee far, and stay away for a long time) that Arnaldus of Villanova, Rhazes, and other medieval medical writers attributed to Galen and Hippocrates, the greatest of all medical authorities.[18]

By the opening of the fifteenth century, the onset of an epidemic was regularly accompanied by the fear of being near those with plague symptoms, the fear of travelers from other potentially infected places, and a suspicion of those crowded public spaces that were the very mainstay of city life. Against the backdrop, we can better understand why Catherine of Siena's (1347–1380) decision to stay in the city to tend to the victims of the plague of 1374 and Bernardino of Siena's similar service during the plague of 1399 were seen by their contemporaries not just as charitable actions but as veritable martyrdoms.[19] We can also better understand the growing need for paintings such as Biondo's in a city like Florence. These vivid pictorial reminders of the miraculous plague-repelling powers of the relics housed in the metropolitan cathedral were the religious and civic leaders' best hope for keeping citizens calm and preventing a fear-driven exodus of the wealthy, a disaster that could potentially devastate the urban poor more than plague itself.

## Pitying Plague Victims in the Age of Renaissance Humanism

Thanks to the recovery and translation of an expanded range of literary and historical sources, artists of the Renaissance period began to represent ancient plagues known only from pagan sources. One of the very earliest of such representations went on to attain such fame that it influenced nearly all subsequent plague imagery, both religious and secular, thus serving as an iconographic watershed for Western plague imagery. That work is Raphael's design for *The Plague of Phrygia*, made around 1512–1513.[20] Although Raphael's original drawing has been lost to us, we know how it appeared thanks to a celebrated and widely reproduced engraving made by Marcantonio Raimondi around 1515, known as the *Morbetto* (Fig. 3.4).[21]

Raphael's image of a plague stands alone as a historical subject, being neither an illustration accompanying a text (although this has been occasionally hypothesised), nor a component of a cycle of images, as in the case of the plague scene in Giovanni del Biondo's altarpiece. Moreover, it may very well be the first visual representation of a plague from pagan literature

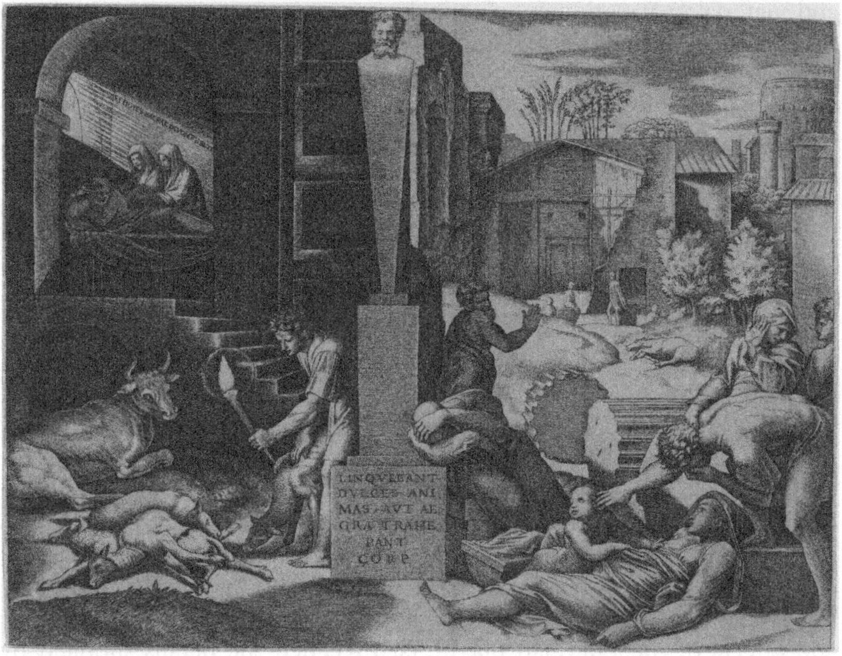


**Fig. 3.4** Marcantonio Raimondi after Raphael, Il Morbetto (The Plague), engraving of ca. 1515 after a design of ca. 1513, 19.5 × 25.2 cm

to have been made by a Christian artist.[22] The new interest in pagan sources for their descriptive accounts of epidemics is of enormous significance. It signals that for certain erudite audiences, plague no longer was solely a protean scourge with which the Judeo-Christian God chastened humans or led them to proper forms of worship. It was now being treated as a non-sectarian phenomenon and as an ontologically specific disease. In effect, epidemic disease had become a subject of discourse in both moral philosophy and natural philosophy, including medicine. In many ways, Raphael's *Morbetto* pertained to the new discourse on epidemic disease within the framework of moral philosophy, as we shall see.

The plague that is represented in the *Morbetto* is an ancient scourge that devastated the Phrygian city of Pergamon, as described in Vergil's *Aeneid*, the epic poem that served as Raphael's primary textual inspiration. Just as Giovanni del Biondo's fourteenth-century image includes both spiritual and material explanations of plague, here, too, two distinct aetiologies of the Phrygian plague are brought into confrontation. In the upper left register, a religious aetiology is indicated by the illustration of the oneiric visit Aeneas received from the household gods known as the Penates. The engraving's erudite audience would have been able to refer to Vergil's epic poem for the account of how the Penates revealed that Italy, not Crete, was the ancestral home to which Aeneas was destined to return his people, with the subtext that the current plague had been sent by the gods in order to chase the Trojans from Phrygia's shores towards the Tyrrhenian Sea. Thus, as Vergil recounted it, the epidemic was an edifying scourge mandated by the Olympian deities. Nevertheless, the poet did not portray its victims as morally repugnant individuals. Rather, Vergil characterised them as being merely hapless, unwitting, and deserving of our pity. Notably, Raphael took great interest in the lines with which Vergil poignantly described the suffering of the Phrygians, inserting the relevant passage into his image, where it is inscribed on the base of a column in the central foreground: 'Linquebant dulces animas aut aegra trahebant corpora' (Men gave up their sweet lives, or dragged enfeebled frames).[23]

Clearly inspired by medical debates on the origin, symptoms, and transmission of contemporary Italian plagues, Raphael's interest in the nature and effects of the Phrygian plague far exceeded the details supplied by Vergil. As imagined by the artist, this plague's material causes (or 'proximate causes' as opposed to divine causes) as well as its observable effects are described in the image's lower and right-hand registers. In this respect, the image provides one of Western art's first visual renderings of the corporal symptoms of plague. In the centre of the image, for instance, we see a woman slumped over the inscription stone: the way she holds her head shows she has a plague fever; the bunching of her garment indicates that she has been writhing in discomfort; and her collapsed position betrays the pain and weakness in her limbs.[24] She is surrounded by a host of evil augurs that coincided with the plague, including ominous formations in the sky, dead livestock, patches of barren earth, a desolate town, and buildings toppled by earthquakes.[25] This constellation of adverse signs makes a clear allusion to the Aristotelian idea of pestilence as being but one facet of a universal process of corruption that initiates in the heavenly spheres before permeating all the lower ones. Also representative of the Aristotelian notion of plague is the detail of a man pinching his nose, a gesture that shows he is trying to protect himself from the miasmas in the breath of plague victims and in the gases released by their putrescent flesh, both of which were regarded as deadly poisons in all Renaissance plague treatises.[26]

Even while alluding to miasmic theory, the *Morbetto* also shows how the Phrygian plague could have reached the city through a contagion spread by animals and the peasants who tend them. Key to this interpretation of Raphael's design is the positioning of a rustic shepherd tending his diseased sheep in a suburban area characterised by liminal architectural elements such as a sewer and a stone wall, and a dead horse in the midst of the city. For these details, Raphael likely turned to the description of another ancient plague, that of BC 464, as described by Livy:[27]

> It was an unhealthy season, and in both town and country there was a great deal of sickness. Cattle suffered as much as men, and the incidence of disease was increased by overcrowding, as farmers together with their livestock had been taken into the city for fear of raids. The smell of the motley collection of animals and men was distressing to city folk...the farmers and yokels, packed as they were into inadequate quarters, suffered no less from the heat and lack of sleep, while attendance upon the sick, or mere contact of any kind, continually spread the infection.[28]

We might speculate about the reason Raphael grafted Livy's description of the plague of BC 464 onto Vergil's account of the Phrygian plague. Although Livy had attributed a divine origin to the plague of BC 464, he also indicated that its toll was exacerbated by the crowding of sick farmers and their livestock in the city. Raphael's placement of a heap of sheep carcasses in the image's left foreground so as to mirror human victims clustered in the image's right foreground encourages speculations regarding a possible zoonotic origin of the human disease. In this scene of cross-species slaughter, the allusion to the nosology of the ancient Roman epidemic would have resonated strongly with Raphael's viewership, particularly since health officials all across Europe regularly clashed over proposals to protect urban populations by sealing cities off from peasants bringing their products to market.[29]

If the religious and material origin of the plague is the theme of the meditative left half of the image, the right half plunges the viewer into a visceral experience of a dystopian place infected not only by plague but also by fear. Like the dying sheep, the sickened humans in Raphael's depiction are also piled near each other in a squalid, extramural location, abandoned to their cruel fate. Much like the bodies thrown out of windows in

Giovanni del Biondo's scene of the Pavian plague in the St. Sebastian altarpiece, the sick Phrygians have been ruthlessly cast out of their city so as not to infect the other citizens with the putrid miasma of their breath and suppurating buboes.[30]

The olfactory and respiratory organs were not the only channel for pestilential contagion according to Renaissance medicine: the organs of sight were also a route of transmission, as we are reminded by the gesture of the woman on the right side of Raphael's scene who holds her hand up to shield her eyes while turning her head away from the direction of Phrygia's agonising plague victims. She enacts a widespread notion about how the plague could spread due to the visibility of plague victims, as voiced by an anonymous physician who wrote in 1349 from Montpellier: 'But the greater strength of this epidemic and, as it were, instantaneous death is when the aerial spirit going out of the eyes of the sick strikes the eyes of the well person standing near and looking at the sick, especially when they are in agony; for then the poisonous nature of that member passes from one to the other, killing the other.'[31]

The visual contagion described in the above passage was a widely circulated medical teaching already during the Black Death of 1348 that held the sight of plague victims to be a cause of sickness and sometimes even death in the viewer.[32] As explained by the Valencian physician Jacme d'Agramont in 1348, 'From imagination alone can come any malady [...] Thus it is evidently very dangerous and perilous in times of pestilence to imagine death and to have fears.'[33] This way of thinking helps explain the longstanding reluctance of artists to represent plague victims realistically and why plague victims—if they were represented at all—were always shown as very small, undistinguished figures, even in the case of artistically sophisticated images with large fields dedicated to the depiction of historical plagues, such as Benedetto Bonfigli's *Plague Banner of San Francesco al Prato* (also known as the *Madonna della Misericordia*) of 1464. Since pictures of plague victims could potentially cause fear in the same way that the sight of real plague victims did, this made pictures of plagues potentially dangerous.

Raphael's *Morbetto* is perhaps the first work of art in modern times as well as ancient times that represents dying and dead plague victims in a realistic manner with symptomatic specificity. Renaissance viewers, whether weaned on superstitious notions of the evil eye or trained in university medicine's theory of visual contagion, would have felt a jolt of terror and instinctively wanted to avert their eyes with an impulse of self-preservation upon seeing plague victims in Raphael's *Morbetto.* However, that perception of danger would have been offset to a degree by the work's magnetic hold on the viewer's eye and heart through feelings of pity and brotherly love for those touched by the scourge, a radically new and unprecedented strategy for an image of plague.

No previous image of plague had ever attempted to do this, certainly not with such deliberate design. Raphael knew that the viewer's eye would initially be attracted to the figures because of the beauty of their proportions, the delightfully lifelike way in which they seem to move, and the fascinating variety of poses that invites the viewer to deduce the story behind these impassioned displays. Once Raphael's audience recognised the tragic story behind the figures' particular poses, their engagement was transformed from superficial pleasure to deeper feelings of empathetic concern reserved for living human beings—a passage surely facilitated by the Christian practice of meditating upon images of Christ's passion.[34] If viewers initially feared for their own safety, prolonged looking resulted in concern for the safety of fictive figures: the wretched old woman who is alone in her pain, the cadaver of a young mother disgracefully abandoned on the ground, and the infant who must either drink her poisonous breastmilk or starve. It was no accident of design that Raphael's composition aligns the images of all these suffering humans with the nearby dead sheep; clearly the parallel was intended to remind Renaissance viewers of the Biblical designation of Jesus Christ as a 'Lamb of God' (John 1:29)—that is, an innocent creature sent to a sacrificial death—and thus to encourage them to see even the plague-ridden populace of a pagan land as deserving of a Christian's pity.

To assist viewers in understanding and controlling their own fearful responses to the plague, Raphael inserted into his scene a number of emotional surrogates who, like the viewer, witness the awful events and react emotionally to the disease's agonised victims. By physically enacting the competing emotional reactions of fear and pity, these witnesses serve as possible models for the viewer's own response. One such witness is the bearded companion of the sickly old woman. Overtaken by fear of possibly being infected, he runs away from her contagious body, even while glancing back at her one last time. A contrasting emotional response is exhibited by a second witness: a handsome, athletically proportioned young man is so deeply moved by pity that he risks his own health in order to prevent an infant from drinking his dead mother's lethal breastmilk. As this man approaches the baby, he pinches his nose to avoid breathing the miasma issuing from the mother's dead flesh, and thus we understand the danger he is in. In Raphael's retelling of this ancient episode from Vergil's famous epic, this anonymous rescuer of the innocent baby exhibits courage and selflessness, and thus appears more heroic than Aeneas himself.

As first noticed by art historian Elizabeth Schröter, Raphael probably based his supine, lifeless mother in twisted drapery upon the roughly similar motif in the foreground of Michelangelo's Sistine Chapel fresco of *The Deluge*, a picture that depicted the terrible process of God's wilful destruction of nearly all of humanity.[35] However, for his portrayal of the infant trying to suckle from that lifeless woman, Raphael drew upon a textual source from classical Antiquity: Pliny the Elder's ekphrastic description (*Natural History* 35.98) of a lost ancient painting by Aristides of Thebes, an artwork which depicted a mother who has been mortally wounded in the chest during a siege, and whose starving infant will die—according to contemporary beliefs—if he drinks her breast milk, since it is tainted with blood.[36] As Raphael would have read in the relevant passages by Pliny the Elder, Aristides was considered 'the first among all painters to paint the soul, and g[i]ve expression to the affections of man [...] and also the emotions'.[37] Thus, what undoubtedly captured Raphael's interest in an ancient picture about war, not plague, was the emotive impact that Aristides' celebrated image derived from its motif of the dead mother and her imperilled child.

By merging Michelangelo's trope for conveying the devastation of the Biblical flood with Aristides' trope for expressing the horror of war, and then by transposing this hybrid form to the context of plague, Raphael was consciously tapping into what Aby Warburg termed a *Pathosformel*, meaning a visual formula for arousing a charged emotional response in the viewer.[38] Raphael further enhanced the poignancy of this *Pathosformel* by having the imperilled infant reach for his deceased mother's breast while looking across the fictive space of the engraving towards the heap of dead lambs, which, as noted above, symbolise an innocence that is defenceless against cruelty.[39] No less evocative than it is poignant, this detail of the infant and its dead mother epitomises Raphael's innovative use of art to raise concern for each individual victim of plague. Whereas previous depictions of plague in Italian painting had emphasised its massive carnage by means of scatterings of tiny, uniform cadavers, Raphael chose instead to portray just a few emblematic plague victims. He endowed each sickly body with a specific age and gender, and in order to evince the peculiar nature of the victim's suffering with a precision that recalls the roughly contemporary emergence of medical case histories, a form of medical writing that followed the course of a disease in a single patient. Individualised and indeed humanised by Raphael's composition, each and every plague victim becomes a heart-rending meditation on the manifold tragedies instigated by a mysterious disease.

An explanation for Raphael's empahtetic portrayal of Phrygia's pagan plague victims is furnished by a growing deontological pressure on Europe's governing class to remain in city during an epidemic. Priests, doctors, and civic leaders all needed to diminish their fear of catching the disease and to overcome their ingrained reluctance to look at those who were afflicted. With the Black Death, a fierce debate had broken out over the ethics and morality of fleeing from a plague-infected city. Humanist critics like Petrarch (1304–1374) and Coluccio Salutati (1331–1406) charged that flight was a shameful and unmanly form of pusillanimity unbefitting of a citizen or a physician; religious critics like Rodrigo Sánchez de Arévalo, Bishop of Zamora (1404–1470) considered flight a sinful lack faith in God, an omission of charity unbefitting of a Christian, and, in the case of disobedient priests, an abandonment of their charge to minister to their flocks.[40] By contrast, those who wished to justify flight cited examples of worthy and saintly men who had chosen self-preservation in times of danger. Jesus (whose family took him to Egypt to escape King Herod's assassins) and St. Paul were among these examples, as was the same widely admired hero who appears in the upper left scene of Raphael's *Morbetto*: Aeneas, who fled the Phrygian outbreak to settle in Italy.[41]

With imagery that compels our sympathy for the dying, the dead, and the helpless, Raphael led the Renaissance viewer to see the mythic hero's decision to flee Phrygia in a newly censorious light. Even while the artist acknowledges the fear that is natural amidst a plague-ridden place, he pleads with the viewer to think with pity of the innocents who still might be saved, and of the sick who could be given some comfort and companionship amidst their suffering.[42] Raphael's critical reframing of Vergil's epic resounds with the Christian Humanist ideologies of his age. By dramatising the plight of nameless plague victims and by substituting Aeneas with a more altruistic and empathetic hero, Raphael offered a means of managing spontaneous emotional responses like fear and repulsion, and he gave plague victims a positive visual image that aligned closely with the recent efforts throughout Renaissance Europe to provide medical care to the needy urban masses, especially during epidemic outbreaks.[43]

## Explicit Symptomology in Late Renaissance Images

The *Morbetto* accommodates readings through multiple lenses, including ethics, religion, medicine, and poetry, but the potency of Raphael's image lies in its ability to trigger deep concern and empathy towards the plague-stricken while acknowledging the viewer's underlying fear of the danger they represent. Many artists in the following centuries sought to imitate Raphael's stirring image of the effects of epidemic disease. Their artworks, which took on great urgency with the devastating outbreaks in 1576 and 1630, followed one or the other of two divergent courses due to differences of purpose and audience: While some picked up on the secular and medical understandings of plague in the *Morbetto* in order to cater to elite audiences, others focused on the religious aetiology of the disease in order to address a more inclusive viewership. Nearly all of them, however, sought to express the touching plight of the plague victim, as it had first been explored by Raphael.

In the sixteenth and seventeenth centuries, the utility of such images was seized upon by charitable institutions that needed to raise money in order to fulfil their mission to care for the victims of the epidemic outbreaks that struck large cities with a frequency proportional to the vitality of their commerce. At this time, plague paintings focusing on the drama of the victims were made for a number of churches attached to hospitals and other charitable institutions in cities throughout Italy. A notable example is Tintoretto's large painting of *Saint Roch Healing Plague Victims* (Fig. 3.5) carried out in 1549 for the apse of San Rocco, the church of the Venetian confraternity known as the Scuola di San Rocco and the centre of an important reliquary cult focused on the saint's body that was housed here since 1485.[44]

The painting's subject, drawn from medieval hagiographic legends, is St. Roch's miraculous healing of plague victims during the fourteenth-century pandemic while he was staying at a refuge for pilgrims in Acquapendente.[45] As the first Venetian painting to represent a scene of plague, Tintoretto's painting sets Roch, a saint invoked in Venice as a protector against plague since 1348, in the middle of a dingy, chaotic lazaretto where he is flanked on both sides by patients in their hospital beds, and surrounded by cadavers on the floor.[46] A male patient is showing Roch his wound, while in the darkness behind Roch are two doctors (one of whom appears to be an Arab) absorbed in a discussion, unable to offer a cure. They, like the plague victims, are ultimately in the hands of God.

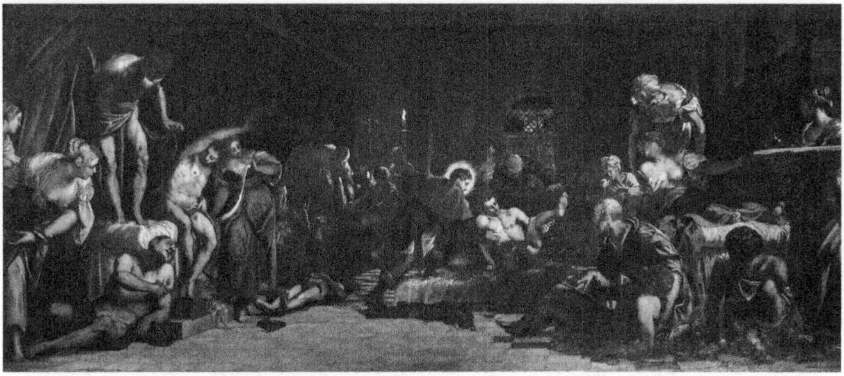

**Fig. 3.5** Tintoretto, 'St. Roch in the Hospital' 1549, oil on canvas, 307 × 673 cm

Although centrally located in the oblong canvas, the representation of Roch's thaumaturgic healings occupies only a small portion of the image. Most of it is devoted instead to the depiction of the caregiving meant to ease the suffering of the plague victims, such as the changing of bandages, the distribution of wine, food and medicine, the ministration of Last Rites, the succour of orphaned children, and the removal of the dead for burial.[47] The doctors may be powerless to offer cures, but even the two humbly dressed old servants are able to offer comfort and care: an elderly male worker and an elderly washer woman on the right side of the canvas are fearlessly alacritous in the provision creature comforts to the sickly young adults in their arms.

Tintoretto has followed Raphael's lead in diversifying and individualising the sick: there are men as well as women (which does not reflect the actual practice of rigorously separating the sexes in lazarettos); one patient is vigorous enough to stand on his bed; another has died and been pushed from the bed to the floor; yet another is a foreigner, as indicated by his turban. There is even a baby in the care of a wet-nurse, apparently taken from the agonised female patient below them.

Although the plague victims vary greatly in terms of their poses, they all share two important characteristics: each exhibits an idealised, heroic nudity, and each is branded by single ruddy aposteme.[48] The athleticism of their physiques is emphasised by the twisting and stretching they do seemingly just to exhibit in the viewer's direction the tell-tale buboes that brought them to this hellish place. Equally idealised is the languishing female plague victim on the image's right side. She is impossible to ignore because of the flame-red blanket wrapped around her feminine form and the bright light on her smooth breast, which presumably she had taken out to nurse a child that is now in the care of the wet-nurse behind her. Even the female attendants are winsome young women with elegant coiffures and colourful dresses. All in all, the combination of athletic nudes, attractive women, and patches of intense, jewel-like colour throughout the darkness succeeds quite well in counterbalancing the array of pitiful and tragic suffering this *locus terribilis*.[49]

Setting buboes and cadavers in plain sight even as it delights the eye with things of beauty, Tintoretto's representation of plague seeks to keep the viewer's opposing reactions in tension in keeping with the peculiar role of this painting. On the one hand, since the commissioning institution had a vested interest in fostering St. Roch's cult as a plague saint, a minatory reminder of the disease was useful in spurring visitors to seek the saint's protection.[50] However even while indicating the intense redness of the buboes, showing the discomfort and feebleness of the patients, and including a corpse, the artist refrained from making the imagery of plague aesthetically unpleasant. By carefully containing the frightening potential of the image, the artist succeeded in keeping the visitors' attention; were they to avert their eyes too quickly, they might miss the important story of Roch's miraculous healings.[51] This need to keep the viewer engaged in a protracted consideration of potentially dangerous subject matter explains why Tintoretto deployed his full armoury of visual delights to ensnare the viewer's eye: nudes with heroic anatomies, youthful beauties, refulgent colour, daring perspectives, chiaroscuro, and figural variety. The result represents an important development in plague imagery. Whereas Raphael had pioneered imagery that stimulated strongly empathetic responses to plague victims, Tintoretto strategically provoked diametrically opposed responses, simultaneously raising the spectre of plague and charming the viewer with beauty.

An understanding of the historical experience of emotions is critical to understanding the profound solace Tintoretto's image offered both male and female viewers in an era of frequent plagues.[52] In Tintoretto's society, emotional responses were given the same weight as good and evil acts; moreover, they were believed to influence the eternal fate of a soul. With so much in the balance, the regime of the emotions became a fundamental focus of Christian practices of confession and prayer, while devotional images and meditational guides were valued as means of modelling emotional reactions and resisting destructive and sinful passions.[53] Within this economy of sentiment and salvation, Tintoretto's image helped viewers to manage their negative responses and to transmute them into spiritually beneficial emotions and possibly even charitable actions.

## Baroque Plague Art to Stay Healthy

The final milestone in this overview is a painting that was arguably designed to excite the emotions of terror and sadness to the point of surfeit: Nicolas Poussin's *Plague of Ashdod* (Fig. 3.6).

This painting was completed in 1631, while much of central Europe and Northern Italy were in the grips of a plague that was headed towards Rome, where Poussin had his studio.[54] Its subject, an unprecedented choice for a stand-alone easel painting, is the same as that of the Biblical illustrations described above. The religious aetiology is alluded to on the left side of the image, where we see the fallen statue of the pagan god

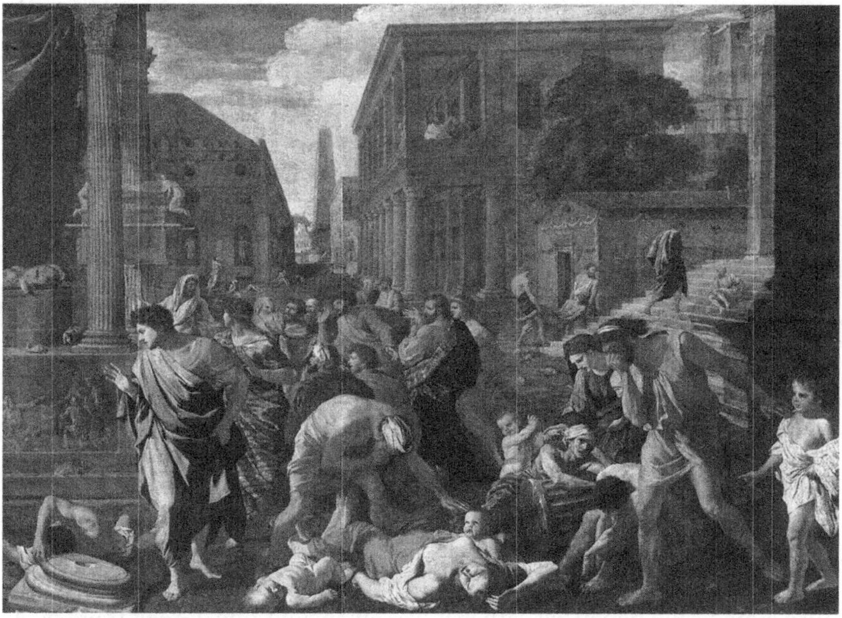

**Fig. 3.6** Nicolas Poussin, 'The Plague of Ashdod' 1630–1631, oil on canvas, 148 × 198 cm

Dagon and the Ark beside it, while the rest of the scene shows the transmission and effects of the epidemic. Any historical plague Poussin might have chosen to depict would have suggested obvious connections with the current calamity, but his unusual choice of the Plague of Ashdod implied that the current scourge was sent by God to punish religious sins like impiety and idolatry. This comparison put the Italians' current situation in the most drastic light possible, suggesting they should have no hope for salvation because God had willed them to suffer as long as they sinned.

In order to show the suffering of the Philistines in a way that would resonate with contemporary fears about the plague, Poussin drew upon the imagery of Raphael's *Morbetto.* Figures that were first debuted in Raphael's image include Poussin's depictions of people trying to avoid plague in various ways: by running from the city, by pinching their noses in the vicinity of victims, and by blocking their eyes to avoid terrifying sights. The Philistine woman slumped over a column was inspired by Raphael's fever sufferer, only now her face is darkened by the disease and locked in a hard, unseeing zombie-like state known as the *facies pestica*, or plague face, which was caused by experiencing too much fear. Above all, we recognise Poussin's adaptation of Raphael's celebrated *Pathosformel* of the infant about to nurse from a dead mother. It is similar in every detail, except that Poussin has intensified its emotive power with the proleptic addition of a dead baby lying on the ground near the same lifeless mother. It is the living baby's unfortunate twin brother, already fallen victim to its mother's milk.

Plague regimens since the fourteenth century had warned readers to avoid sadness and distress, and in ancient and modern accounts alike, it was recorded that some people had died of 'mortal panic', amounting to apoplexy at the sight of, or even the mere thought of, the disease's horrors. Poussin's picture seems at first to fly in the face of the predominant medical teachings of his age, the same teachings that had led Tintoretto to temper the frightening aspects of his lazaretto scene with pleasant details. Rather than counterbalancing his image's perturbing effects, however, Poussin's radical solution was to heighten them as much as possible, engendering so much painful emotion that the surfeit would force a purging effect on the body's humours.

The model for Poussin's use of an artfully simulated cataclysm to produce a medically beneficial catharsis was an ancient one: Greek tragedy. As Aristotle argued in his literary treatise known as the *Poetics*, a well composed tragic play relieves the audience of their unhealthy backlogs of painful emotions and the attendant humoral excesses. Poussin surely came upon this idea while studying Lodovico Castelvetro's treatise on theatre, written in 1570.[55] Here Aristotle's insistence on the health benefits of tragedy is explained through the specific analogy to the plague:

> Tragic drama is the means by which our compassion and fear are diminished, and so it is to our benefit that we experience these passions in diverse situations as the theatre permits. The proof of this is in epidemics, at the beginning of which, when three or four people die, we are gripped by suffering and fear, but then once we have witnessed hundreds and then thousands die, these reactions of compassion and fear cease.[56]

If Poussin and his contemporaries believed that the artist's *Plague of Ashdod* could achieve the important psychosomatic benefits that Aristotle attributed to tragic theatre, they must have taken it for granted that within a person's imagination, artistic representations of plague could attain a certain limited equivalency with real plague. The particular efficacy of Poussin's rhetorical formulas for capturing the potency of real plague and transforming it into a safe 'visual vaccine' would certainly help explain why the painting enjoyed such enormous success. Not only was Poussin's painting resold for an astonishing price to the French crown, but it was also copied in oil, reproduced in engravings, and endlessly imitated.

## Conclusion

The human imagination was a perilous place, above all during times of plague. For early modern Europeans, it was an area of concern for both private and public medicine. Priests tried to console it with sermons and Masses; physicians advised cheering it through diet and pleasant pastimes like music; and governments endeavoured to protect it by keeping the plague's devastation out of sight (and thus out of mind) to the degree it was possible. However, because early modern Europeans believed that the imagination's processes were preponderantly oriented towards visuality, artists were recognised to possess a particularly potent means of reaching the human imagination. From this perspective, then, artists—no different than priests, physicians, and government officials—played a role in protecting their society from the devastating effects of epidemic disease.

We have seen how, even before the second pandemic, artists were charged with the making of religious images of plague so that future populations could learn from the tragedies of the past and remain mindful of the punitive power of their divine creator. Once the pandemic reached Europe in the mid-fourteenth century, images of epidemics occurred in new devotional contexts. They included representations of Christian-era outbreaks for the first time, and they began to incorporate secular notions of disease and its cure. Art in this period rarely showed plague's impact on the body; instead, art described the plague's impact on the city, devastated by a high rate of mortality as well as by the flight of its healthy citizens for safer places.

Beginning with Raphael, artists of the sixteenth century aimed at ever more descriptive renderings of the plague's many different effects on its human victims, keying their depictions to excite not only the standard emotional responses of fear and revulsion, but also the newly important reaction of empathy. The resulting images helped foster concern for the painted sufferers as well as for their real-life counterparts in plague hospitals and lazarettos. By the seventeenth century, representations of plague were more abundant than ever, appearing not only in devotional imagery but also in secular paintings with literary and historical themes. These images of plague continued to inspire emotions in the viewer, but now they were designed to inure the viewer to the much stronger emotional reactions triggered by real plagues. This was achieved through the viewer's repeated looking at, and empathetic engagement with, representations that functioned like gymnasiums for the eye and heart. The artist aspired to do what the physician could not: to inoculate the imagination against despair, fear, and mortal panic, and to help contemporaries face head-on the plagues that by then were part of the predictable pattern of life.

## Notes

1. The study of coeval plague imagery to gain insight into the history of epidemics was pioneered by Louise Marshall, 'Manipulating the Sacred: Image and Plague in Renaissance Italy, *Renaissance Quarterly* 47, no. 3 (1994): 485–532. Here, orienting our inquiry in a subtly different direction, we will interrogate images to reveal pervasive mentalities and the social imaginary over the longue durée, rather than to draw inferences about cultic and religious response to specific plague events.
2. The geographical limitation of this chapter to Western Europe and the focus on Italy roughly reflects the historical conditions of the production

of plague imagery. As affirmed by Dionysios Stathakopoulos, 'Crime and Punishment: The Plague in the Byzantine Empire, 541–749', in Lester K. Little (ed.) *Plague and the End of Antiquity. The Pandemic of 541–750*, pp. 99–117 (Cambridge: Cambridge University Press, 2007), p. 111: 'There is no trace of the disease and its impact in any Byzantine work of art.'

3. Jeffrey H. Tigay, 'Exodus. Introduction and Annotations'. In Adele Berlin, Marc Zvi Brettler, and Michael A. Fishbane (eds.), *The Jewish Study Bible*, pp. 102–210 (Oxford: Oxford University Press, 2004).
4. Bucher, *Pamplona Bibles* 1:3–4, 9–10. Kurt Weitzmann, 'Die Illustration des Septuagint', *Münchner Jahrbuch der Bildenden Kunst* 3–4 (1952–1953): 96–120, pp. 116–117.
5. On the description of the disease, see Lawrence I. Conrad, 'The Biblical Tradition for the Plague of the Philistines', *Journal of the American Oriental Society* 104, no. 2 (1984): 281–287, p. 285.
6. Compare the use of dark windows here to the open doors in later paintings to symbolise the decimation of the urban population, as described in: Louise Marshall, 'Plague in the City: Identifying the Subject of Giovanni di Paolo's Vienna Miracle of Saint Nicholas of Tolentino', *Renaissance Studies* 27, no. 5 (2012): 654–680, pp. 674–675. For the reference to rats in the biblical source, see: P. Berger, 'Mice, Arrows, and Tumors: Medieval Plague Iconography North of the Alps'. In Franco Morando and Thomas Worcester (eds.), *Piety and Plague from Byzantium to the Baroque*, pp. 23–63 (Kirksville, MO: Truman State University Press, 2007), p. 24.
7. Kay P. Jankrift, *Brände, Stürme, Hungersnöte. Katastrophen in der mittelalterlichen Lebenswelt* (Darmstadt, 2003) pp. 151–166.
8. In regard to Italian cities, see: Roberto Greci, 'Il controllo della città. L'ufficio dei fanghi e strade a Bologna nel XIII secolo', Nuova Rivista Storica 75 (1991), 650–651; Duccio Balestracci, 'The Regulation of Public Health in Italian Medieval Towns' In Helmut Hundsbichler, Helmut, Gerhard Jaritz, and Thomas Kühtreiber (eds.), *Die Vielfalt der Dinge: Neue Wege zur Analyse mittelalterlicher Sachkultur*, Forschungen des Instituts für Realienkunde des Mittelalters und der Frühen Neuzeit: Diskussionen und Materialien 3, pp. 345–357 (Vienna: Verlag der österreichischen Akademie der Wissenschaften, 1998); Roberta Magnusson and Paolo Squatriti, 'The Technologies of Water in Medieval Italy'. In Paolo Squatriti (ed.), *Working with Water in Medieval Europe. Technology and Resource-Use*, Technology and Change in History 3, pp. 217–266 (Leiden: Brill, 2000), pp. 253, 256. For French examples, see: Jean-Pierre Leguay, *L'eau dans la ville au moyen âge* (Rennes: Presses universitaires de Rennes, 2002), pp. 117, 119, 136 ff. For Spanish ones, see: L. García-Ballester et al. (eds.), *Practical Medicine from Salerno to the Black Death* (Cambridge: Cambridge University Press, 1994).

9. The presence of a peasant was first noted in Berger, 'Mice, Arrows, and Tumors', p. 24.
10. On the intersection between communal memory and plague, see: Ann G. Carmichael, 'The Last Past Plague: The Uses of Memory in Renaissance Epidemics', *Journal of the History of Medicine and Allied Sciences* 53, no. 2 (1998): 132–160.
11. The circumstances of its commission are not known. It replaced an earlier altarpiece for the same altar, which had been commissioned by Florence's bishop, Filippo dell'Antella, in 1362 in order to observe the cult of the saint whom he credited with his salvation from case of plague he contracted while in Avignon in 1348. See: Detlev von Hadeln, *Die wichtigsten Darstellungsformen*, pp. 8–9 and Sheila Barker, 'The Making of a Plague Saint. Saint Sebastian's Imagery and Cult Before the Counter-Reformation', in Franco Mormando and Thomas Worcester (eds.), *Piety and Plague from Byzantium to the Baroque*, pp. 90–131 (Kirksville, MO: Truman State University Press, 2007), pp. 99–100.
12. On the mortality rates in Florence during the Black Death, see, for example: D. Cesana, O. J. Benedictow, and R. Bianucci, 'The Origin and Early Spread of the Black Death in Italy: First Evidence of Plague Victims from 14th-Century Liguria (Northern Italy)', *Anthropological Science* 125, no. 1 (2017): 15–24, p. 17. Given the tremendous mortality caused by these late fourteenth-century plagues, it is not surprising that some societies at this very same time began developing record-keeping practices that allowed for the analysis of mortality rates. According to Ann G. Carmichael, 'Contagion Theory and Contagion Practice in Fifteenth-Century Milan', *Renaissance Quarterly* 44, no. 2 (1991): 213–256, p. 214, 'death itself became a meaningful social event to Renaissance Italians, who first began to keep records of mortality during the late fourteenth century'.
13. On the irregularity of this burial and this theme as a trope for other medieval plague imagery, see: Marshall, 'Plague in the City', pp. 669–674.
14. Riccardo Simonini, 'Il codice di Mariano di Ser Jacopo sopra "Rimedi alibi nel tempo di pestilenza"', *Bolletino del Istituto Storico Italiano dell'Arte Sanitaria* 9 (1929): 161–169.
15. See: Luciano Patetta, 'Nuove ipotesi sul lazzaretto Quattrocentesco di Milano', *Bollettino d'arte* 71, no. 35/36 (1986), p. 25.
16. Ann G. Carmichael, 'Plague Legislation in the Italian Renaissance', *Bulletin of the History of Medicine* 57, no. 4 (1983): 508–525, p. 512. Carmichael shows that whereas the medical community preferred corruption theory over contagion theory, Visconti and Gonzaga were able through their autocratic control of the state to impose measures based on the theory of human-to-human contagion.
17. Carmichael, 'Plague Legislation', pp. 512–513.

18. The plague-time Latin dictum '[Fuge] cito, [vade] longe, [rede] sero' was also expressed, 'Haec tria tabificam tollunt adverbia pestem: mox, longe, tarde, cede, recede, redi'.
19. Bertrand Lançon, 'Maladie et médecine dans la correspondance de Jérôme', in Y.-M. Duval (ed.), *Jérôme entre l'Occident et l'Orient: VIe centenaire du départ de saint Jérôme de Rome et de son installation à Bethléem. Actes du colloque de Chantilly*, pp. 355–366 (Paris: Brepols, 1988), p. 361. The unwillingness of both family members and priests to give succor to plague victims because of their fear of catching the disease is discussed in relation to art and literature of the 14th and 15th centuries in Louise Marshall, 'Affected Bodies and Bodily Affect: Visualizing Emotion in Renaissance Plague Images', in P. Maddern, J. McEwan, amd A. Scott, Performing Emotions in Early Europe, pp. 73–103 (Turnhout: Brepols, 2018), pp. 81–82.
20. The related extant drawings are in the Gabinetto dei Disegni e Stampe degli Uffizi, inv. nos. 1348 F and 525 E, and in the Royal Collection of Her Majesty the Queen, RCIN 990117.
21. Raimondi's engraving has been analysed in relation to plague imagery in: Gauvin Alexander Bailey, Pamela M. Jones, Franco Mormando, and Thomas W. Worcester (eds.) *Hope and Healing. Painting in Italy in a Time of Plague, 1500–1800*, (2005), cat. no. 5, pp. 186–187.
22. It has been noted that Raphael's design was partly inspired in part by an illustrated late Antique codex of the *Aeneid* that was at that time in the possession of Pietro Bembo, and which is now in the Vatican Library (cod. Vat. Lat. 3225); see: David Herndon Wright, *The Vatican Vergil: A Masterpiece of Late Antique Art* (Berkeley CA: University of California Press, 1993), pp. 110–114. Regarding the passages describing the plague of Phrygia, the relative illustration in this manuscript is one showing Aeneas receiving the visit from the Penates at night (cod. Vat. Lat. 3225, fol. 28r), which Raphael imitated closely. The immediately preceding illustration (cod. Vat. Lat. 3225, fol. 27r) shows Aeneas' ships docked on the shores of the city they founded, Pergamon. In my view, no signs of famine and pestilence are indicated in this illustration of Pergamon, and indeed, plague itself is not represented visually anywhere in the Vatican Vergil.
23. Vergil, *Eclogues, Georgics, Aeneid I–VI*, trans. H. Rushton Fairclough (New York: G. P. Putnam's Sons, 1916), vol. I, book III, ll. 140–141, p. 359. See also note 15 for the late Antique codex that Raphael consulted.
24. Plague fever and other medical symptoms of pestilence, largely drawn from the writings of Avicenna, are described in plague treatises such as that by Petrus Pintor, *Aggregator sententiarum doctorum de praeservatione curationeque pestilentiae*, dedicated to Pope Alexander VI and published in Rome in 1499.

25. Only the first three of these are mentioned in Vergil's account (bk. III, ll. 137–142): '[…] when on a sudden, from a tainted quarter of the sky, came a pestilence and season of death, to the wasting of our bodies and the piteous ruin of trees and crops. Men gave up their sweet lives, or dragged enfeebled frames; Sirius, too, scorched the fields with drought; the grass withered, and the sickly crop denied sustenance' (Vergil, *The Aeneid*, trans. Fairclough, p. 359).
26. On the pinched nose as an indication of the theory of miasmic vapours, see: Katrin Achilles-Syndram, '"So macht nun Abbilder eurer Beulen und eurer Mäuse": Die Pest als Thema der bildenden Kunst'. In Hans Wilderotter (ed.) *Das grosse Sterben: Seuchen machen Geschichte*, exh. cat., pp. 94–121 (Berlin: Jovis, 1995), pp. 100–101. For Raphael's derivation of these symptoms and causes of plague from ancient medical treatises, see: Stefania Mason Rainaldi, 'Le immagini della peste nella cultura figurativa veneziana'. In *Venezia e la Peste: 1348–1797,* pp. 209–286 (Venice: Marsilio, 1980), cat. no. a12, pp. 238–239. Raphael could have also found most of these ideas in contemporary treatises as well; for the miasmatic theory in medieval and early Renaissance medicine, see: Carmichael, 'Plague Legislation'.
27. On Livy's description of this and other plagues, see: Susan Satterfield, 'Livy and the Pax Deum', *Classical Philology* 111, no. 2 (April 2016): 165–176.
28. Livy, *The Early History of Rome*, translated by Aubrey de Selincourt (New York: 1984), bk.3.6, p.189. The imagery of dying livestock also recalls the ancient belief that pestilence among humans was generally preceded by a die-off of animals. There are many sources for this notion, from Homer until the Middle Ages, as demonstrated by Lodovico Antonio Muratori, *Del governo della peste* (Milan: 1832), pp. 3–6.
29. On the association between rotting flesh and plague, see the sources in note 9 above. On the growing interest of sixteenth-century physicians in the possible role of contagion in the transmission of plague, see: Carmichael, 'Contagion Theory and Contagion Practice'.
30. On the longstanding medical concern about the role of corpses and even the breath of living plague victims in the generation of pestilence, see note 9 above. Closer to the era in which Raphael made his print, Pope Leo X was concerned about the danger to public health posed by plague victims and thus granted syndics the power to remove the *incurabili* to special hospitals, even against their will, with a 1515 bull; see: Jon Arrizabalaga, 'The French Disease and the Hospitals for Incurables in Italy until 1530'. In Jon Arrizabalaga, John Henderson, and Roger French, *The Great Pox. The French Disease in Renaissance Europe* (New Haven and London: Yale University Press 2004), p. 157. It is possible that Raphael's comparison of the dead sheep and dead humans was inspired by Boccaccio's statement in

the *Decameron* that plague victims were treated no better than dead goats; for a discussion of Boccaccio's observation in relation to an earlier artwork, see Marshall, 'Affected Bodies and Bodily Affect', p. 84.

31. Anonymous practitioner of Montpellier, translation by Justin K. Stearns, *Infectious Ideas: Contagion in Premodern Islamic and Christian Thought in the Western Mediterranean* (Baltimore MD: Johns Hopkins University Press, 2011), p. 95.
32. Justin K. Stearns, *Infectious Ideas: Contagion in Premodern Islamic and Christian Thought in the Western Mediterranean* (Baltimore MD: Johns Hopkins University Press, 2011), pp. 95–96. Horrible sights were thought to materially enter the body as visible 'species' carried by rays into the onlooker's eyes, and from there into the faculty of the imagination where the images excited the passions of the soul and in turn caused deleterious perturbances in the body's humoral balance. For the medical texts describing the danger of fearful images in the imagination, see: Martin Marafioti, *Storytelling as Plague Prevention in Medieval and Early Modern Italy. The Decameron Tradition* (London: Routledge, 2017), pp. 11–15.
33. Charles Edward Armory Winslow and M. L. Duran-Reynals, 'Jacme d'Agramont and the First of the Plague Tractates', *Bulletin of the History of Medicine* 22, no. 6 (1948): 747–765, p. 761.
34. An early, fundamental study of the use of images in Christian devotions is Sixten Ringbom, 'Devotional Images and Imaginative Devotions: Notes on the Place of Art in Late Medieval Private Piety', *Gazette des Beaux-Arts* 73 (1969): 159–170.
35. Elizabeth Schröter, 'Raffaels Madonna', *Zeitschrift für Kunstgeschichte* 50, no. 1 (1987): 46–87, p. 71.
36. Elizabeth Cropper, 'Marino's "Strage degli Innocenti", Poussin, Rubens, and Guido Reni', *Studi Secenteschi* 33 (1992): 153; Sebastian Schütze, 'Aristide de Thèbes, Raphäel et Poussin: La représentation des affetti dans les grands tableaux d'histoire de Poussin des années 1620–1630'. In ed. Alain Mérot (ed.) *Nicolas Poussin (1594–1665): Actes du colloque organisé au Musée du Louvre*, pp. 571–601 (Paris: Documentation Française, 1996), pp. 576–578.
37. *The Elder Pliny's Chapters on the History of Art*, translated by K. Jex-Blake (Chicago IL: Ares Publishers, 1976), bk. 35.98: pp. 133–135.
38. On this term, coined by Aby Warburg in his *Das Nachleben der Antike*, see: Colleen Becker, 'Aby Warburg's Pathosformel as Methodological Paradigm', *Journal of Art Historiography* 9 (2013): 1–25.
39. It is no coincidence probably that the drawing of Raphael's that Raimondi had engraved just a couple years earlier depicted the Massacre of the Innocents, a subject that also reveals an interest in the related *Pathosformel* involving infants, their mothers, and tragic death on a calamitous scale.

40. Criticism of those who fled from cities during the plague was as old as the history of this advice to flee, as shown in the important study by Heinrich Dormeier, 'Religiös motiviertes Verhalten von Laien und Klerikern in Grenz- und Krisensituationen: die Pest als "Testfall wahrer Frömmigkeit"'. In Klaus Schreiner (ed.), *Laienfrömmigkeit im späten Mittelalter Formen, Funktionen, politisch-soziale Zusammenhänge*, pp. 331–392 (Munich: De Gruyter, 1992). Dormeier's collection of source material suggests that the debate became particularly tense and politically charged among the leadership of the Roman Catholic Church in the 1480s.
41. Aeneas' example was invoked as an argument by Giovanni da Bologna, a physician in Muggia, in 1395. See: Dormeier, 'Religiös motiviertes Verhalten', p. 350.
42. In her study of the image, Christine M. Boeckl, *Images of Plague and Pestilence: Iconography and Iconology* (University Park, PN: Pennsylvania State University Press, 2001), pp. 92–96, also found this image to be noteworthy for its emphasis on the victims of plague.
43. In 1514, for example, Rome's first hospital expressly for plague victims, and open to men of all trades, was established by a confraternity devoted to Roch, the saint famous for having tended to plague victims at the expense of his own health. See: Fiorani, 369; and Maroni Lumbruso, 343–345. Around 1475, Pope Sixtus IV had the hospital of Santo Spirito in Sassia rebuilt. Alexander VI, however, was the first pope to establish an institution for the care of plague victims; see: E. Schröter 'Raffaels Madonna: ein Pestbild?', *Zeitschrift für Kunstgeschichte* 50, no. 1 (1987): 46–87, p. 68 n.117.
44. Maria Agnese Chiari and Moretto Wiel, 'Il culto di san Rocco a Venezia: la Scuola Grade, la sua chiesa, il suo tesoro'. In Massimo Tirotti and Claudia Rossi (eds.), *San Rocco nell'arte: Un pellegrino sulla Via Francigena*, pp. 67–70 (Milan: Electa, 2000).
45. Louise Marshall, 'A Plague Saint for Venice: Tintoretto at the Chiesa di San Rocco', *Artibus et Historiae* 33.66 (2012): 153–187, p. 170.
46. Ibid., p. 173.
47. Although it is tempting to imagine that Tintoretto was illustrating remedies that his patrons at the Scuola di San Rocco offered to plague victims, this is not the case; nor did they operate a plague hospital; ibid., p. 184, n. 36. In ibid., p. 173, it is suggested that the image responds to criticisms levelled against the confraternity for not doing more for the city's poor.
48. Ibid., p. 173.
49. For a discussion of plague imagery that features only idealised beauty and avoids almost altogether the disturbing signs of the disease, see: Sheila Barker, 'The Making of a Plague Saint. Saint Sebastian's Imagery and Cult Before the Counter-Reformation'. In Franco Mormando and Thomas

Worcester (eds.) *Piety and Plague from Byzantium to the Baroque*, pp. 90–127 (Kirksville MO: Truman State University Press, 2007), pp. 114–119, 123–127.

50. The association of Roch's cult with protection from plague was encouraged by the wide circulation of printed pamphlets with prayers. See: Pierre Bolle, 'Saint Roch de Montpellier, doublet hagiographique de saint Raco d'Autun. Un apport décisif de l'examen approfondi des incunables et imprimés anciens'. In E. Rénard et al. (eds.) *Scribere sanctorum gesta Recueil d'études d'hagiographie médiévale offert à Guy Philippart*, pp. 525–572 (Turnhout: Brepols, 2005). On the use of the imagery of St Sebastian by hospitals and confraternities to incite fears of sudden death, ultimately for the purpose of inciting charitable donations, see: Elvio Lunghi, *Il* Martirio di San Sebastiano *di Pietro Perugino a Panicale* (Perugia: Fabrizio Fabbri Editore, 2005), pp. 135–138. More generally on the topic, see: Barker, 'The Making of a Plague Saint', p. 102.
51. On the attitudes towards terrifying imagery in the Renaissance, see Sheila Barker, 'Poussin, Plague, and Early Modern Medicine', *The Art Bulletin* 86, no. 4 (2004), 659–689; Elisabeth Hipp, 'Poussin's The Plague at Ashdod. A Work of Art in Multiple Contexts'. In Franco Mormando and Thomas Worcester (eds.), *Piety and Plague from Byzantium to the Baroque*, pp. 177–223 (Kirksville MO: Truman State University Press, 2007).
52. It is well established that theoretical medical treatises of the Renaissance distinguished between male and female humoral constitutions, with implications for the passions. Nevertheless, this chapter, in describing the use of art to manage emotional responses to plague, has not considered women as a separate category of viewer in light of recent findings that early physicians did not treat women's emotions differently in their practice of medicine. See: Na'ama Cohen-Hanegbi, 'The Emotional Body of Women. Medical Practice between the 13th and 15th Centuries'. In Pioska Nagy and Damien Boquet, (eds.), *Le Sujet des émotions au Moyen Âge*, pp. 465–482 (Paris: Beauchesne, 2009).
53. Na'ama Cohen-Hanegbi, 'The Matter of Emotions: Priests and Physicians on the Movements of the Soul', *Poetica* 72 (2009): 21–42, pp. 28–31.
54. The following paragraphs summarise the ideas first put forth in: Barker, 'Poussin, Plague, and Early Modern Medicine'.
55. Poussin referred to the work in his 'Observations on Painting'. See: Barker, 'Poussin, Plague, and Early Modern Medicine', p. 667.
56. Lodovico Castelvetro, *Poetica d'Aristotele volgarizzata e sposta*, ed. Werther Romani, 2 vols. (Rome: Giuseppe Laterza e Figli, 1978–1977), vol. 1, 161.

CHAPTER 4

# Pesthouse Imaginaries

*Ann G. Carmichael*

Verona's lazaretto, the longest surviving of Italy's large plague hospitals, was reduced to rubble at the end of World War 2. As Allied forces pressed north, retreating Germans tried to ignite stored munitions. Irreparable destruction to the former hospital structure came later, on 20 May 1945, with massive explosions that killed at least 30 persons and destroyed buildings within a 500-metre radius, rattling houses in the city five kilometres away. Of the structure itself, only the bottom part of a beautiful open-air chapel facing the Adige River and the remnant of a ruined loggia survived. The calamity was initially blamed on careless squatters and looters, but by late July a different story had taken shape: the German attempt to raze the pesthouse had been sabotaged by a pro-Resistance *ragazzo*, who supposedly cut a trip wire.[1]

At several points over the post-war decades, partial rehabilitation of the lazaretto seemed possible. Then in 2007 a documentary filmmaker, Mauro Vittorio Quattrina, created new emotional connections to the ruins by

A. G. Carmichael (✉)
Indiana University, Bloomington, IN, USA
e-mail: carmicha@indiana.edu



C. Lynteris (ed.), *Plague Image and Imagination from Medieval to Modern Times*, Medicine and Biomedical Sciences in Modern History, https://doi.org/10.1007/978-3-030-72304-0_4

using overhead camera drones to lead viewers downriver, inviting them to imagine anew the feared last passage that many victims of the 1630 plague made.[2] By 2014, Verona had secured restoration funding from Italy's cultural trust, the FAI (*Fondo Ambiente Italiano*). Only the damaged chapel designed by renown Veronese architect Michele Sanmicheli in the early 1540s was ever a priority, and rebuilding just its cupola proved prohibitively expensive. Workers were laid off.

The whole area may now return to silent abandon, although not all agree that the space should preserve any of its historical connections.[3] Once deliberately sited well away from the city in a flood-prone area that limited development, today residential expansion encroaches. Some want to rehabilitate the acoustically sophisticated chapel for modern open-air concerts; some mourn the impending loss of nearby green space; still others want a large park and sports centre, fully erasing the area's link to its premodern past. Proponents of preservation emphasise the serenity of the area, the natural beauty, the almost 'religious' sense of walking around the ruins.[4] It is an interesting dilemma: what to do with non-curated reminders of a plague-connected past, spilling across landscapes that are not demarcated as archaeological sites, or encountered as artefacts safely segregated in a museum? For centuries plague history and literature—and to a surprising extent even modern plague science—have advanced the notion that plague could be walled off, contained through human interventions. Instead, humans have largely imagined the security of built boundaries.

The filmmaker's establishing shot did not feature typical plague imagery. By training an overhead camera on wild, untended landscapes and hovering above Verona's ruined lazaretto area, the opening scene unsettles a longstanding narrative that plague could be contained, whether or not that was Quattrina intention. Much plague history to this day conveys the comforting assurance that locking up affected households and/or warehousing the sick and suspect in pesthouses mitigated overall mortality, even if deaths inside sequestered areas soared. Here the camera hovered over the river, its banks and a series of unbuilt areas leading to the lazaretto. The opening scene thereby invites us to envision these spaces as relevant terrain, materially involved in the epidemic cycle: a useful perspectival shift. Historians can interrogate plague's spatial and temporal silences to see something other than our traditional storied narratives, where human actions lead to plague outcomes and thus write a new plague history more in line with recent scientific knowledge.

Quattrina evocative film invited viewers to make a mental picture of a long-ago plague from moving overhead images of a landscape devoid of humans, to see what was no longer there. The salience of the terrain to the history of Verona's great plague can be difficult to appreciate. Yet it is fundamentally relevant to our new scientific understanding of these great, recurring epidemics in world history. For a full century plague science has shifted towards study of non-human events and species that allow the persistence of the pathogen that causes plague. Meanwhile the tug of plague history tends towards human actions and interactions. New 'facts' now destabilise some established plague narratives. Rapidly advancing scientific technologies in the 2010s have facilitated the recovery and speciation of ancient pathogen DNA (aDNA) from human remains.[5] Since 2011 conclusive evidence exists that *Yersinia pestis* infections contributed to human morbidity and mortality experience in late medieval and early modern Europe.[6] This is a consequential finding because *Y. pestis* is not a classic 'contagious' disease, spread directly from one human to another. This archaeogenetic research sits uncomfortably alongside the experiential testimony of historical actors, especially when they proclaim plague a contagion.

## Plague, Re-seen from Above

Some brief consideration of overhead photography related to plague history seems useful, in keeping with this volume's primary focus on plague's 'Third Pandemic'. Quattrina opening to the documentary uses film devices so familiar to twenty-first-century audiences that we forget both cinematography and aerial photography developed contemporary with plague outbreaks of the 1890s to the 1920s. According to Teresa Castro, early cinematographers deliberately thrilled movie-goers by fusing their sense of 'I am flying' with the sense of 'I am seeing'.[7] The Quattrina film merely harnesses this now commonplace film experience when it documents a present-day view above a languid stretch of the Adige River to the ruins of San Pancrazio. It is thereby 'cinematic', in that it moves us as viewers to, through and around the pesthouse ruins, while the film's disembodied narrator urges us to connect plague narration to a contemporary environmental setting. The camera hovers over a weedy wasteland along the river's banks, then sends us around the remnants of walls. We are to imagine a tragic past of this place beneath what is actually viewable.

This cinematographic viewer-in-motion developed alongside some techniques of overhead photography that had practical, military-operational importance. Deciphering agrarian and rural landscapes along the Western Front advanced significantly during World War 1, requiring new visual and interpretive skills to analyse aerial surveillance photographs. As the camera moved higher, away from its targets, the human presence and built environment were dwarfed and flattened by vertical, high-altitude perspective.[8] From a balloon or airplane familiar structures and landmarks were often unrecognisable, confounding military recognisance efforts. Deciphering enemy troop movements from above further necessitated images of the same terrain at different points in time. The French military thus created teams of experts in visual analysis (artists, architects, geographers) to read sequenced aerial photographs, one moment in time compared to a prior set of images: an inherently historical perspective. Operationally these overhead images needed to be coordinated with traditional on-the-ground military intelligence.

Aerial photography, both cinematic and surveillant, developed contemporary to the Third Pandemic and its germ-theory causal model, but interconnections remain elusive. F. W. Murnau's *Nosferatu*, the famous 1922 German expressionist film, brought moviegoers, and plague, into this new perspectival context. Count Orlock, known as the 'Bird of Death', was land-bound, a vampire-plague menace who passed at night and in stillness across temporarily un-minded spaces. This film did not use aerial photography to achieve its haunts, however. Meanwhile, scientific understanding of the plague replaced the human-centred, contagion model of plague transmission with evidence of plague as a zoonosis, a pathogen primarily affecting rodents and transmitted by their fleas. The new plague paradigm was established by 1928, with Ricardo Jorge's synthesis of global research on rodent plague reservoirs.[9] For the most part, historians of plague's past did not fall in line with scientific consensus, instead training the camera on persons and spaces where 'contagion' lurked.

Social scientists (historians, anthropologists, sociologists) during the Interwar decades harnessed the analytical view from above to new objectives, including some studies actually relevant to past plagues. Aerial photography was first directly applied to the historical study of deserted late medieval villages in the 1920s, particularly in England, though subsequent documentary research ruled out an initial claim that villages had disappeared with Black Death demographic collapse.[10] Aerial photography used to surveil ongoing plague outbreaks in wild rodents occurred only after

World War 2. During the 1940s in both the US and the USSR, public health control of plague shifted to monitoring the disease in rodent colonies. The earliest US published proposal for the use of aerial photography related to the public health control of plague in rodents dates from the early 1970s.[11]

Given that for nearly a century scientific models of plague have described the ecology and epidemiology of plague as a zoonosis primarily involving non-commensal rodents and their fleas, it is surprising that the study of historical plagues continues to advance a contagionist model of plague's premodern ravages. Plague's early photographic canon reinforced the human-centred, contagionist perspectives on past plagues during outbreaks from the 1890s to 1920s, before plague scientists shifted to the new ecological model of plague.

Made during the first decade of the current century, Quattrina river journey to the pesthouse appeared during a time when the relationships between plague history and plague science shifted significantly. By the late 1990s an influential subset of prominent medievalists offered vigorous resistance to scientists' claims about past plagues in Europe. With close re-reading and analysis of historical sources they concluded that the great pestilences could not have been caused by the same bacterium responsible for modern plague epidemics. Meanwhile, newer archaeological evidence and molecular laboratory assays of human remains associated with mass burial sites drove a view-from-below, enlarging what could not be seen without the visualisation that molecular scientific technologies could provide.[12] Plague killed its victims too quickly to leave traces on bones, which precluded use of paleopathology to assess causes of deaths in mass burial sites before the era of nanotechnologies.[13]

A decade later, in 2010, *Yersinia pestis* was recovered from human remains a London cemetery used only during 1349 and 1350, at the time of the Black Death.[14] Today we are certain that the same rodent pathogen that causes cases of human bubonic plague today was present and responsible for some portion of the human deaths during many premodern epidemic outbreaks.[15] The status of the contagion-model of premodern plague dynamics, significantly refined by dedicated historical investigations, is no longer fully supportable. We are therefore invited to re-see plague history, shifting attention to the landscapes and environmental conditions that supported human plague outbreak.

I argue that with new attention to written sources and the contexts in which they were generated, historians are in position question the assumptions conveyed by historical actors who proclaim that the great pestilences in late medieval and early modern Europe were 'contagions'. Boundary technologies invented centuries ago to contain the spread of this disease can also be historicised. Anthropogenic landscapes and terrains that accelerate early transfer of plague to human communities are parallel, interacting worlds that matter to plague history. But the ecological parameters governing plague spillovers do not necessarily operate at the same temporal and spatial scales that frame plague histories, presenting considerable methodological challenges if addressed head-on. Sustained interdisciplinary work is thus required for historians, social scientists and scientists to re-see how earlier plagues were supported by environmental factors. For example, traditional plague histories do not problematise non-human aspects of a given city's outskirts and rural hinterlands.[16] At the same time historical climatology and landscape archaeology do not quite convey recognisable historical perspectives, which are usually governed by particulars of time and space and cultural production.

The wider panoramic view of plague-related landscapes that Quattrina film trailer inadvertently featured can lead us to see the spaces of former pesthouses and other structures and activities used in the control of early modern plagues. Situated in semi-rural, often agrarian, landscapes that humans shared with small, unnoticed animals such as rodents, such landscapes materially participated in what we now know with certainty: that a flea-borne rodent pathogen played a role in many outbreaks.

## Verona's Ill-fated Lazaretto

How did lazaretto landscapes participate in the events of Verona's great plague?[17] Seen from above Quattrina San Pancrazio nestles in overgrown surroundings, in a sizeable 'empty' space encompassed by the ruins. The overhead camera meanwhile moves around, denying the viewer any static location or group of people from which to understand a long-ago plague in this place. It is useful to see the *longue durée* of Verona's lazaretto as a cyclical story of destruction and abandonment, followed by intervals of renovation. The 1944 explosion that inspired Quattrina 2007 documentary was merely the last of these cycles.

Before the sixteenth century, Verona handled plague and famine emergencies on an *ad hoc* basis, an approach typical of late medieval municipal reforms and the shifting objectives of charitable institutions.[18] The indigent and the plague-stricken poor were usually sheltered in temporary huts and derelict church properties outside urban walls, while inside the city emergency needs created an endless series of skirmishes between municipal and ecclesiastical authorities over funding.

In the early fifteenth century, Venice incorporated Verona and many other cities into its terraferma, mostly in a non-violent takeover that did not interfere with most local practices, including plague control.[19] All that changed following a series of devastating famines, epidemics, and military setbacks that Venice suffered during the war of the League of Cambrai (1508–1516), and up to the disastrous 1527–1529 plague across north and central Italy. During these decades Venice re-conceptualised a much wider territorial management strategy to address combined military and plague threats, seeing lazarettos as multifunctional defensive installations important to the state's security and expansion, both overseas and on the mainland.[20] From this point on, Venice defended its lazaretto complexes in the lagoon, the Adriatic, and the terraferma as rational anti-plague policy.[21] Venice's subject cities and most of its maritime colonies built fortress-like bulwarks that conflated differing security functions: quarantine and disinfection stations, occasional use as a military garrison, staging areas for the deportation of unwanted immigrants, and hospital-enhanced isolation facilities to use in epidemic years.

Sixteenth-century plague management involved reorganisation of suburban landscapes and rural areas just beyond the urban periphery. For example, when Venetian troops wrested Verona from occupying imperial troops in 1517, bureaucrats immediately orchestrated the demolition of suburban structures outside the city's eastern walls. The destroyed extramural area became known as the '*spianata*': a flat, cleared expanse that permanently redefined Verona's territorial orientation.[22] Verona's previous pesthouse terrains were clumped in just this area. Verona's well-designed defensive gates and impregnable walls reflected another commonality of Venice's border-defence policy. A recently built Misericordia hospital survived, because it mirrored ongoing Venetian hospital reforms to create charitable safety-nets for the city's own 'deserving poor'. But this newer public hospital had no responsibilities for plague control or for plague-stricken patients.

Relocation and construction of a more permanent lazaretto in line with the Venetian defensive plan became necessary.[23] Verona oligarchs needed to acquire the land; the Venetian Senate had to approve their decisions at each stage. Finally, in January 1539, Verona's city council approved plans for a grand and permanent pesthouse complex designed by Verona native

Michele Sanmicheli, then Venice's premier military architect. He earlier designed Verona's new walls as well as a few palaces for prominent Veronese aristocrats, but he became better known for his various fortress installations around the Venice-controlled Adriatic. Sanmicheli's design for a monumental stronghold-as-lazaretto echoed the architecture of his Renaissance city gates. The remote complex was to be bigger, better, safer, and more beautiful than any prior plague isolation area.

Drawing on both Milanese and Venetian concepts, Verona's new pesthouse hoped to fuse two different best-practice models of plague isolation facilities. Sanmicheli's personal familiarity with state-of-the art lazaretto facilities in both Venice and Milan informed his design of San Pancrazio.[24] Traditional Venetian plague control required boats to transport plague victims and their to contacts islands in the lagoon, first the old lazaretto (*Lazzaretto Vecchio*), then by the 1460s to a second lazaretto island (the *Lazzaretto Nuovo*). By the 1480s, both islands were managed by a new Health Magistracy, and both soon doubled as customs barriers and ammunition depositories in non-plague years.[25] Verona's lazaretto was meant to serve territorial defence in similar fashion, including the use of boats to transport victims from the city to the pesthouse.[26] Sanmicheli's first-hand study of Milan's quite different suburban pesthouse may have been facilitated by one of the other men on Verona's initial lazaretto committee: Francesco Bevilacqua. Francesco was the nephew of Count Galeotto Bevilacqua, famous for donating the land for the Milanese lazaretto, *Santa Maria alla Sanità*. The cornerstone of this iconic extramural pesthouse was laid in 1488, after the catastrophic plague of 1483–1486, but it was not built or used until the 1510s. Treated in more detail later in this chapter, Milan's lazaretto became as imposing architecturally as the two Sforza structures that survive today: the *Ospedale Maggiore* and the ducal *Castello*.[27]

During the decades of urban redesign, Girolamo Fracastoro was a member of Verona's College of Physicians, and served as its Prior from time to time. He would have been an ardent voice in support of a lazaretto, because he fiercely believed that the plague was contagious. Indeed his 1546 medical treatise *On Contagion and Contagious Diseases*, was written in Verona during these same years.[28] But for the most part pesthouses were not promoted by medical authorities.

The actual construction of Verona's San Pancrazio lazaretto began in the late 1540s, when Sanmicheli and his local supporters were no longer involved. Work then stopped within a year or so, and despite recurring epidemics the project lapsed for decades, meeting ecclesiastical opposition to civic funding strategies. The sixteenth-century structure was far less imposing than typical Sanmicheli complexes, probably reflecting cost-cutting decisions of architect Giangiacomo Sanguinetto, who also served as auditor of the hospital of (Saints) Giacomo and Lazzaro alla Tomba. That institution's patrimony was to support the lazaretto's construction.[29] Whatever structures actually existed, they played only a supporting role during the pestilence of the 1570s. The very scale of mortality in Verona's 1575–1576 plague may have spurred new attention to San Pancrazio. Though we have no documentation of their motives, the Veronese returned in the 1590s to execute the expansive original plan, or some semblance of it, finishing the massive enclosure by 1628.[30] Just in time, or so one would have supposed.

Any optimism that remote, improved lazaretto facilities might actually mitigate human losses from devastating plague was demolished with the 1630 epidemic. The running census that Verona physician Francesco Pona provided in his 1631 narrative of the epidemic, *Il Gran Contagio*, puts Verona death rates at Black Death levels. Pona's description is laced with lazaretto horrors, conveying all the characteristic frights that we find in similar early modern plague histories. Unfortunately, he did not differentiate deaths in the pesthouse from deaths in the city, choosing to emphasise instead the sufferings that his fellow (privileged) citizens endured along their fateful last journey: stumbling out of their houses, hovering on the squalid piers, scrunched into the boats along with the dead and the dying, touching persons of very different social station. Pona emphasised the singular plight of those unaccustomed to confinement, to bodily indignities and the rough care by servants, having to eat unfamiliar foods, listening to screams and moans both day and night. Understandably, he argued, some despairing citizens took their own lives by breaching the guarded areas; others died because they were dragged away from their houses in night-clothes, or because their bare feet touched infected boats. Deaths in the lazaretto became 'like leaves falling in a great forest'.[31] All of this came from hearsay, research in city documents, and his imagination, for Pona was not an eyewitness to any events at San Pancrazio. He elected household isolation early on.[32]

The finished lazaretto had only 152 patient rooms or cells, surrounding an area divided into four rhomboid sectors intended to separate the plague-ill from merely suspected plague cases, from contacts and from convalescents. During the peak weeks of the plague, the complex held over 4000 persons at any one time. Clearly the built fabric was insufficient to the scale of the emergency. Sailcloth tent shelters were used in the open space of each quadrilateral area of the enclosure. The diagonal design of the dividing walls inside the compound was intended to allow each patient from his/her cell to participate in the Mass at the central chapel, while preventing the occupant of a room to see his or her neighbour, or to look into another sector.

Pona's treatise was similar in purpose to Alessandro Tadino's in Milan: to debrief survivors of an unmitigated catastrophe by narrating what happened, what went wrong.[33] Pona claimed that two-thirds of those stricken by 'contagion' died. He also named more than a dozen physicians who died within hours to days of falling ill. Thus, the evidence for contagion was clear. But was the inflexible approach of the Venetian general sent to impose order wise? Pona rocked no political boats, so to speak, in his assessment: the general arrived when order in the city had collapsed, thus his iron-fisted approach was justified. He denied the wealthy the choice to remain in their houses should they fall ill. No assistance or support would be forthcoming if they refused to go to the lazaretto; they would die alone. Orchestrated convoys of boatmen ferried the ill, the well, and the already deceased downriver. Pona provided nauseating detail of the administrators' attempts to deal with long-unburied cadavers. When gravediggers and boatmen could no longer be found, in desperation they consigned the dead to the Adige River.[34]

In many other cities and towns, human losses during the plague of 1629–1633 were stunningly high, though few at the level that Verona suffered.[35] The pre-plague census taken in 1627 numbered 53,285 citizens; only 20,630 were counted in January 1631. In other words, assuming that those who fled the city early on had returned and were therefore included in the second census, Verona had lost around 60 percent of its resident population despite having a fine and fully functioning isolation hospital facility situated at a safe distance from the city.[36] It is difficult to imagine that Verona's 1630 plague could have been even worse had it lacked this available, well-planned lazaretto.

Our uncertainty about either the urban population actually at risk during the epidemic's peak or the number who died at San Pancrazio makes it difficult to calculate the risk of dying in a pesthouse versus the risk of dying if one remained in the city. At a distance of centuries, the ability to relocate and confine many thousands of plague-stricken citizens does not seem to have mitigated Verona's mortality effectively, whatever psychological or managerial benefits the grand structure may have delivered. By 1600 premodern administrators never questioned the premise that plague was contagious and could be controlled by rigorous separation of all ill, suspect or convalescent victims from those who were still well. To this day early modern plague testimonials such as Pona's are often accepted as empirical evidence that plague behaved like a contagious infectious disease in the past. These testimonials are better understood with the scepticism that historians routinely apply when working with documents.

San Pancrazio, having failed its first and only test as an effective plague containment facility, for two centuries stood guard at Venice's southwestern frontier. As was true of plague fortresses elsewhere in northern Italy, for more than a century it functioned intermittently to garrison troops or as an isolation facility during non-plague epidemics. It also performed as a customs and/or disinfection station.[37] None of these old contagion warehouses remained serviceable as Italy modernised. Poor people with little entitlement to city resources typically inherited them.

Given how long San Pancrazio survived intact, images of this Verona bulwark are exceedingly rare. A bit of it can be seen in the background of an etching by Joseph de Montalegre, found in botanist-physician Johan Christoph Volckamer's 1714 *Continuation of the Nuremberg Hesperides.* We glimpse the cupola of Sanmicheli's central chapel peeking above the fortress walls, un-altered before the mid-eighteenth century. The crenelated walls, likely not in Sanmicheli's design, mask the chimneys of each patient cell. Volckamer's popular work meanwhile chose this and other scenes around Verona as backdrop to his celebration of exotic citruses (Fig. 4.1). In this plate an enormous *cedrato di Fiorenza*—produced by a fragrant tree, shown at scale in the bottom right corner—dominates the image foreground. The tree's bitter blood-orange fruit purges the landscape of its lugubrious plague legacy.

## Lazaretto Landscapes, Plagued Terrains

Lori Jones first used the term 'plaguescape' to add temporal and disease specificity to millennia-old ideas about the relationship between pestilence and contaminated airs, waters and places.[38] As she shows, authors of plague treatises steadily refined their discussions of environments, human frailty, and the remediable causes of plague-causing environments, all while implementing a series of plague controls based on the less traditional

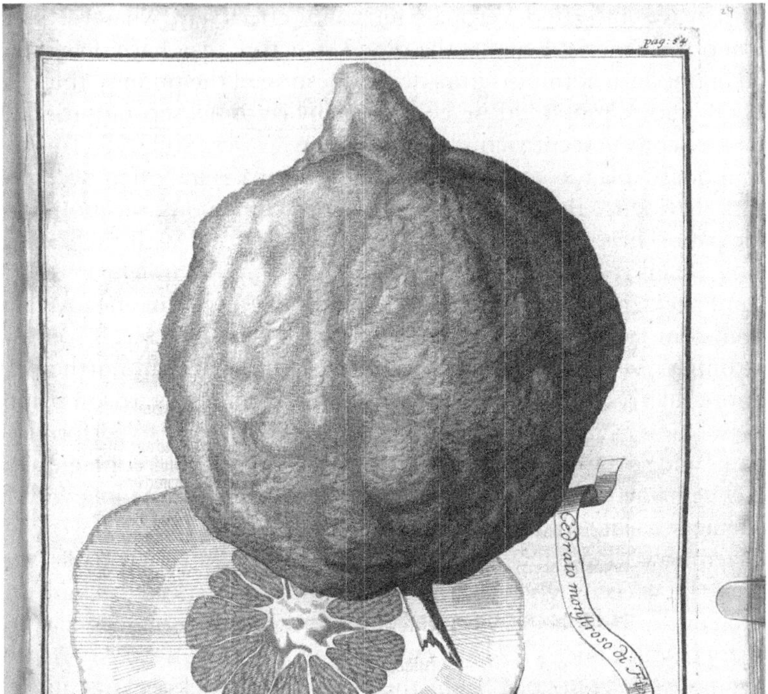


**Fig. 4.1** Johann Christoph Volckamer: 'Bitter Orange and The Lazaretto of Verona, 2 Miles from the City', 1714

notion of 'contagion'. Their discussions pertained to ground-level plaguescapes and what could be sensed directly, thus differ from the aerial perspective invoked here.[39] My reprise of Jones's vivid term offers a rather different imaginary of plaguescapes, seen from above to pull our attention to the wider geographical scale of lazaretto settings and the metastatic dispersal of plague-control activities.[40]

Widely used visual representations of lazarettos hone our gaze to human-filled clinical spaces, as well as situate individual lazarettos within a built urban fabric. Little comparable attention turns to rural plague-control spaces. Building plans reveal the size and functions of hospital rooms, thus can show how indoor lazaretto areas performed as medicalised spaces. Jane Stevens Crawshaw and John Henderson both provide haunting analysis of health office staffing and extramural plague workers, particularly in the 1630 plagues of Venice and Florence, respectively. Both offer extended and comparative examination of architectural plans, translations of survivor memoirs, and glimpses at patient graffiti etched into the hospital's walls.[41] Both show how great commissioned paintings that feature the direct interventions of ecclesiastical figures can reveal precious details of wards and alcoves, including the devotional art at each bedside, fundamental to early modern clinical spaces. Attention to the artistic and material culture of pesthouse experience thus allows us to understand better traditionally used eyewitness documents and narratives.

Nelli-Elena Vanzan Marchini's work on Venetian-inspired lazarettos instead compares architectural sketches of the overall building design to understand the lazaretto as stand-alone structures that imposed trade protocols for centuries, in a specialised Mediterranean-wide mode of plague control.[42] These maritime lazarettos were initially built in the midst of recurring plagues, when control of persons deemed contagious dominated architectural and spatial planning, and so have strong commonalities in the ways they locate kitchens and staff housing. Later-built island lazarettos served mainly as quarantine stations, and thus assigned larger facilities for the disinfection of merchandise, retaining the spaces where grain and other foodstuffs were stored, or where those who died in confinement were buried.[43]

In other words, much that is written about lazarettos and quarantine stations addresses space-planning and the experiences of persons detained or serving with walls or on islands. How, then, does a view from above add to study of these spaces? We first see that contagion-based plague controls of the early modern era sprawled over far more terrain than one might conjure when focused on intramural spaces of isolation and quarantine, even with the Venetian island version of a lazarettos. Walls shrink in height, doorways disappear as barriers, docks become tiny hubs where flotillas of small boats deposit or collect plague victims.[44] Aerial photographs of Venice's *Lazzaretto Nuovo* illustrate the island's spatial magnitude. This island's six-hectare footprint featured a main building, the sixteenth-century *Tezon Grande*, where the still-healthy contacts of plague victims were sent to perform prophylactic quarantine.[45] Disinfection of merchandise from overseas took place elsewhere on the island. By the late 1500s, many spots in and around Venice were used by the city's plague personnel, all of which we should see as plague-involved environments. Thus, the building itself and the other designated island for use in plagues, the *Lazzaretto Vecchio*, were only two component parts of an evolving surveillance-containment system deployed across the city and its extramural periphery.

Venice's plague defences in the terraferma similarly envisioned generous extraurban space would be involved in plague control. The Verona lazaretto area enclosed 4.08 hectares, which did not include the banks of the river-as-moat that surrounded it on three sides, nor the separate cemetery and specialised disinfection areas outside the walls.[46] Plague-related activities in Verona simultaneously involved other areas in the city and suburban zone. In the 1576 plague Verona still used temporary wooden huts as additional isolation facilities, occupying the no-man's-land (the *spianata*) that Venice created in the 1510s.[47] By 1630 that same suburban zone served as a staging ground for transport operations, housing for petty officials, and the isolation of some well contacts of prominent plague victims, and so we still must include these areas as plague-relevant landscapes where new exposures to rodent fleas could occur.[48]

The enclosed area of Padua's lazaretto was smaller than Verona's, covering around 1.4 hectares; each long exterior façade measured 120 meters. From the 1510s to 1533, the decision was taken to replace Padua's earlier structures for isolating plague patients, a timetable similar to Verona's planning, and here, too the actual building began later, in 1555.[49] The main lazaretto was only one component of a local, geographically distributive control of plague, which involved the various staging areas for disinfection of goods, isolation of convalescents, housing for support staff, practices based on the revolutionary 1576 plague treatise of Giovanni Filippo Ingrassia, a manual widely used by the seventeenth century.[50]

At the dawn of a profound late sixteenth-century shift in plague management practices, Padua's pesthouse in 1576 offered more available patient cells than did Milan's model lazaretto (discussed below), even as enterprising farmers nevertheless created additional villages of huts for the use of city folk fleeing plague. Bergamo's new lazaretto, decided upon in 1503, begun in 1504, expanded modestly in 1576, then finally enclosed by 1581, was a 1.628-hectare compound, with many other linked units of extra-urban plague control 'immersed in the open countryside'.[51] Brescia's lazaretto took its name from a nearby chapel of San Bartolomeo *in clausuris*, a reference its agricultural surroundings and legal status, not to a walled or fenced enclosure.[52] Temporary huts were replaced by a permanent hospital structure in the late fifteenth century, with walled space and elegant porticos, a chaplain's apartment with his private orchard and vegetable garden, an enlarged church and a vast walled cemetery stretching before it, much like Padua's new lazaretto. Brescia's governors also commandeered other houses for plague control, spread across an area with radius of several kilometres.[53] Surrounding walls or bends in rivers were trusted to contain plague and to impair escape of plague suspects as well. The division of unbuilt land into differently used spaces for plague control reflected the newer objectives in social and public health management of epidemic crises: landscape segmentation. Significantly, such approach translated contagion ideas about plague control to reconfiguration of spaces often used for production, whether agricultural or animal husbandry.

In an interesting inversion of the theme of 'inside' and 'outside' spaces, Susan Einbinder's study of Jewish literature produced during the 1630 plague examines the especially informative account of Abraham Catalano. Padua's Ghetto was already segregated when the plague arrived, but not all Jews lived in it. Catalano, for example, lived in the city and was well-connected to businessmen and administrators, but in the plague was forced to move inside the Ghetto. He imposed order, found a protected outside warehouse space to prevent goods from their destruction as 'infected', and secured use of a property to serve as the community's lazaretto outside the Ghetto. The community supplied food, medicine, gravediggers, and transport personnel to their own, separate pesthouse. In this case Catalano kept two different daily death lists, one for the Ghetto, one for the lazaretto, naming each victim.[54] But the divisions made little difference. Only 72 of 721 members of the community remained well during the epidemic.

Similarly, among Christian communities, transporting persons to vacant, private structures and monastery out-buildings severed many victims and suspects from their familiar surroundings, a disorienting experience even for the well. Women and children were quite often separated from the men of their household, further depriving them of accustomed social identities and status. Relocation was as important to early modern administrators as was isolation and other separations. Clustered by criteria other than their pre-existing ties to family and neighbourhood, some plague-touched residents were moved to and through these outdoor spaces in ways that we do not typically associate with care of the gravely ill. For example, Florence's *Sanità* undertook logistically complicated, costly removal of any poor person tainted by the merest whiff of plague. Carts laden with ill and suspects, lumbered up the surrounding hills where they were sorted by gender and age into different fortresses, monasteries, churches.[55] Each location required continual transports of food, medicine, personnel, and all needs for patients' bedding. Lucca selected isolated farmsteads only for the city's victims; every outlying village and hamlet was to use a house or cabin, if possible near running water, exposed to the north wind, and not close to any inhabited area, all on the assumption that rural areas were less deadly.[56]

Like Florence and Lucca, most Italian cities and towns did not invest in permanent, purpose-built lazarettos. During plagues they instead rented or appropriated little-used and abandoned extramural buildings, spaces filled with evacuees once a plague took hold. Many designated lazaretto areas of seventeenth-century Italy thus did not rely on walls, towers, and moats to control movement into and within these spaces.[57] They were wild or semi-wild areas at the city's agricultural periphery, usually situated near natural water boundaries, open cemetery areas or untended land. Some even used palisade fencing and guards to segregate those who had touched plague from those presumed well. In common with internees' experience inside the large, built-enclosure lazarettos and barrier islands, few of these more open pesthouse areas would have been familiar spaces to the persons removed from their urban parish neighbourhoods. The uncertainties of plague turned city dwellers into temporary emigrants, disorientated to place, living among ill and unfamiliar persons.[58]

Our gaze retrained to the broad geographical footprint of plague control, it is useful to return to Milan's great inland pesthouse with detail, because we have some more direct evidence that persons could contract plague once confined. The complex is as iconic as Venice's island lazarettos in premodern plague history, largely because Alessandro Manzoni featured it in his great 1820s novel, *I Promessi Sposi* (*The Betrothed*). The history of *San Gregorio*, as it was called locally, offers a comparable chronology to the Venetian saga. In the fifteenth century, lazaretto pre-history involved the legalities and logistics of attempts to find ample extramural land, not merely abandoned buildings, for use in plague control. In 1448–1449, Milanese oligarchs re-purposed a wooded 10.2-hectare ducal villa in Cusago initially hoping to clear the streets of hungry beggars, and thus avert the environmental alteration believed to cause plague.[59] But plague cases erupted faster than homeless refugees could be removed. Ferrying the ill down the city's Great Canal was furthermore difficult to orchestrate and many died en route. The Milanese thus resorted to temporary use of a wide array pre-existing peri-urban structures and spaces, while implementing rudimentary attempts to separate the ill, the suspect, the convalescents and the dead. After the 1450–1452 plague, the successor Sforza dynasty (1450–1535) reclaimed the Cusago villa as a hunting park and the city no longer had a future option to use that remote refuge.

During the next plague of 1468–1469, administrators of the new Sforza public hospital, the *Ospedale Maggiore*, still under construction, debated the cost and benefits of an additional Venice-like lazaretto complex. The original plan offered by notary Lazzaro Cairati called for a massive, water-rich 26-hectare location at Cresenzago, over ten kilometres to the north and east of the city, was rejected because its use would compromise lucrative and privately owned farmland. It was furthermore difficult to reach because stretches of the intended transport canal were not yet fully navigable. The eventual site selected for Milan's historically famous lazaretto was much closer to the city, at the old church and cemetery of San Gregorio, and had a considerably smaller footprint, at around 14 hectares.[60]

Count Galeotto Bevilacqua's will in 1486 made his bequest of property near San Gregorio conditional on the *Ospedale Maggiore's* use of this particular suburban zone. He also did not allow hospital administrators to delay: building had to be underway within two years after his death, else the land reverted to his surviving heirs (it is quite possible that not even the absolutely devastating plague cycle of 1483–1486 would have convinced wealthy aristocrats of its need). And so, the lazaretto's cornerstone was laid in 1488, but its actual construction never really got started until after 1499, when the French were in charge. The facility was partially in use by late 1512 or 1513, but its first true test came with the catastrophic plague in 1524, dubbed the 'plague of Charles V'. Haunting accounts of that outbreak nevertheless do not even mention the lazaretto, which was, after all, only one of several areas receiving those who were removed from the city. Each governing sector of the city secured its own suburban isolation areas, sometimes to keep families together and in clusters with others from their parish churches, sometimes separating the well from the sick.[61] Many of these areas were little more than open-air campgrounds with temporary isolation huts, places where families within the same quarter of the city could undergo isolation together.

By the 1576 plague, called the 'plague of St Charles [Borromeo]', separation of the ill from those who were suspect or who had recovered was standard practice, and to some limited extent distinguished clean and contaminated personnel. San Gregorio does not appear to have been fully overwhelmed, though we do not actually know whether cabins or tents were erected in the central 13-hectare 'prairie' with the hospital's walls, as happened later. As was the case in 1524, temporary cabins and campgrounds for the still-well and the recovering also sprawled outside four of the city's six main gates, and at least two other extramural lazarettos served the southern and western suburban periphery.[62]

Complacency and confidence may have set in after 1575–1577—their experience was different than Verona's in this particular plague wave. Prepared or not, Milan was, like Verona, crippled by the great plague of 1630 and also had a large and competent lazaretto. The prior year had funnelled massive numbers of upland refugees towards the city, hungry persons seeking relief from dire famine as well as safety to those terrified by advancing German and French troops. The health office cleared the streets of these migrants, taking them to San Gregorio. Conditions worsened considerably inside the pesthouse when plague arrived. Those interned in San Gregorio alone reached 12,000–15,000 during the 2–3 peak weeks, and the facility was used only to isolate the ill. The enclosed pesthouse became comparable to modern-era refugee camps where basic services collapse under the massive needs of daily-arriving destitute migrants.

Rare documentation from inside this lazaretto complex during Milan's catastrophic 1630 plague further allows us some evidence about the ecological conditions inside pesthouse walls. Writing in 1646 an anonymous Capuchin memoirist described in vivid detail the heroic deeds of friars inside the lazaretto. He claimed that just three of his fellow monks survived, of the 20 who went in, and he offered uncommon detail to illustrate their sacrifices: 'the poor friars gave up their own [blankets and straw mattresses] to minister to the sick poor. When necessary they slept on the bare ground, at other times on raised grates, using them because of the humidity of the rooms and the abundance of rats, fleas and other troublesome animals.'[63] He would not have needed to point out to anyone who then lived in Milan that the terrain inside the pesthouse was rich in numerous freshwater springs. Usually a life-supporting feature of the lazaretto, the interior meadow (*prato*) also turned into a lake when a torrential downpour began on the night of 23 July 1630. Devastating loss of life followed, even before plague deaths skyrocketed during the early autumn months. About this incident Giuseppe Ripamonti wrote that Father Felice Casati rushed to rescue drowning plague victims: 'like a fishermen pulling the little fish out of the net, … [he] untangled the children and passed them to helpers, transferring them from hand to hand from the [flooded] meadow to the portico [of the built lazaretto] and from there to rooms.'[64]

These two environment-centred stories provide circumstantial evidence to my argument, that even walled pesthouses could support ongoing *Y. pestis* infection, areas humans would have shared with rodents and fleas. But we have some surviving physical evidence that the pathogen was also present. A recent study of the protein residues left on the pages of death registers that were kept inside Milan's large lazaretto during the 1630 plague found rat dermal tissue, showing that rats scampered over paper registers when the clerks were not around. Moreover, the investigators also found presumptive evidence of *Y. pestis* surface proteins in their analysis of residual plant, animal and human proteins, further material evidence of the pathogen's likely presence inside the lazaretto.[65] Just as would have occurred in more rural confinement areas, in suburban pesthouses, temporary occupation entailed human immersion into terrain co-occupied by field rodents and their fleas.

## Painting Pesthouses and Plaguescapes

Some pesthouse images painted between the 1570s and the 1650s include background scenes of the expansive peri-urban terrain given over to plague control. Wider, not-quite-overhead cityscapes were then popular, and in some of these we glimpse confinement zones with the thicket of tents and huts that they contained. In others outdoor, still largely urban, plague scenes take place at street level, as in artistic depictions of Milan's

1576–1577 plague. The campaign to sanctify Archbishop Charles Borromeo (d. 1586) involved commissioned representations of his life, including direct acts penitence and charity during a dangerous epidemic. Twenty years after Borromeo's death, Gian Battista Crespi (called 'il Cerano') in 1602 made a cycle of 20 very large paintings, each to be stretched between pillars inside Milan's cathedral. All four panels that dealt directly with Borromeo's interventions in the plague of 1576–1577 show events that occurred outdoors. Two of them illustrate tents, huts, the burial of corpses, another provides a glimpse into the lazaretto itself from the city-side entry into the complex.[66]

The bottom third of a very different kind of public art, a large votive painting made in Milan during the 1630 plague, instead depicts tents inside the San Gregorio lazaretto. Those actually at street level would have seen this part of the painting from an oblique overhead perspective, seen from a location south and slightly east of the city centre. A Madonna of Mercy dominates the top two-thirds of the image with plague saints Roch, Sebastian and Charles Borromeo and the artist himself, prior Bernardo Catoni, under her sheltering cape. At this outdoor, inner-city shrine the devout would first be immersed in a view of plague victims' last passage, because that very different representation occupied the entire bottom third of the painting. This section of the painting is shown in Fig. 4.2.

The movement of persons is read from the right (city) side of the vignette to the left, beginning with bodies on carts as they crossed a bridge over the canal encircling the city walls, and then over the smaller bridge into the lazaretto's entrance. A three-meter-wide moat supposedly offered

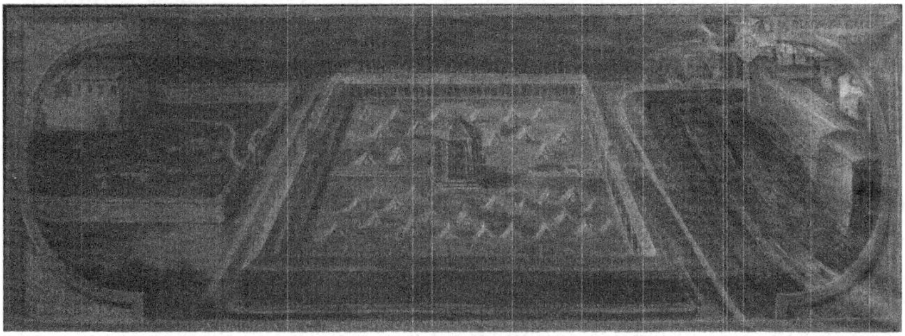

**Fig. 4.2** 'The Lazaretto of Milan': original by Prior Bernardo Catoni, 1630; repainted in 1890 by Giovan Battista Rastellini

a further measure of security, although through the lens I imagine, we can further see a supportive habitat for rodents. In the central part of the view the lazaretto's exterior walls are not re-scaled their imposing height and breadth, which actually framed a two-story monastery-like enclosure surrounding the preserve. Indeed they would appear diminished in a view from above. But the entrances and exits are given prominence in this adjusted overhead view, likely because they were meaningful liminal spaces to surviving parishioners. Many would further find a sort of map, how to walk to the great cemetery where their loved ones were buried.

The structure proper had only 288 rooms, not all of which served as patient cells. Here, however, the viewed area inside the lazaretto is shown filled with an orderly array of tents, each with a tiny occupant. Meanwhile the image shows nothing of the actual, four-quadrant segmentation that parsed evacuees into different areas within the campground—infected from the recovering who were not yet well enough to move to different convalescent areas in the countryside; men from women and children. Running waterways and palisade fencing, patrolled fiercely by some of the clerical staff, were also dominant aspects of the lived experience in San Gregorio that the images do not capture.[67] Through the lazaretto's backdoor, away from the city and on the left side of the image, we can just see where the dead were transported under an arch into what the Milanese called the *foppone*—a great public cemetery—situated around the old church of San Gregorio.[68] Close up, the image's cemetery portion shows other small human figures, some the dead being buried, some still-living gravediggers.

The image reproduced here is an 1890 copy by Giovan Battista Rastellini, done from prior Bernardo Catoni's 1630–1631 original. Called the 'Madonna di Tencitt',[69] Rastellini's image was made at a very historical different watershed. Residents of the parish in 1890 could not have walked along their ancestors' route to the pesthouse because Milan's ongoing modernisation program had dismantled the lazaretto. Furthermore, significant hydrological renovations in the northeast sector had effectively erased the image's orienting structures.[70] The original ex-voto painting once reassured a working-class neighbourhood of *scaricatori* or dock unloaders that plague-stricken friends and neighbours would stay together in life and in death, that the pesthouse portals were patent and accessible. The copy's purpose was instead to preserve icons of popular piety at the neighbourhood level as its own face and purpose changed rapidly.[71] Thus the copy was a comforting fiction. By the 1890s the lazaretto had already been dismantled and the old canal system mostly covered over, renovations that partly suppressed the neighbourhood, its livelihoods, its local language.

In both the Crespi plaguescapes and Catoni's view of Milan's lazaretto our gaze fixes on entrance and exit points, but passageways into and out of plague-control areas were not ubiquitous details in paintings that include lazarettos. For example, Carlo Coppola's 'The Pestilence of 1656 in Naples' offers a striking composite view of a borderless pesthouse, borderless in this case because any possible enclosing structures lie out of the frame. We see up close Coppola's horrifying array of cadavers being handled by still-living plague workers in an undifferentiated, unbuilt landscape. It is a scene with no clear boundaries, offering the viewer no safe distance, no separation. We are inside a space of undetermined expanse, where body-handlers (also called body-clearers) wear red vest-like garments over their clothed torsos; the gravediggers are shirtless; the dead tumble into the foreground. Coppola's disturbing image stands in dramatic contrast to another contemporary painting of this particular and consequential plague, Domenico Gargiulo's 'Largo del Mercatello During the Plague in Naples'. Gargiuolo allows us to peer into an artificially enlarged city piazza, where various plague scenes are represented simultaneously.[72]

Some contemporary seventeenth-century paintings depict more discernibly open lazaretto spaces. The young Flemish artist Anthony Van Dyck was present during the horrific 1624 plague in Palermo, captivated by a legend important to this plague. He made two versions of his 'Saint Rosalie, Interceding for the Plague-Stricken of Palermo' before he returned home in late summer 1625, both featuring the teenage saint as she gestured to a sprawling extramural lazaretto area near the harbour. The location had salience for plague control.[73] From an anonymous memoir we know that the sea was situated an 'arquebus-length' away from the pesthouse area, and that the city's Senate commandeered seaside huts for use in a final cleansing of patients who had recovered and completed a lengthy quarantine. Survivors were required to bathe in the sea before they could be issued new clothing and allowed to return home.[74] The entire area was rather like a lazaretto without walls. The enclosed areas were instead the city and its aristocratic abodes.

The memoirist's testimony captures another sense of working within an expansive plague-control terrain. The explosive rise in plague cases during the late spring, 1624, overwhelmed existing facilities, compromising the government's ability to confine suspect and convalescent cases within designated isolation areas and structures. Palermo administrators thus converted an entire suburban zone contiguous with the sea into an isolation zone.[75] By late July 1624, the city Senate had to recruit a new chief physician, hiring a Milanese friar who boasted prior plague experience. He negotiated the use of a 'small animal', probably a donkey, to reach all who needed him.[76] Even so, his activities were limited to the ill, rather than to all those dislodged from their neighbourhoods (i.e. contacts, suspects, convalescents, and the post-convalescent in yet another quarantine). These other evacuees occupied various areas in the extramural suburbs, tended or guarded by other health office subalterns.

Even in very small communities, the land surrendered to plague service is vaster that one might imagine. When thinking of temporary pesthouse facilities, at a minimum we should include all spaces involved in the plague-related functions that early modern towns and cities required. A few places that served as these lazarettos without walls can still be seen in the Italian countryside, such as the '*campo della peste*' marked with a cross in San Giorgio Lomellina, or that in Botticino Mattia, near Brescia.[77] Many more such sites once existed.

For example, Chieri's experience during the 1629–1631 plague was dire enough to prompt annual commemoration for over 200 years.[78] This small town, 18 kilometres southeast of Turin, was initially a plague refuge for the larger city's aristocrats but was itself overwhelmed during the summer months of 1630. Nineteenth-century historian Giachino Montù searched public and private archives to find any records related to this annually remembered plague in Chieri, as buildings, grave markers and ecclesiastical records were all fast disappearing. Through surviving records of plague deaths that he pieced together, we glimpse the expanse of this plague-relevant zone. Because many records had been deliberately destroyed in the wake of the Napoleonic wars, especially with the suppression of monasteries in 1802, Montù tracked down the minutiae, noting the names of prominent Chieri plague victims in death registers, the former and current names of dilapidated or suppressed monastic properties, the private orchards and other surrounding structures where each had died or been buried.[79] The records that interested Montù were fragmented in content and not held in a centralised location, thus necessitating his trek through some of the tiny confinement areas sprinkled across Italy's Piedmont. For example, there was the farm in Vallèro that the town rented as its pesthouse: this structure lay close to a Capuchin monastery where friars were willing to provide care and comfort to the ill.[80] Chieri also used chapels outside various city gates, other farm properties, vineyards, and meadows, all situated at a suitable distance from the town. Some properties were used for 'suspect' cases, other specific localities were named in the surviving records because they had been deemed safe from peril. Later remembrances, in other words, recorded the precise locations of deaths, burials or where a testament was made. Plaques posted on doors survived generations, identifying where the occupants had died in the 1630 plague. These were some of the many measures used to connect families over time, because those dead from plague were not permitted burial in parish church cemeteries.[81]

## Writing Contagion into Plague History

By the later 1500s the words *contagio* and *peste* were used interchangeably. An unshakeable belief in the contact-borne spread of plague governed Italian control policies by the later sixteenth century and justified the elaboration of isolation practices and boundary-establishing protocols. Dedicated attention to the minutiae of separations, disinfection, and quarantine procedures occupied administrators after the first cases of *morbi contagiosi* were found anywhere within a city or its suburbs. Plague management began by categorising persons according to their degree of contact with putative plague victims or connections to known plague hotspots. During the early stages when a town or city struggled with the first recognised cases of a plague outbreak, as each new victim was identified, the house itself, the family, the servants, and most of furnishings were subjected to a bewildering array of containment and purification procedures. Regulations not only specified which persons were allowed to be out-and-about during the epidemic, but also involved increasingly rigid, ritualistic separations within categories of city workers, whether public health staff or the small armies of persons managing animals and transport vehicles, such as wagons and carts and enclosed sedan chairs. Lengthy quarantines were summarily imposed on those who came from regions where plague reigned as well as on those who had merely casual contact with someone later diagnosed with plague. Civic authorities also quarantined for weeks to months any who recovered from a plague-like illness, a policy that related more to fear than to reason or general contagion-based rationale.

Samuel K. Cohn, Jr., in his 2010 *Cultures of Plague*, convincingly documented the appearance a new kind of plague treatise or genre, structured by a conviction of plague's contagious nature.[82] He argued that histories and medical treatises written during and after a peninsula-wide wave of plague outbreaks in the 1570s wrote about plague epidemics in a new, narrative way, breaking with a centuries-old approach in plague treatises.[83] In a tripartite fashion the latter summarised the remote, or celestial causes of plagues, the proximate (terrestrial) causes, and a final section devoted to prevention and cures. Treatises composed after this watershed decade were organised chronologically, providing evidence about geographically discrete outbreaks and the authors' personal experience in a plague. Providing enhanced coverage of places near the intended audience, a storied sequence tended to convince readers that humans spread plague as they travelled from one place to the next. These new-style plague treatises, mostly written after the threat had subsided, at the same time offered selective accounts of plague's impact on a particular city or region, bound into a tight and authoritative narrative that excluded rumours or evidence that did not fit neatly within it. Authors of these new-style treatises also collected local archival evidence and survivor stories, providing authenticity to the account. These treatises further exonerated city authorities of mismanagement, and often vilified subalterns and medical authorities who thwarted right-minded attempts to interrupt the epidemic's spread.

This early modern genre of plague literature typically placed the epidemic's origins well outside the terrain or social groups most familiar to the intended audience. Each writer steered the story towards local landmarks and traditions, noting the plague's effects on people and places of primary concern, lingering on details of the city's early contagion-based interventions; most of these led to a briefly happy reprieve. However, after weeks to months of contained, aggressively managed individual cases, plague commonly escaped official control. At the midpoint of the drama and the reader's city descended into horrors and cruelties of all sorts, now viewed at a temporally safe distance. The plague's worst weeks permitted readers' immersion into lurid local particulars, as the drama unfolded. With the first hints of winter the epidemic's grip relaxed, generating hope and celebrations if not the play's denouement. Plague sometimes rekindled when trade resumed and fields thawed, befitting literary tragedy. Identifying the week or day on which no new cases were discovered opened the way to the plague's fitting finale: proper and permanent rituals to thank God and the saints for deliverance, with unalloyed praise for the actions of heroic local leaders, even when the outcome was disastrous.

In none of the plague narratives that Cohn assembled and analysed did the authors ever question plague's contagious nature. We, now centuries later, can no longer accept their premises, assumptions and manipulation of evidence as proof that early modern plagues actually did spread along person-to-person pathways. In other words, contagion is a plague trope. Storied plague itineraries often fail to include evidence from localities and areas along plague's supposed path that did not experience an outbreak at all, or that suffered plague later, after most of their neighbours were in recovery.[84] Modern historians who have worked with archival documentation alongside these official narratives can demonstrate that during the crisis itself, before some stories and rumours achieved quasi-factual status, local officials typically struggled to determine the directional flow of possible threats.

Aggressive contagion-based policies ultimately failed to avert an epidemic altogether, else there would have been no drama to tell.[85] After plague receded, early cases of plague were an oracle's warning, prelude to a local catastrophe. Literate survivors learned to attribute the calamity to breaches of regulations, carelessness in the performance of purification protocols, medical equivocation, or a popular that denied early cases were instances of 'true' plague. The deliberate spread of plague-causing substances, an extreme view of plague's contagion, generated rioting, arrests and gruesome public executions, to an extent not seen since the Black Death. In a subsequent work Cohn shows how this kind of plague-specific predicament, a reaction to the evident failure of many expensive policy choices, led to equally typical eruptions of violence that has not been typical of comparable non-plague epidemic outbreaks.[86]

One of the most elusive subsets of available historical data concerns the causes of deaths in lazarettos. Even the overall mortality among persons sent to the pesthouses is difficult to estimate. Stevens Crawshaw's analysis of lazaretto-associated deaths in Venice here, too, is quite valuable, because to some extent her extensive archival research dispels any notion that lazarettos could contain plague cases. Could removal of the ill have provided overall public health benefits to those allowed to remain in the city? Stevens Crawshaw interestingly suggests, providing evidence to the point, that refinements to surveillance, isolation, and separation protocols likely reduced the risk of death in lazarettos during Venice's 1630 plague, but not in the city and its suburban zones. No more than 40 percent of recorded deaths in Venice's devastating plague of 1575–1577 occurred in the lazarettos. In the 1630–1631 Venetian plague pesthouse deaths comprised an even lower proportion of the overall mortality, achieved, she argues, through more rigorous separation of the well and the ill.[87] John Henderson's evidence from Florence tells an opposite story of lazaretto mortality, thus very high deaths in these spaces, and an overall lower mortality for Florence (around 12 percent), compared to other Italian cities.

In other published literature, I have thus far found only one similarly detailed comparison of pesthouse vs. urban deaths in a great plague that Stevens Crawshaw did not include, that for Palermo in 1624–1625. A cleric residing in the lazaretto copied the city's official daily death tallies from June 1624 through June 1625, as part of his reports to his superiors in Rome. Overall, he claimed that 6088 persons died in Palermo's lazaretto areas, while 6298 died in the city. Seductively thorough, the evidence at the same time frustrates any hope for definitive evidence. From reported daily numbers the persons sent to the lazaretto totalled 6015, which means that 73 fewer persons were sent to these areas than died there. City authorities furthermore made no count of those who recovered, nor did they note how many remained well after being confined as suspects or contacts.[88] The pestilential degradation of lazaretto landscapes in the premodern world escapes notice until we look (metaphorically) from above, overhead.[89]

## Conclusion

With new visualisation technologies, new problems emerge that were before unseen.[90] The interlinked assumptions we have long made about what caused very high, supposedly contagion-driven plagues of the 'Second Pandemic' now require new analysis. What we are able to, or choose to see, then depict or photograph or describe in words, profoundly shapes our understanding of historical plagues.

A view of plague as contagious emerged during the era of recurrent plagues in Italy, forming a significant component of premodern experience and responses to great epidemic crises. However, seeing plague as a human-spread contagious disease leaves us less able to interrogate the spatial-temporal distances between the local and the regional, the regional and the global, which now matter to analysis of the wider ecological and abiotic conditions that supported plague spread and recurrence over time. The horizons and frames that plague historians have long used are oriented to the activities of persons in a theatrical series of vignettes constrain the boundaries of a broader environmental gaze—we zero in on the stricken city's streets; inside the pesthouse; at a patient's bedside; in the graveyard, and so forth.

We must entertain doubts about the supposedly robust evidence for a contagion model of plague's temporal-geographical spread, because the retrieval and identification of *Y. pestis* in mass burial sites from the sixteenth through the eighteenth century confirms persistence of the organism that causes modern bubonic plague in Western Europe. Moreover, the modern pathogen is a direct descendant of that found in human remains of the Black Death era.[91] We thus cannot presume that *Y. pestis* infection led to a clinically or ecologically different disease half a millennium ago.[92] In other words, we have two broad but separate phenomena to which available historical evidence can speak, phenomena that have heretofore been conflated with the routine characterisation of plague as contagious.[93] Explaining the pathogen's historical ecology—thus its spread and diffusion geographically and over time—requires a different stance towards evidence and different methodological approaches than does analysis of human responses to catastrophic surges in human deaths in a particular time or place. My imaginary—to see pesthouses and plaguescapes from above, without respect for walls and doors, or for the separation of persons into clusters of well, ill, suspect and rehabilitating—is a claim that none of these structures or practices effectively interrupted the interannual spread of this particular pathogen or the conditions that led to outbreaks in one local versus another. These interventions invented in response to recurrent plagues were not inconsequential to human experience during crises, but in actuality did not determine urban mortality outcomes, thus whether losses were exacerbated or mitigated by the use of lazarettos.

Why were plagues before the modern, germ-theory era so costly in human lives? It seems most likely that plagues opened the way to a lethal combination of mortality/morbidity from other infectious diseases and crisis supermortality, as many in a population lost access to the basic support and care they needed to survive.[94] Deaths among young children, the elderly, and the disabled soared. For many others, recovery from bubonic plague offered no protection against protracted hunger, the hardships of confinement, new exposures to other infections, or the worsening of preexisting afflictions, such as tuberculosis. We have long understood all this about premodern pestilences.

Contagion practices developed in response to recurring plague have a clear legacy of public health successes in the control of infectious diseases that do spread person-to-person, such as the ongoing SARS-CoV2 pandemic. Lazarettos are gone, but non-pharmaceutical management of COVID-19 is clearly useful. We have no firm supportive evidence that social distancing, masking, or attempts to alter the air of urban spaces produced similar effects to brake premodern plagues. Rigorous adherence to the elaborate protocols of contagion control in the early modern era separated 'clean' from 'dirty', but did little to protect persons unable to flee to safer locales or unable to avoid routine contamination of stored food. Nothing in the armoury of plague control protected well persons, plague workers, or plague victims from repeated and inescapable exposures to waterborne pathogens.

At a distance of centuries the spread of plague by contact between persons (as in a household cluster or a closed, squalid pesthouse), is not distinguishable from infections due to a shared environmental risk (e.g. rodent colonies near a city mill or grain storage facility). We cannot determine how plague spread, whether by contagion or through contamination from a common source, from available evidence. Such issues become difficult to untangle with either the narrative accounts of the time or the archival evidence that they selectively collect. Knowing that *Y. pestis* was a recurring pathogenic presence in Europe invites us to re-evaluate some empirical claims we have long made about plague's earlier *modus operandi*.

**Acknowledgements** My students in a senior history seminar at Wofford College (Spartanburg, South Carolina, 2018) initially inspired aspects of this enquiry, and led to my efforts at the St Andrews' conference and another on 'Cities in Crisis', summer 2018 and held at the Biblioteca Herziana (Rome, Italy). I am particularly grateful to Daniele Abdon and Margaret Bell, organisers of the Rome conference; to Daniele in particular for sharing her photographs of the *Lazaretto Nuovo* and bibliography she shared on recent architectural perspectives on urban crises; to Lori Jones, Nükhet Varlik and Hendrik Poinar for advice at multiple junctures; to the anonymous referee and his/her suggestions; and above all to Christos Lynteris for redirecting my historical gaze.

## Notes

1. G. P. Marchi, 'Il lazzaretto di Verona'. In Francesco Pona, *Il gran contagio di Verona nel milleseicento, e trenta*, Photostatic edition, pp. vii-viii (Verona: Centro per la formazione professionale grafica, 1972 [1631]). Francesco Pellegrini provides lengthy quotes from the contemporary newspaper accounts in his 'Il lazzaretto di Verona', *Studi Storici Veronesi* 2, no. 2 (1949/50): 143–191.

2. M. V. Quattrina, 'La peste. La guerra. L'abbandono', with images of the crew and the camera drones used in filming: https://lazzarettovr.jimdo.com/documentario-sulla-peste-al-lazzaretto-di-verona/
3. See the blog of Alessandro Boggian, November 12, 2012, on the first public assembly held for saving the lazaretto: www.aleboggian.altervista.org/salviamo-il-lazzaretto-di-verona/. On work stop: L. Costantino, 'Raccolta di fondi per salvare il Lazzaretto', *L'Arena* (3 May 2016).
4. A. Braioni and A. Conforti Calcagni, 'Il lazzaretto di Verona, da luogo di sanità a luogo per la cultura della salute'. In *Lazzaretti europei: Da luoghi di sanità a rete di rapporti internazionali*, pp. 15–17 (Atti del Convegno, Venice, 2013). The authors argue that 18,000 cubic metres of concrete will destroy the 'incantata belezza' (enchanted beauty) of the location.
5. S. Duchêne, S. Y.W. Ho, et al. 'The Recovery, Interpretation and Use of Ancient Pathogen Genomes', *Current Biology* 30 (5 October 2020): R1215-R1231; M. A. Spyrou, K. I. Bos, et al. 'Ancient Pathogen Genomics as an Emerging Tool for Infectious Disease Research', *Nature Reviews: Genetics* 20 (2019): 323–340.
6. K. I. Bos, et al. 'A Draft Genome of *Yersinia pestis* from Victims of the Black Death', *Nature* 478 (2011): 506–510; S. Haensch, et al., 'Distinct Clones of *Yersinia pestis* Caused the Black Death', *PLoS Pathogens* 6:10 (2010): e1001134; L. Little, 'Plague Historians in Lab Coats', *Past and Present* 213 (November 2011): 267–290; M. A. Spyrou, et al. 'Historical *Y. pestis* Genomes Reveal the European Black Death as the Source of Ancient and Modern Plague Pandemics', *Cell Host & Microbe* 19 (2016): 874–881.
7. T. Castro, 'Aerial Views and Cinematism, 1898–1939'. In Mark Dorrian and Frédéric Pousin (eds.), *Seeing from Above: The Aerial View in Visual Culture*, pp. 118–33 (London: IB Tauris, 2013), pp. 120–21.
8. J. Haffner, *The View from Above: The Science of Social Space* (Cambridge, MA: MIT Press, 2013), pp. 7–33.
9. R. Jorge, ed., *Les Faunes régionales des rongeurs et des puces dans leurs rapports avec la peste: résultats de l'enquête de comité permanent de l'Office International d'Hygiène publique, 1924–1927* (Paris: Masson, 1928).
10. C. Dyer and R. Jones, *Deserted Villages Revisited: Explorations in Local and Regional History* v. 3 (Hatfield: University of Hertfordshire Press, 2010); R. Glasscock (ed.), *Historic Landscapes of Britain from the Air* (New York: Cambridge University Press, 1992), pp. 103–124.
11. Final Report, 'Public Health Applications of Remote Sensing of the Environment (An Evaluation)', Contract number NAS-9-11522, Manned Spacecraft Center, NASA, pp. 80–83, retrievable at https://ntrs.nasa.gov/search.jsp?R=19730005342. A path-breaking survey of twentieth-century Soviet plague surveillance and control makes no specific mention

of aerial photography assisting ground surveillance. See: S. D. Jones, et al., 'Living with Plague: Lessons from the Soviet Union's Antiplague System', *PNAS* 116:19 (2019): 9155–9163.

12. For important new archaeological work on the impact of the Black Death in rural England see C. Lewis, 'Disaster Recovery: New Archaeological Evidence for the Long-Term Impact of the 'Calamitous' Fourteenth Century', *Antiquity* 90, no. 351 (2016): 777–797; S. Kacki, 'Digging up the Victims of the Black Death: A Bioarchaeological Perspective on the Second Plague Pandemic'. In C. M Gerrard, P. Forlin, and P. J. Brown (eds.), *Waiting for the End of the World? New Perspectives on Natural Disasters in Medieval Europe*, pp. 232–247 (London and New York: Routledge, 2021).
13. M. Achtman, et al. 'Microevolution and History of the Plague Bacillus, *Yersinia pestis*', *Proceedings of the National Academy of Science, USA* 101 (2004): 17837–17842; Little, 'Plague Historians in Lab Coats'.
14. On the archaeogenomics related to the breakthrough study of London's East Smithfield Black Death cemetery see: K. Bos, et al. 'A Draft Genome'; Little, 'Plague Historians in Lab Coats'; M. H. Green, 'The Globalisations of Disease'. In N. Boivin, R. Crassard, and M.D. Petraglia (eds.), *Human Dispersal and Species Movement: From Prehistory to the Present*, pp. 494–520 (Cambridge: Cambridge University Press, 2017).
15. Spyrou et al., 'Historical *Y. pestis* Genomes'.
16. Two exceptions should be noted. J. L. Bolton, 'Looking for *Yersinia pestis:* Scientists, Historians and the Black Death'. In L. Clark and C. Rawcliffe (eds.), *Society in an Age of Plague*, pp. 15–38. Fifteenth Century Studies, XII (Woodbridge: Boydell Press, 2013); N. Varlık. *Plague and Empire in the Early Modern Mediterranean World: the Ottoman Experience, 1347–1600* (Cambridge, UK: Cambridge University Press, 2015).
17. Camera drones have been commercially available only since the 2010s; see B. Rao, A. Goutham Gopi, and R. Maione, 'The Societal Impact of Commercial Drones', *Technology in Society* 45 (2016): 83–90. Such devices have become fundamental to the study of archaeological sites and landscapes: S. Campana, 'Drones in Archaeology: State-of-the-Art and Future Perspectives', *Archaeological Prospection* 24 (2017): 275–296. How they will affect consumer (and student) expectations for new and interactive ways of exploring past urban spaces and earlier human experience is not yet clear.
18. G. M. Varanini, 'La carità del municipio: Gli ospedali veronesi nel Quattro e nel primo Cinquecento'. In A. Pastore, et al. (eds.), *L'Ospedale e la città: cinquecento anni d'arte a Verona*, pp. 14–41 (Verona: Cierre, 1996).

19. M. Knapton, 'The *Terraferma* State'. In Eric Dursteler (ed.) *A Companion to Venetian History, 1400–1797*, pp. 85–108 (Leiden and Boston: Brill, 2019).
20. See L. Camerlingo, 'Il lazzaretto a San Pancrazio e l'ospedale della Misericordia in Bra: Le forme dell'architettura'. In *L'Ospedale e la città*, 179. On this era in Verona's new approaches to charitable assistance with the Misericordia (in contrast to new approach to territorial state building) see: P. Lanaro, 'Carità e assistenza, paura e segregazione: Le istituzioni ospedaliere veronesi nel Cinque e Seicento verso la specializzazione'. In A. Pastore, et al. (eds.), *L'Ospedale e la città: cinquecento anni d'arte a Verona*, pp. 43–55 (Verona: Cierre, 1996).
21. N.-E. Vanzan Marchini (ed.), *Rotte mediterranee e baluardi di sanità: Venezia e i lazzaretti mediterranei* (Venice: Skira, 2004); Camerlingo, 'San Pancrazio', pp. 182–183. In Padua, the old lazaretto was demolished in 1509; construction of the new one, situated on the left bank (Venice-facing) of the Brenta canal, did not begin until 1555, and was completed in 1576. See C. Ferrari, *Il lazzaretto di Padova durante la peste del 1630–3* (Padua: Società Cooperativa, 1905).
22. A. Conforti Calcagni, *Le Mura di Verona: La città e le sue difese dalla fondazione romana all'unità d'Italia* (Verona: Cierre, 1999), pp. 82–83. Destruction of pre-existing walls and a suburban isolation hospital happened in nearly identical fashion at Brescia, during the first two decades of the sixteenth century. Venice's Senate dictated that new lazarettos be built a distance of 1600 *passi* from the urban walls. See Pellegrini, 'Il lazzaretto di Verona', p. 150; and on Brescia, see P. Guerrini, 'Miscellanea Bresciana di studi, appunti e documenti con la bibliografia giubilare dell'autore (1903–1953). *Memorie Storiche della Diocesi di Brescia* 21 (1954): 97–98.
23. Camerlingo, 'Il lazzaretto a San Pancrazio', p. 179.
24. For details of the stages of the lazaretto's construction and Sanmicheli's involvement and designs, see Camerlingo, 'Il lazzaretto a San Pancrazio', pp. 183–188. Sanmicheli's designs did not survive; on the interim structure and full description of the building of the structures that survived into the twentieth century see Marchi, 'Introduction', x-xiii; and P. Davies and D. Hemsoll, *Michele Sanmicheli* (Milan: Electa, 2004), pp. 114–121.
25. N. E. Vanzan Marchini, curator. *Venice and the Mediterranean Lazarettos* (Venice: Edizioni della Laguna, 2004), pp. 22–26.
26. Vanzan Marchini, 'Introduction'. In Vanzan Marchini (ed.) *Rotte mediterranee e baluardi di sanità: Venezia e i lazzaretti mediterranei*, pp. 12–13 (Venice: Skira, 2004).
27. L. Beltrami, *Il lazzaretto di Milano' (1488–1882)* (Milan: Umberto Allegretti, 1899); L. Patetta, 'I temi nuovi dell'architettura Milanese del Quattrocento e il Lazzaretto'. *Arte Lombarda*, n.s. 79:4 (1986): 75–84.

28. Pellegrini, 'Il lazzaretto di Verona', p. 157, believes that the initial planning committee would have solicited Fracastoro's advice in 1539, though there is no firm evidence to the point. Fracastoro himself blamed Germans for an earlier plague in Verona: 'We saw, in the year 1511, when Verona was in the hands of the Germans, plague break out, of which about ten thousand persons died. One coat made of skins caused the death of no fewer than twenty-five Teutons…'. See Fracastoro, *De Contagione et Contagiosis Morbis et Eorum Curatione*, Libri III, trans. and ed. W. C. Wright (New York: G.P. Putnam's sons, 1930), p. 239.
29. Officially the transfer of the leprosarium's management to the Health Magistracy came in 1451, but the effective use of these resources was not ordered until 1576. At that point wooden isolation huts (called *casotti*) in the city were no longer acceptable; mandatory removal of some plague victims occurred at this point. See: Pellegrini, 'Il lazzaretto di Verona', pp. 146–147. Thus, before the disastrous outcomes of the 1576 epidemic, it is also possible that many in Verona believed traditional, temporary isolation camps would serve well enough in a crisis.
30. Flooding in 1591 undermined the earlier structure: Camerlingo, 'San Pancrazio', p. 179. Pellegrini, 'Il lazzaretto di Verona', pp. 151–153, and pp. 163–170, instead argues that official assessment of mortality during the 1575–76 plague was worse because recovering plague patients were housed too close to the city.
31. Though he said that it 'horrifies the pen' (*inorridisce la penna*) to write of tormented noble victims, Pona, *Il gran contagio*, p. 27, relished these details. Some were 'amorously delirious' others escaped guarded areas and threw themselves into the river, and some even beat their heads on stones, splattering brains, gushing blood: '*Alcuni, rotte le custodie, con precipitoso corso, si lanciavano nel fiume; altri, con feroce moto, percuoteano del capo alle pietre; e con larga effusion di sangue seminavano le cervella*'.
32. Pona, *Il gran contagio*, p. 32, where the author describes his resolution to despise the profits plague would bring him and preserve himself from danger 'for the sake of the Patria'.
33. A. Tadino, *Raguaglio dell'origine et giornali successi della gran peste contaiosa, venefica, & malefica seguita nella Città di Milano, & suo Ducato dell'anno 1629 sino all'anno 1632* (Milan, 1648), discussed in L. di Giammatteo and J. A. Mendelsohn, 'Reporting for Action: Forms of Writing between Medicine and Polity in Milan, 1580–1650'. In J.A. Mendelsohn, A. Kinzelbach and R. Schilling (eds.) *Civic Medicine: Physician, Polity and Pen in Early Modern Europe*, pp. 152–156 (London and New York: Routledge, 2020).
34. Pona, *Il gran contagio*, p. 26.

35. G. Alfani and M. Bonetti, 'A Survival Analysis of the Last Great European Plagues: The Case of Nonantola (Northern Italy) in 1630', *Population Studies* 73:1 (2019): 101–118.
36. Marchi, 'Introduction', xlii, repeating numbers that Pona supplied, *Il gran contagio*, pp. 60–61. On the demographic scale of this epidemic wave in Italy see: G. Alfani and T. E. Murphy, 'Plague and Lethal Epidemics in the Pre-Industrial World', *Journal of Economic History* 77:1 (2017): 314–343. And still useful: L. Del Panta, *Le epidemie nella storia demografica italiana (secoli XIV-XIX)* (Turin: Loescher, 1980), pp. 157–163.
37. Vanzan Marchini, 'Introduction', pp. 12–13, and Camerlingo, 'San Pancrazio', pp. 182–184, discussing the lazaretto in the context of Venetian territorial state commercial policy. Milan's more famous lazaretto was used as a jail, as soldiers' quarters, as shops and businesses, as a Veterinary School, as a municipal workhouse for the poor, and as a festival site—see: Beltrami, 'Il lazzaretto di Milano', p. 56. Contemporary with Volckamer's publication, in 1715 the lazaretto was used as a prophylactic quarantine site for part of the entourage of Maria Teresa Grandduchess of Tuscany and her consort Duke Lorenz; the duchess and duke instead performed quarantine in the sumptuous Villa Burri just across the river; see Pellegrini, 'Il lazzaretto di Verona', p. 177.
38. Lori Jones, 'The Diseased Landscape: Medieval and Early Modern Plaguescapes', *Landscapes* 17, no. 2 (2016): 108–123.
39. For example, Florence's survey of living conditions in Florence, on the eve of the 1630 plague: J. Henderson, *Florence under Siege: Surviving Plague in an Early Modern City* (New Haven and London: Yale University Press, 2019), pp. 51–70.
40. The best and most detailed description of this deliberate approach to plague control is R. Cancila, 'Salute pubblica e governo dell emergenza: la peste del 1575 a Palermo', *Mediterranea: Ricerche storiche* 13, no. 37 (2016): 231–72. See esp. pp. 248–256: 'The nine *lazzaretti* of Palermo'. Also very useful is Jane Stevens Crawshaw's discussion of the evolution of Genoa and Ligurian confinement zones, though her primary interest is 'quarantine hospitals': J. Stevens Crawshaw, 'The Places and Spaces of Early Modern Quarantine'. In Allison Bashford (ed.), *Quarantine: Local & Global Histories*, pp. 15–34 (London: Palgrave Macmillan, 2016).
41. J. Stevens Crawshaw, *Plague Hospitals: Public Health for the City in Early Modern Venice* (Farnham, Surrey: Ashgate Publishing, 2012); J. Henderson, *Florence under Siege*, pp. 181–216.
42. Vanzan Marchini, 'Introduction', *Venice and the Mediterranean Lazarettos.*
43. A. Bashford, 'Maritime Quarantine: Linking Old World and New World Histories'. In Bashford (ed.), *Quarantine: Local & Global Histories*, pp. 1–12 (London: Palgrave Macmillan, 2016); A. Chase Levinson, 'Early

Nineteenth-Century Mediterranean Quarantine as a European system'. In Bashford (ed), *Quarantine: Local & Global Histories*, pp. 35–53 (London: Palgrave Macmillan, 2016).

44. Stevens Crawshaw, *Plague hospitals*, pp. 53–54 and 88–91.
45. Ibid., pp. 67–70; Vanzan Marchini, 'Venezia e l'invenzione del lazzaretto'. In Vanzan Marchini (ed.), *Rotte mediterranee e baluardi di sanità: Venezia e i lazzaretti mediterranei*, pp. 25–28 (Venice: Skira, 2004). Poveglia was in service from the 1790s as Venice's a third island isolation area, called *lazzaretto nuovissimo*; see Vanzan Marchini, 'Venezia'. In Vanzan Marchini (ed.), *Rotte mediterranee e baluardi di sanità: Venezia e i lazzaretti mediterranei*, pp. 201–202 (Venice: Skira, 2004).
46. I calculated the area from Pellegrini's measurements: a rectangular space of 238.67 x 171.11 meters; 'Il lazzaretto di Verona', p. 158. The FAI consigned only three hectares to renovation: www.veronaeconomia.it/leggi-notizia/argomenti/economia-veronese/articolo/fai-fondo-ambiente-italiano-comune-di-verona-e-leccezionale-e-storica-zona-lazzaretto-da-dicemb.html
47. Pellegrini, 'Il lazaretto di Verona', pp. 161–170.
48. Ibid., pp. 164–170. At the time, Veronese would also have seen the Adige River as a plague consignment area, for when burial space and/or personnel grew scarce, the dead were consigned to the river; Pona, *Il gran contagio*, pp. 45–47.
49. Padua's lazaretto had 300 patient cells, and in this respect was more capacious than Verona's complex. It was authorised by Venice in the 1530s, building begun in 1555, and in use for the 1575–76 plague. Not as architecturally impressive as Verona's structure, it was demolished in 1819, after long disuse. See V. Marzocca, *La peste a Padova nel 1575–1576: cause e meccanismi di diffusione dell'epidemia*, Università degli Studi di Padova, masters' thesis in *Scienze Statistiche*, Academic Year 2017/2018, pp. 43–49.
50. R. Cancila, 'Salute pubblica'.
51. A. Melchioni, 'Il lazzaretto di Bergamo'. In *Lazzaretti europei: Da luoghi di sanità a rete di rapporti internazionali.* Atti del Convegno, Venice, Isola del Lazzaretto Nuovo, 14 September 2013, (Venice: Archeoclub d'Italia, 2015), pp. 13–14. Stable url: www.archeomedia.net/wp-content/uploads/2015/10/Lazzaretti-europei.pdf. The quote is from Melchioni's description of the lazaretto zone: 'immersa in aperta campagna'.
52. The English medieval equivalent was a 'close', describing field and forest belonging to a great lord, once the Bishop of Brescia, and farmed by unfree peasants, but by the mid-fifteenth century the commune claimed it as a civic lazaretto because it had been long abandoned. See: P. Guerrini, 'S. Bartolomeo al Lazzaretto', *Memorie Storiche della Diocesi di Brescia* 15,

no. 1 (1948): 64–67. I have not been able to determine how many hectares the *chiusura* included.

53. Guerrini, 'S. Bartolomeo', pp. 66–67: the chaplain assigned to the lazaretto had duties beyond its supposedly closed perimeter: at Costalunga [a few hundred meters distant], and even at Casadesimo [today's Chiesa della Stocchetta, 4.5 kilometres away]. And the houses proximate to, but outside, the lazaretto were considered infected (*sporche*), so the chaplain sent to the pesthouse was to minister to those persons as well as the plague stricken. An image of part of the remaining loggia can be found in A. Vaglia, 'La peste del 1478 a Brescia dal diario di Giacomo Melga', *Le Infezioni in Medicina* 2 (2011): 131.
54. Similar to the lugubrious evidence from Verona, partitioning spaces and categorising seems in retrospect a failed plague policy. Ninety percent of the Ghetto residents (numbering 721 before the plague) fell ill; 58.4 percent died. See S. Einbinder, 'Poetry, Prose and Pestilence: Joseph Concio and Jewish Responses to the 1630 Italian Plague'. In Haviva Yishai (ed.), *Shirat Dvorah: Essays in Honor of Dvorah Bregman*, pp. 77–78 (Be'er Sheva: Ben Gurion University and Mossad Bialik, 2018).
55. Henderson, *Florence under Siege*, pp. 195ff. Putting a lazaretto at San Miniato allowed a whole-hill isolation zone, which may have reduced overall mortality. Upland regions and other areas not converted fully to agricultural production are usually more biodiverse, with plague circulating among a greater variety of rodents and their fleas. Spillover events, producing human plague cases, are less common than in farmland, even though a greater proportion of rodents are affected by plague. See R. J. Eisen, K. MacMillan, et al. 'Identification of Risk Factors for Plague in the West Nile Region of Uganda', *American Journal of Tropical Medicine and Hygiene* 90, no. 6 (2014): 1047–1058.
56. G. Sforza, *F.M. Fiorentini ed i suoi contemporanei lucchesi. Saggio di storia letteraria del secolo xvii* (Florence: F. Menozzi, 1879), pp. 66–67.
57. Ibid., pp. 41–57.
58. See here D. Zanetti, 'A Milano nel 1524: i sospetti del Gentilino'. In C.M. Cipolla and R.S. Lopez (eds.), *Fatti e idee di storia economica nei secoli xii-xx. Studi dedicati a Franco Borlandi*, pp. 313–137 (Bologna: Il Mulino, 1976). Zanetti draws on the fortuitous survival of a register of plague evacuees from the southwestern, porta Ticinese district of Milan, an administrative technique that kept families and neighbourhoods together, if only on paper. In this devastating 1524 plague, Milan was still using extramural campgrounds—the Gentilino was a cemetery area for that city district—for plague control, despite the availability of the San Gregorio lazaretto.

59. Cusago was a Visconti villa and hunting park in the fourteenth century, but the property was seized by the Ambrosian Republic oligarchy after Filippo Maria Visconti died in 1447. C. Decio, *La peste in Milano nell'anno 1451 e il primo lazzaretto a Cusago* (Milan 1900), pp. 14–16, describes uses of Cusago and its dimensions. Beltrami, 'Il lazzaretto di Milano', pp. 63–64, reprints the original donation: '*pertiche 1500 di bosco nel territorio di Cusago*', around 900 hectares. After the Sforza came to power in 1450, Cusago was again made private hunting park; see C. A. Martelli, *Ducks and Deer, Profit and Pleasure: Hunters, Game and the Natural Landscapes of Medieval Italy*, Graduate Program in History, Ph.D. thesis (Toronto, York University, 2015), chapter 3. On plague control within Milan, 1450–52, see G. Albini, *Guerra, fame, peste: Crisi di mortalità e sistema sanitario nella Lombardia tardomedioevale* (Bologna: Capelli, 1982), pp. 121–132; A. Bottero, 'La peste in Milano nel 1399–1400 e l'opera di Gian Galeazzo Visconti', *Atti e Memorie dell'Accademia di Storia dell'Arte Sanitaria*, ser. 2, 8 (1942): 17–28; A. Pasi Testa, 'Alle origini dell'Ufficio di Sanità nel Ducato di Milano e Principato di Pavia', *Archivio storico lombardo*, ser. 10, 102 (1976): 376–386.
60. For the various Milanese lazaretto dimensions, which I interpolated from the data he provides, see: Beltrami, 'Il lazzaretto di Milano', pp. 18–19, 22, 49. Estimated a little differently, with Crescenzago's expanse at 260,000 $m^2$ and the eventual San Gregorio location, 150,000 $m^2$. V. Cavenago, *Il lazzaretto: storia di un quartiere di Milano*, 2nd ed. (Milan: Nuove Edizioni Duomo, 1989), pp. 49–51.
61. See Zanetti, 'I sospetti del Gentilino'. And on the long-lasting political effects of this plague, see: S. D'Amico, *Spanish Milan: A city within the Empire, 1535–1706* (New York: Palgrave Macmillan, 2012), p. 11, and pp. 61–62.
62. Cavenago, *Il lazzaretto*, 61–62.
63. Ibid., 24–27, provided the text of the Capuchin manuscript. Felice Casati was still alive in 1646 and may have been the other Capuchin informant.
64. G. Ripamonti, *La Peste di Milano del 1630 Libri Cinque, Cavati dagli annali della Città*, trans. F. Cusani (Milan: Perotta, 1841), Book II, pp. 115–116, notes the Capuchin manuscript but provides only the testimonial to its authenticity, not the detail about rodents and fleas all over the meadow.
65. A. D'Amato, et al., 'Of Mice and Men: Traces of Life in the Death Registries of the 1630 Plague in Milano', *Journal of Proteomics* 180 (2018): 128–137.
66. P. M. Jones, 'San Carlo Borromeo and Plague Imagery in Milan and Rome'. In Gauvin Alexander Bailey, Pamela M. Jones, Franco Mormando and Thomas W. Worcester (eds.) *Hope and Healing: Painting in Italy in a*

*Time of Plague, 1500–1800*, pp. 65–96 (Worcester, MA: Clark University; College of Holy Cross, and Worcester Art Museum, distributed by the University of Chicago Press, 2005). The plague painting done by Cerano himself depicts Borromeo distributing his own clothing and household furnishings to the still quarantined in a tent area outside the city. I have nonetheless found no direct evidence of tents inside the San Gregorio lazaretto during the 1575–77 plague.

67. See here the commonly reproduced sketch of the lazaretto and the surrounding suburban zone, made by an artist-inmate during 1630: G. P. Bonetti, 'Il lazzaretto di Milano e la peste del 1630, a proposito di un nuovo document iconografico', *Archivio Storico Lombardo* 50, fasc. 3/4 (1923): 388–442.
68. Actually there were three adjoining *foppone* there, two of which were dug to receive the bodies of victims at the lazaretto and some of those who instead died in the city or in transit, See Cavenago, *Il Lazzaretto*, 190–200; and Redemagni, https://storiedicimiteri.wordpress.com/2018/11/18/i-fopponi-di-milano/
69. Bevacqua, 'Peste manzoniana'. In Milanese dialect the dock workers were the men with dirty faces, called *tencia* or *tencitt.*
70. V. Bevacqua, 'Peste manzoniana e 'Madonna di tencitt', initially published in *La Ca' Granda*', accessed at: www.formazione.eu.com/_documents/cagranda/articoli/2005/0117.pdf. See also the blog of Sergio Codazzi, http://vecchiamilano.wordpress.com/2011/02/10/la-madonna-dei-tencitt, which provides different particulars about the painting, its fate, and the prior who painted it. On the hydrological modernisation of Milan see P. Redondi, M. Richiuti and G. Tartari, 'Milan, sa nappe, e le Lambro: la quantité au détriment de la qualité'. In Laurence Lestel and Catherine Carré (eds.) *Les rivières urbaines et leur pollution*, pp. 123–172 (Versailles: Quae, 2003). And see: Jones, 'San Carlo Borromeo', pp. 82–84.
71. A useful overview of Milanese cemeteries and modernisation can be found at www.storiadimilano.it/citta/cimiteri/cimiteri_milanesi.htm. See also P. Redemagni, *I cimiteri* (Milan: M&B publishing, 2004). The painting survived Allied bombing in 1943, which levelled much of the nearby *Ospedale Maggiore.*
72. The Coppola painting is reproduced in *Hope and Healing* (above note 66), catalogue number 26, and is held by the Princeton Art Museum. James Clifton, 'Art and Plague in Naples', pp. 97–117, discusses what is more typical of the era: Gargiulo's manipulation of the piazza's scale in order to represent all city plague activities. Clifton usefully invokes Roland Barthes's distinction between conveying a sense of 'having been there' (the Gargiulo painting) and 'being there' (the Coppola canvas), p. 114. On the Naples

plague generally see I. Fusco, 'La peste del 1656–58 nel Regno di Napoli: diffusione e mortalità', *SIDeS, Popolazione e Storia* 1 (2009): 115–138.

73. Cancila, 'Salute pubblica'.
74. G. A. Bailey, 'Anthony van Dyck, the Cult of Saint Rosalie and the 1624 Plague in Palermo'. In Gauvin Alexander Bailey, Pamela M. Jones, Franco Mormando and Thomas W. Worcester (eds.) *Hope and Healing: Painting in Italy in a Time of Plague, 1500–1800*, pp. 118–133 (Worcester, MA: Clark University; College of Holy Cross, and Worcester Art Museum, distributed by the University of Chicago Press, 2005). Van Dyck left Sicily in September 1625; we can date the image more closely, through events described in an anonymous report, by a prelate present throughout the plague: S. Salomone-Marino (ed.), 'La peste in Palermo negli anni 1624–1626: Relatione di anonimo', *Archivio Storico Siciliano* 30 (1905): 221–289, here p. 232. An arquebus shot could reach a distance of around 1000 meters. On the discovery of St Rosalie's remains see E. Frasca and M. Costa, 'La peste in Sicilia in età moderna: tra scienza e devozione', *Annali della facoltà della formazione Università degli studi di Catania* 19 (2020): 106–107.
75. The city council on 23 June decided to hire 150 builders (*operarij*) to create a capacious lazaretto in Borgo Santa Lucia. It was operational by the night of 24 June, when 66 patients from the previous site, called the *Spasimo*, were transported there. See Salomone-Marino (ed.), 'La peste in Palermo', pp. 231 and 263.
76. Ibid., p. 246: '*per essere detto luogo molto spatioso, per poter attendere a tutti i bisogni con sollecitudine, cavalcava un animaletto...*'.
77. For images of these fields and historical markers, see: https://ilmestiere.wordpress.com/2014/09/06/san-giorgio-lomellina-e-il-campo-della-peste/ and M. Tedeschi, *Il bosto, la vigna, la pietra: Botticino nella storia* (Brescia: Comune di Botticino, 1988), p. 80.
78. G. Montù, *Memorie storiche del gran contagio in Piemonte negli anni 1630 e 31, e specialmente del medesimo in Chieri e ne' suoi contorni raccolte e pubblicate per la solennità secolare celebrata in questa città in onor di Maria SS.ma delle grazie, 1, 2 & 3 Sett. 1830* (Turin: Giacinto Marietti, 1830), pp. 3–5 on the 200-year-old tradition. I am especially grateful to Susan Einbinder for calling my attention to this work.
79. For examples, a Barnabite friar and layman who died 22 and 28 September were buried '*in curtili apud murum ecclesiae*' (Montú p. 35); a nobleman's widow who dictated her will '*alla cascina detta il Pescòre*' (p. 58); Stefano Piovano made his will at the farm called *tetto de' Piovani*, which Montù located precisely between two rural chapels, and hired guards to block the road to the north (pp. 60 and 118). On the choice of lazaretto location,

the regulated dress and comportment of hired *monati* (body-clearers), surgeons, and physicians, see pp. 86–87. The contaminated (*brutti*) and 'clean' (*netti*) were differently dressed, providing another way for citizens to read the wall-free plague landscapes. Many plague plaques that commemorated the 1630 epidemic were mounted much later, between the 1790s and the 1820s; see Montù, *Memorie storiche*, 104–106.

80. Montú, *Memorie storiche*, 39. It was also called the 'Lazzaretto di Valle in Chieri'. The Capuchin monastery was popularly called S. Anna del Lazzaretto, but Montù, p. 49, claims that it was field used as a burial site for plague victims, and belonged to a different monastery. There was also a 'lazzaretto di Baldissero' closer to Turin than Vallèro. The town of Moncucco used various lazarettos; Moriondo instead sent and buried victims at the '*cascina delli Zucca*'. In other words, town victims were seeded across the rural landscape.
81. Similar improvised burial plots have been excavated in Provence. See, for example, B. Bizot, P. Reynaud, and P.-H. Rigaud, 'Le cimetière de l'infirmerie des Fédons (Lambesc, Bouches du Rhône) avril—septembre 1590'. In M. Signoli, et al. (eds.), *Peste: entre épidemies et societies*, pp. 67–70 (Florence: Florence University Press, 2007).
82. S. K. Cohn, Jr. *Cultures of Plague: Medical Thinking at the End of the Renaissance* (New York and Oxford: Oxford University Press, 2010), pp. 77–139.
83. Ibid., pp. 77–94. The single most important treatise appears to be Giovanni Filippo Ingrassia's massive tome on the plague in Palermo, *Informatione del pestifero et contagioso morbo del protomedico siciliano* (1576). Cohn finds it a medically traditional treatise that adds a narrative retelling of the recent plague in Palermo. Cancila, 'Salute pubblica', instead finds it a novel management-oriented plague treatise, given over to Ingrassia's recommended procedures and protocols for mitigating a contagious epidemic. The work was soon translated into Latin becoming widely known outside Italy, as well as widely used in north Italy.
84. For example, the contemporary account of G. M. Montini, *Della peste in Bassano nel 1631* (Bassano: Baseggio, 1855). Bassano not only was a refuge for Venetian and Veronese merchants and noblemen fleeing plague in 1630, but the whole region was located along thoroughfares used by the German army. Udine, to the north and east, also suffered a severe plague in 1629–1630, as did the whole of Lombardy to the west and southwest.
85. One of England's main places for plague tourism, Eyam, in the Peak District, has long shaped a grim story to accentuate plague's contagion and the heroism of villagers who isolated themselves, sparing their neighbours through their sacrifice. But it was mostly a story 'established, manipulated

and reshaped to fit changing literary and historical fashions': see P. Wallis, 'A Dreadful Heritage: Interpreting Epidemic Disease at Eyam, 1666–2000', *History Workshop Journal* 61, no. 1 (2006): 31–56. Of 330 villagers 259 died, but those deaths were spread out across a 14-month interval.

86. Samuel K. Cohn, Jr., 'Plague spreaders'. In S. K. Cohn, *Epidemics: Hate and Compassion from the Plague of Athens to AIDS*, pp. 153–93 (New York: Oxford University Press, 2018).
87. Stevens Crawshaw, *Plague Hospitals*, pp. 184–193. She further estimated that pesthouse mortality accounted for a significantly lower proportion of overall (or plague only?) deaths in 1630–1631, and she meticulously parsed the evidence on which her numbers were calculated. However, for evidence about the devastating 1483–1486 plague in Milan she should have relied on G. Albini's 'La mortalità in un grande centro urbano nel '400: il caso di Milano'. In R. Comba, G. Piccinni, G. Pinto (eds.), *Strutture familiare: epidemie, migrazione nell'Italia medieval*, pp. 117–134 (Naples: Edizioni Scientifiche Italiani, 1984). This was the epidemic that persuaded Milanese authorities to invest in a grand lazaretto.
88. Salomone-Marino, 'La peste in Palermo', pp. 263–76.
89. Stevens Crawshaw, *Plague Hospitals*, argues that unburied bodies may have contributed to excess pesthouse mortality. Such sights clearly increased terror, but cadavers do not usually pose a direct threat because most infectious pathogens require a living host to survive and replicate. So, too, hungry fleas require a living host. Still the best single introduction to the role that ectoparasites of all sorts played in shift of disease experience, premodern to modern, is J. C. Riley, 'Insects and the European Mortality Decline', *The American Historical Review* 91, no. 4 (1986): 833–858.
90. C. Lynteris and R. Stasch, 'Photography and the Unseen', *Visual Anthropology Review* 35, no. 1 (2019): 5–9.
91. K. Eaton, 'Historical Insights from 'Modern' Plague: plague from the 19th to 21st Century', working paper, Department of Anthropology: McMaster Ancient DNA Centre, 20 June 2018; and M. A. Spyrou, et al., 'Phylogeography of the Second Plague Pandemic Revealed through Analysis of Historical *Yersinia pestis* Genomes', *Nature Communications* 10, no. 4470 (2019). https://doi.org/10.1038/s41467-019-12154-0
92. Unfortunately, Samuel K. Cohn, Jr. makes this assertion in his otherwise valuable *Cultures of Plague*, pp. 39–76.
93. The thoughtful, COVID-19-era study by Frasca and Costa, 'La peste in Sicilia', is interesting here. The authors summarise the Ingrassia treatise composed in Palermo, 1575, (see above note 80); his costly, sustained attention to the minutiae of innovative contagion control is usually held responsible for the much-reduced mortality the city faced when compared

to its neighbours. While Frasca and Costa cite traditional blame for this plague (importations by ship from areas of the Afro-Mediterranean), they themselves neither support nor reject such conclusion. As they show, strong evidence of ongoing plague spread in the Sicilian hinterlands supports the retrospective view that plague could have been endemic in some areas of the island. I find their biomedically agnostic approach to the historical evidence quite useful.

94. Stuart Borsch and Tarek Sabraa, 'Refugees of the Black Death: Quantifying rural migration for plague and other environmental disasters', *Annales de Démographie Historique* 134, no. 2 (2017): 63–93.

CHAPTER 5

# Picturing Plague: Photography, Pestilence and Cremation in Late Nineteenth- and Early Twentieth-Century India

*David Arnold*

Of all the societies struck by bubonic plague during the third pandemic, India was by far the greatest sufferer, experiencing more than ninety percent of all fatalities worldwide.[1] Between 1896, when the pandemic first reached Indian shores, and the 1920s (by which time the disease had largely subsided), plague caused an estimated twelve million deaths. Within India, the city of Bombay (now known as Mumbai) and the surrounding Bombay Presidency were both the earliest to be affected (from mid-1896 onwards) and among the areas of the subcontinent worst affected.[2] Bombay in the 1890s was an administrative capital, an industrial and commercial hub, and the most technologically sophisticated city in British India. It is not therefore surprising that Bombay, more than anywhere else in India, gave rise to a very extensive photographic archive,

D. Arnold (✉)
University of Warwick, Warwick, UK
e-mail: D.Arnold@warwick.ac.uk



C. Lynteris (ed.), *Plague Image and Imagination from Medieval to Modern Times*, Medicine and Biomedical Sciences in Modern History, https://doi.org/10.1007/978-3-030-72304-0_5

detailing many aspects of the disease and, more especially, the remedial measures taken against it. In recent years photography has been one of the most widely discussed visual technologies of the colonial era in India and regarded as one of the principal manifestations of colonialism's far-ranging powers of persuasion, command and self-legitimation. But, subtle and extensive though this literature now is, there is still much to be said about how photography was deployed, by whom and for what purpose: in this respect the plague photography of the 1890s and 1900s can provide critical insights into the wider nature and purposes of India's colonial-era photography.[3]

Although in India (in common with much of the colonised world) image-making of the human body was one of photograph's most obvious and significant functions, much of the debate has focused elsewhere—on the iconography of architecture and landscape, on the self-portraiture of Europeans and Indians at home or at leisure, on ceremonial durbars, public works and political events. Where the body has received attention, it has often been in the limited context of the ethnography of castes, tribes and social 'types'.[4] By contrast with this tableau of the living, the photographic representation of the dead has seldom been considered with respect to India, though it has, there as elsewhere, a complex and distinctive genealogy of its own. Photographing the dead might have several different functions—in relation to forensics and police investigations (where the body became both an exhibit and a source of evidence), in connection with anatomy, surgery and epidemiology (where it was used to show the physiological effects of disease or to illustrate pathological states), as a means of memorialising celebrated public figures, and as part of mortuary rites and funerary procedures.[5] It is on the last of these genres that this chapter will concentrate.

Photography in India had a particularly close and complex relationship with the third plague pandemic, as it did in many other parts of the world.[6] A large number of photographs, running into hundreds, were taken there during the plague, especially in its early and most traumatic years. This substantial photographic archive, far removed from the familiar world of black-and-white studio photography and sepia self-portraiture, is, in Indian terms, exceptional and yet not without precedent. For example, India's destitute and famine-stricken poor were a subject of photography as early as the 1860s, in part because these recurrent crises of subsistence and survival attracted international attention and charitable relief (and so required photographic testimony) but also because the effects of starvation were so graphically inscribed on the emaciated, barely clad body of the non-Western subject.[7] Indeed, the use of photography to illustrate 'unpleasant' or 'horrific' scenes in India was by no means as uncommon as is sometimes assumed.[8] Like the plague corpses to which we will shortly turn, these famine 'victims', as they were collectively portrayed, were not only near-naked; they were also nameless and stripped of all other marks of individual identification. As Western images of Indian famine they functioned as stark evocations of poverty, dehumanisation and imminent death; they were also calculated to show Indians' moral and physical weakness and their dependency on the relief and remedial support provided by the state, by Christian missionaries and philanthropic agencies.[9] But while major epidemics—cholera, malaria, smallpox—followed in the wake of famine, and were the subject of numerous reports, statistical surveys and exercises in disease mapping, few photographs exist to record their passage or the suffering they caused. The most striking absence of all is the apparent lack of photographic images relating to the influenza epidemic of 1918–1919, which in the space of a few months caused between twelve and eighteen million deaths.[10] Clearly, the scale of epidemic mortality in India was no indication of the volume of photographic evidence likely to be produced. Millions could die without leaving any enduring image. But in this respect plague was strikingly different.

## Plague as Image

Although plague photographs were produced in other parts of western India, in Poona (Pune), Karachi, Dharwar and elsewhere, by far the greatest number pertain to Bombay, the city where bubonic plague first struck and where drastic, indeed draconian, anti-plague measures were first introduced. Many of the Bombay plague photographs were incorporated into an album with the title *Plague Visitation, Bombay, 1896–97*. Of the 142 images in the album the majority are attributed to Captain C. Moss of the Gloucestershire Regiment with a smaller number, eight in all, attributed to a professional 'photo artist', F. B. Stewart of Poona, who is also credited with having complied the volume.[11] Compared to Stewart's technically more accomplished and carefully composed albumen prints, which each occupy a whole page of the album, Moss's gelatin silver images look more amateurish, are less carefully composed, and appear as smaller prints in the volume, mostly grouped together as five or six to a page.[12] The description of Moss's work as 'amateurish' calls for further explication. By the 1890s most Indian cities had professional photographers and photographic studios: some, like Bombay and Calcutta (Kolkata), had a dozen or more: these included studios run by Indians as well as Europeans and intended for both an Indian and a European clientele.[13] For instance, one of the most successful photographic enterprises in Bombay in the 1880s and 1890s was that of E. Taurines, whose studio produced a wide range of images including 'views and native characters', buildings, public works and domestic interiors: some of these images were official commissions.[14] It is the more remarkable, therefore, that Moss, an otherwise obscure army officer with no obvious technical credentials, should have been chosen to make a photographic record of the Bombay plague, especially when a professional, Stewart, appears to have taken a number of more carefully crafted images.[15]

Perhaps Stewart and his photographers was unavailable to follow the Bombay Plague Committee on its almost daily excursions around the city, including sometimes impromptu visits to burial grounds and cremation sites.[16] I have been unable to find any official correspondence relating to the commissioning of these photographs, and it is possible that Moss was paid privately by the committee's chair, Brigadier-General W. F. Gatacre, to provide a personal record of his activities and achievements. Indeed, Gatacre may have been made aware of the value of photography as a means of visual documentation from the Chitral campaign of 1895 on which he had submitted a report illustrated with photographs by Sergeant-Major Devlin of the Royal Engineers. Soldiers, most often engineers or artillery officers, had a long association with photography in India.[17] As the *Times of India* observed of Devlin's images showing troops crossing the Lowan Pass: 'With these specimens of the photographer's art before one, it is impossible to doubt that the lens will play an important part in the campaigns of the future.'[18] Combatting plague was another kind of 'campaign' and one in which photographs—and soldier photographers—might have their uses.

The motives behind the compilation of the *Plague Visitation* album are made clear from its preface, written by Gatacre as chair of the four-man Bombay Plague Committee appointed by the Government of Bombay in March 1897 (the other members—all Europeans—were an official from the Indian Civil Service, a physician from the Indian Medical Service and a civilian sanitary engineer). Gatacre remarks that at the time of his appointment plague had already caused nearly 20,000 fatalities in Bombay since its first appearance in August 1896: plague deaths were numbering more than 150 a day. But by June 1897, when the committee was wound up, the disease had 'practically disappeared'. Although this claim proved wildly over-optimistic, it is evident that Gatacre intended the album to record and celebrate the work of his committee and the 'preventive measures [taken] for its extinction'.[19] The album opens with a studio portrait of the Governor of Bombay, Lord Sandhurst, accompanied by a photograph of the members of the committee and its secretary. The images that follow then document the work of Gatacre's committee in implementing the brief given to it by Sandhurst in March 1897. The general was instructed to pursue three objectives—to discover all plague cases, to ensure the treatment in hospital of all plague cases, and to oversee the segregation of all individuals suspected of having plague. But, in the light of intense local opposition to earlier anti-plague measures, Sandhurst also urged the committee to adopt 'persuasion and gentleness', to combine 'firmness' with 'explanation', and to avoid creating a 'feeling of distrust' among Bombay's Indian population.[20]

We can, therefore, understand the Gatacre album as providing a semi-official narrative not of the epidemic itself (which was already well established by the time the committee began its work and was anyway well documented through published sources) but of the vigorous, yet 'humane', remedial measures taken to contain and suppress plague in Bombay and the surrounding countryside. The images duly show the committee's travels, the inspection of villages and houses, the role of disinfection squads, the hosing, cleansing, limewashing and demolishing of infected buildings, the outside and interior of plague hospitals (as in Fig. 5.1), the segregation camps to which plague suspects were evacuated, the administration of Waldemar Haffkine's anti-plague prophylactic, and, occasionally, the victims of the disease and their widowed or orphaned relatives. European doctors and nurses feature prominently in these images, as do, to a lesser extent, the Indians who volunteered to help implement anti-plague measures: here was evidence of 'native' collaboration rather than resistance. These photographic images of ameliorative (or, following Christopher Pinney, 'curative') anti-plague action have much in common with the genre of nineteenth-century famine photography which similarly showed relief works, food distribution, road- and rail-construction and famine camps in British India or the princely states, thereby suggesting the 'humane' benefits of state intervention or European agency rather than the unsettling reality of mass suffering and catastrophic deaths.[21] Even relatively small government or missionary hospitals embellished the record

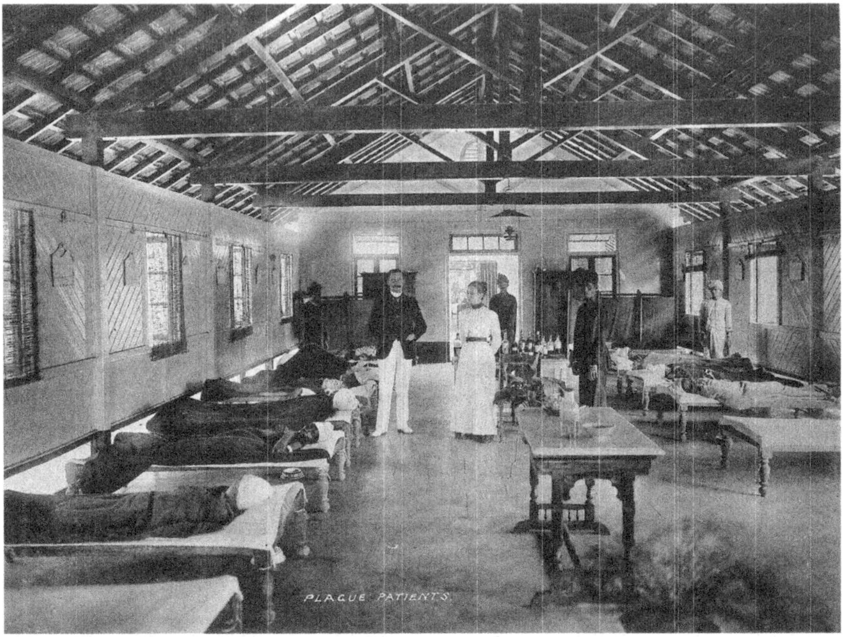


**Fig. 5.1** 'Interior of Plague Hospital, Bombay', c. 1897. (Courtesy of the Wellcome Collection)

of their anti-plague activities with photographs of white doctors and nurses tending to the sick or injecting Indians with Haffkine's anti-plague vaccine, one of the most visible signs of preventive intervention.[22] Hence, there is less evidence in the Gatacre album or elsewhere for India in 1896–1897 of the iconography of 'ruination' Robert Peckham has described with respect to the plague photographs taken in Hong Kong a few years earlier.[23]

Relatively few of the images taken by Moss and Stewart or indeed by other contemporary photographers show actual plague corpses, though some do—with deceased victims lying on mats, being transported by cart to graveyards or carried by bearers to cremation grounds. The transportation of the dead, and hence the risk posed by the removal and disposal of infected bodies, is one of the recurrent themes in the Gatacre album as it was in the medical pronouncements of the time.[24] As with India's famine photography, there seems to have been no taboo on photographing the plague dead. Both, after all, deal with the anonymous Indian dead: neither famine nor plague counted many white 'victims'. Just as such direct corporeal references served to illustrate physical suffering and dehumanisation in famine, so in plague they helped to articulate the sanitary view that 'a plague corpse is a focus for infection' and so needed careful treatment and destruction (ideally by fire) to preclude onward transmission of the disease.[25]

Among the images from the plague pandemic in India that do show human corpses perhaps the most striking are those relating to the cremation of plague victims. Some of these, appearing in British and American illustrated newspapers, were not photographs but artists' sketches. They show plague corpses being carried into cremation or burial grounds or depict blazing funeral pyres being stoked by attendants (Doms) at the burning *ghats* (cremation grounds). Sketches were used, in preference to photography, to suggest movement, sensation and affect in ways that contemporary photographs might fail to capture. They also allow (as, of course, do photographs) for the use of text to explain and amplify the scene. One such sketch appeared in a London illustrated weekly, *The Graphic*, in January 1897. Titled 'The Plague at Bombay: Hindoos Disposing of Their Dead at the Burning Ghauts', it was drawn by Sydney P. Hall, a well-known illustrator, from a sketch made locally by Frederick Airey. As birds (kites or vultures) gather menacingly above the half-open gate of the cremation ground, two parties of turbaned men bearing near-naked bodies on bamboo stretchers arrive to deliver their dead. The funeral scene within is only distantly and hazily glimpsed, as if veiled from closer scrutiny. The caption reads:

> The plague is still carrying off hundreds in this city. The number of deaths reported each day is sixty or seventy, but there is not much doubt that in reality there are many times as many. It is a common sight to see funerals passing through the town—Parsees wending their way with their sad burden to the Towers of Silence on Malabar Hill, the Mahommedans with theirs to their cemetery, and the Hindoos with their dead lying exposed on an ordinary stretcher to their Burning Ghauts in Queen's Road, which at night are lit up with the lurid glare of the many funeral pyres.[26]

Another sketch, which appeared in New York's *Harper's Weekly* in June 1899, gives evidence of the international concern India's plague outbreak had aroused. Bearing the caption 'Incinerating the bodies of victims of plague in Bombay', it gives a dramatic, if still sanitised, version of plague cremation in Bombay (Fig. 5.2). It shows a turbaned, dhoti-wearing Dom stoking one of several visible funeral pyres (the feet of a corpse can just be seen projecting from the nearest pyre), while other attendants sit or stand, exhausted or despondent, against the enclosure's perimeter fence. The

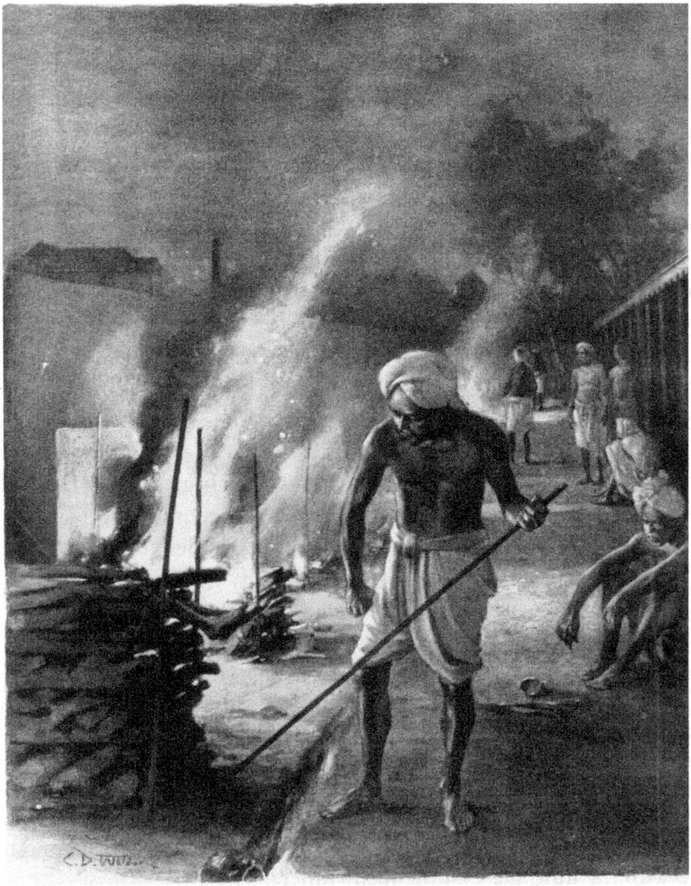

INCINERATING THE BODIES OF VICTIMS OF THE PLAGUE IN BOMBAY.

**Fig. 5.2** 'Incinerating the Bodies of Victims of the Plague in Bombay', *Harper's Weekly*, 3 June 1899. (Courtesy of U.S. National Library of Medicine, Digital Collections)

thick smoke billowing up into the night sky (a brooding scene contemporary photography might not easily have captured) compounds the overall sense of gloom and gestures to the abomination—whether of plague and the menace it posed to the West or the barbarity of burning bodies in such a crude and visible manner.[27]

Turning to photographs rather than sketches, there are several of scenes taken at cremation grounds in Bombay, which show, often more starkly than the sketches, bodies about to be burned or in the process of being consumed by fire. The Gatacre album contains a number of these images, mostly taken at Bombay's principal Hindu cremation ground at Sonapur, on Queen's Road, close to the western shore of the island.[28] Two of the Bombay photographs merit particular attention (Figs. 5.3 and 5.4) and are central to the discussion that follows. They were taken in 1897 and show the interior of the cremation ground, surrounded by a wall and fence roughly ten feet high, with bodies being burned or about to be burned,

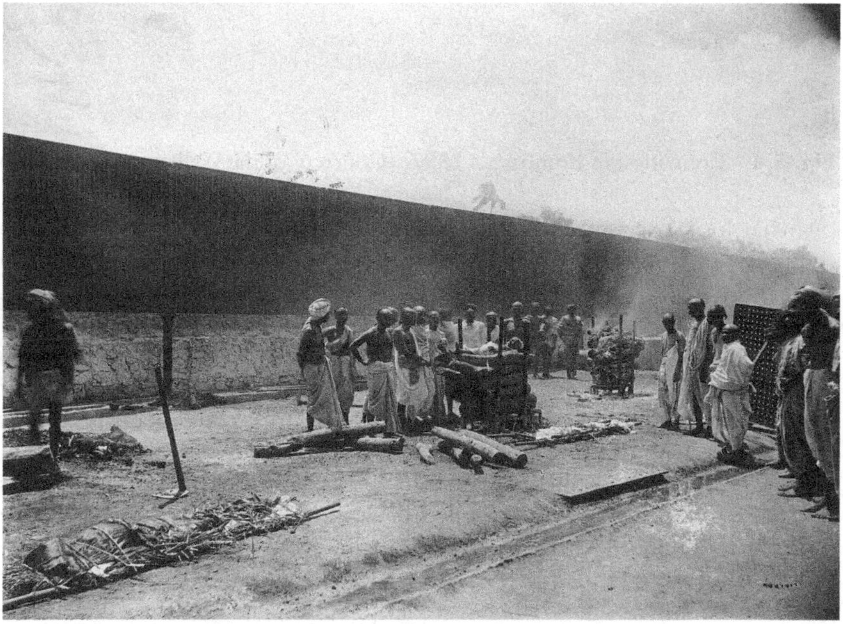

**Fig. 5.3** Cremation in Bombay, c. 1897. (Courtesy of the Wellcome Collection)

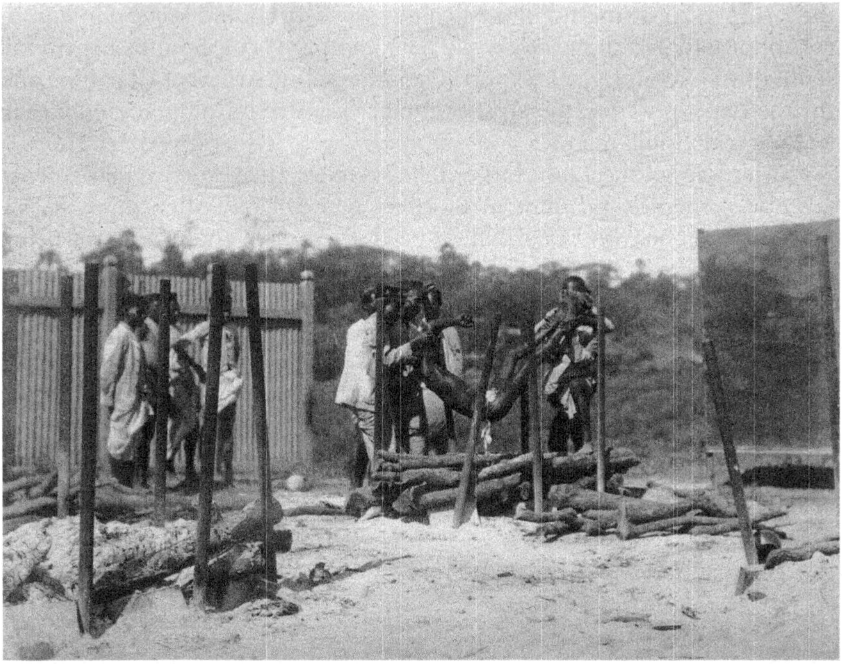

**Fig. 5.4** Cremation in Bombay, c. 1897. (Courtesy of the Wellcome Collection)

on what look like minimal open-air pyres. In the first of these (Fig. 5.3) the body—significantly the *clothed* body—of an old man has been laid on a funeral pyre which is about to be lit: his upturned face is clearly visible. A second pyre can be seen smouldering behind it to the right with a third just visible on the left-hand margin of the photograph. Three groups of men are shown standing around (women did not at this time normally attend Hindu cremations), in rather fixed postures; many peer directly at the camera. Among those close to the main pyre are two or three Brahmin priests, naked to the waist, wearing sacred threads, the front of their heads tonsured, the rest of the hair pulled back in a pigtail. Here then is some indication of religious ceremony. This is significant in that no other plague cremation photographs (or sketches) I have encountered show any religious rites being observed. The scene implicitly challenges the official view that, since the plague body was a 'focus for infection', 'all religious rites and ceremonies should … be curtailed as much as possible'.[29] Conceivably some of the men standing close by are family members and mourners, but they present no obvious expressions of grief, and, aside from the priests, they are likely to be cremation-ground attendants, drawn from an untouchable caste of Doms. Nearby lie the objects and detritus of a Hindu cremation—more logs for the pyre, mattocks for removing the ashes, a few discarded water-pots, and a bamboo litter with another, still shrouded, corpse awaiting cremation.

Although this cremation-ground scene appears in the Gatacre album with the caption 'Sonapore: Interior of Burning Ground with Corpse on Fire', it was also published in September 1897 in *The Graphic* in London, alongside three other images of Bombay plague scenes (see below).[30] There it bore the caption:

> Cremating the Bodies of Victims in the Hindu Burial Ground: While the plague was at its height the bodies of the Hindu victims were taken to the burial ground to be cremated. The pile of wood is supported by two iron stakes driven into the ground on each side. Five carts of wood were used for each adult body. The ashes were afterwards collected and thrown into the sea.[31]

In the second photograph (Fig. 5.4), which did not appear in either the Gatacre album or *The Graphic*, a body, clad only in a loincloth, is being manhandled onto an awaiting pyre consisting of barely a dozen logs; another pyre, already reduced to ashes, can be seen in the foreground.[32] In this scene there are no priests, no hint of any religious rites, and the dark body is almost naked. The two groups of attendants appear from their caps and jackets to be municipal workers or the Doms employed to prepare and maintain the funeral pyre. In both of these photographs (Figs. 5.3 and 5.4) the surroundings appear stark and unadorned: in the first trees can vaguely be seen beyond the surrounding wall and fence; in the second a gap in the fence shows rough ground with trees beyond.

## Burning the Dead

What, then, do these plague images signify, and how do we assign them a context within the wider visual (and textual) representation of death, disease and cremation in late nineteenth- and early twentieth-century India? We might first of all take these images of burning bodies and cremation fires at their face value, as offering photographic insight into the grim reality of mass mortality during India's plague epidemic and of a fate (cremation) that awaited many millions of Indians who died from the disease. They could be understood as registering the impact in particular of the first phase of the epidemic, as glimpsed at its Indian epicentre and before such scenes had become all too familiar (and so less worthy of photographic record) across great swathes of western and northern India. Although there were many other ways in which the epidemic was recorded and given visual expression (through statistics, maps, research papers, official reports), the images present a more immediate vision of epidemic mortality and convey something of the alarm and dread that the arrival of the epidemic created in Indian as well as European minds. At the same time, as the preface to the Gatacre album indicates, many of these images were made when plague was still a novelty and there remained a high degree of optimism that it could be contained, even eliminated, if appropriate measures were adopted.

The Bombay photographs also indicate the importance of one specific means of disposing of the plague dead by illustrating the extraordinary number of Indians who were cremated during the epidemic. However it might strike the European eye, the scene revealed in these photographs is not that of a conventional open-air Hindu cremation. In Fig. 5.3 the logs on the pyre are held in place by four iron posts and supported by a metal frame: a perforated iron screen, to allow the ashes to pass onto the ground underneath the pyre, can also be seen behind the figures on the right. Such iron structures and supports had no place in conventional Hindu cremation. Although the cremations take place in the open air they are not, as custom would dictate, alongside a river or the sea: instead they are within a strictly delimited and enclosed space. The walls and fences so evident in these scenes were by the late nineteenth century a common requirement of municipal authorities, intended to shield the sight of cremation from passers-by and to dissipate the smoke and the smell of burning flesh.[33]

Those cremated in this often peremptory fashion, as suggested by the near-naked figure in Fig. 5.4, were generally the poor (by far the main victims of Bombay's plague), and so were cremated not by their families and attended by priests but by (or on behalf of) the hard-pressed municipal authorities.[34] This was cremation at its most economical, the utilitarian 'disposal of the dead', as it had been adopted in Bombay ever since the famine of the 1876–1878 left thousands of paupers dead on its streets. There was something raw and unseemly about this narrowly functional and unconventionally modern cremation. In Bombay the number of bodies cremated rose sharply in the early plague years, reaching more than 35,000 by 1900 (see Table 5.1). This represents at its peak the burning of nearly 100 bodies a day, or close to half of all those who died in the city. By 1901 fewer bodies were being burned but those cremated still represented more than half of the total number of deaths. Although the photographs, with several bodies on display at the same time, only hint at it, Bombay's burning grounds were almost overwhelmed by the number of the dead awaiting cremation. Such was the demand that wood for pyres was in short supply. With so many bodies there were long delays at the burning *ghats*, as corpses waited in turn to be cremated: in the cities, unlike in the countryside, bodies could only be burned at approved cremation sites and, despite their number, corpses were still burned individually and not on collective pyres. At 11 o'clock on a January morning in 1897 nineteen corpses were undergoing cremation at Sonapur; six more were waiting. By 12:30 another eleven had arrived. One body had barely been reduced to ashes before another was put in its place.[35] This remorseless processing of the human dead created a lasting visual impression on

**Table 5.1** Deaths and cremations in Bombay City, 1873–1903

| *Year* | *Total deaths recorded* | *Number cremated* | *Percentage of cremations to all deaths (with Plague deaths from 1897)* |
|---|---|---|---|
| 1873 | 16,450 | 4786 | 29.09 |
| 1880 | 22,399 | 5569 | 24.86 |
| 1886 | 21,232 | 6354 | 29.93 |
| 1891 | 25,241 | 8040 | 31.85 |
| 1893 | 24,652 | 7987 | 32.40 |
| **1897** | **49,264** ***(11,003)*** | **22,818** | **46.32** |
| **1900** | **81,412** ***(13,285)*** | **35,480** | **43.58** |
| **1901** | **61,648** ***(18,736)*** | **33,672** | **54.62** |
| **1902** | **50,431** ***(13,820)*** | **22,108** | **43.58** |
| **1903** | **52,331** ***(20,788)*** | **23,623** | **45.14** |

*Source*: Annual Report of the Municipal Commissioner of Bombay and Administration Report of the Municipal Commissioner for the City of Bombay for the years stated.

Note: Cells with bold values relate to the plague years, 1897 onwards with actual plague deaths in italics those who witnessed it, one the camera could only partly convey. One European resident described seeing 'day by day and night after night … the sky above Queen's Road crimson with the glow of funeral-pyres'.[36] Another, a missionary, wrote that the burial or burning of the dead became 'a hideous and nauseating spectacle. The smoke of the funereal pyre ascended day and night … Bombay was as a city of the dead.'[37]

In appraising these photographic images of the plague dead, we need to consider two further factors. Firstly, many Indians were not cremated. It remained a practice largely confined to (upper-caste) Hindus, Sikhs and Jains, and as such its photographic representation might serve to inform outside observers of the funerary rites peculiar to those specific communities. Muslims, Christians and many low-caste Hindus were normally buried. Parsis from Bombay's small but influential Zoroastrian population exposed their dead on the Towers of Silence. One of the photographs in the Gatacre album shows the arched entrance to the Haines Road burial ground used by Hindus; a second depicts a huddle of onlookers watching the burial of a low-caste Hindu at the same location.[38] However, contrary to custom, many low-caste Hindus, especially the unclaimed dead, were cremated and not buried during the plague years, from sanitary necessity and municipal expediency, as a ready and effective way of disposing of deceased paupers and to prevent their corpses from infecting others.

Cremation was not only a religious rite: it was also, in colonial India, an approved sanitary procedure. From at least the 1840s, cremation was used by the colonial medical and sanitary authorities to destroy the bodies of Indians who had died in epidemics, such as the Pali plague in northern India in 1850 and during outbreaks of cholera, in the belief that incineration was the most efficient means of curbing contagion.[39] According to the *Indian Medical Gazette* in 1887, 'destruction by fire, or cremation, which entirely abolishes all the offensive and perilous stages of decomposition, is infinitely preferable to any other mode of disposing of the dead'. It was, besides, the established custom of 'a large majority of the population' and so more acceptable in India than it currently was in the West.[40] In the early plague years miasmatic theory still lingered on in India: it was thought better to burn the dead (along with their clothing and belongings) rather than burying them, which might allow their decomposing corpses to release toxic miasmas into the air, water and soil.[41] The Bombay plague photographs touch upon the moral ambiguities of funerary fire. Was cremation a means of honouring the dead as Hindu tradition would suggest, or was it a mere matter of sanitary convenience, facilitating the speedy disposal of the unwanted dead? Figure 5.3 suggests the former; Fig. 5.4 hints more at the latter.

Secondly, we should note that by contrast with India, where a million or more corpses were burned each year, the number cremated in Britain, continental Europe and the United States in this period was extremely small. In 1896 there were 733 cremations in the whole of Britain and by 1899 there had only been 1523 cremations since the Woking crematorium in Surrey (the first in Britain) began its operations in 1885.[42] Moreover, in the West the dead were burned in modern cremation furnaces where their bodies, doubly secluded from sight inside closed caskets within purpose-built crematoria, were invisible to mourners and onlookers, not consumed by fire on open-air pyres and exposed to the public gaze as in India.[43] India's were a highly visible dead, and so gave photography a rare opportunity to observe and record the destruction of a human body by funerary fire. This exceptional visibility and the supposed primitivism of Hindu cremation compared to Western crematory techniques, meant that the cremations depicted in the Bombay photographs and sketches might speak, intentionally or otherwise, to the unsettling (but also alluring) spectacle of 'Oriental' exoticism, of bizarre 'Eastern' bodily practices and strange funeral rites far removed from Victorian convention.[44] The photographs of Sonapur in the Gatacre album, especially those attributed to Moss, suggest the primitive, even haphazard, nature of the scene as seen through Western eyes, though ironically (as already indicated) they also showed how cremation grounds in urban India had been reformed over the previous half-century—with their iron stakes and screens, their entrance gates and perimeter walls, their gas or electric lamps, and benches for mourners to rest on.[45]

If cremation photography in India was an ancient rite seen through the modern lens, one seemingly untouched by the industrial modernity of the cremation furnace, in the West cremation in any form was still a novelty (or, at most, a distant echo of pre-Christian practice). But the growth of the cremation movement in the West from the early 1870s made the example of India the more pertinent—as a living illustration of how to (or, more commonly, how not to) cremate the dead.[46] Cremation was proposed by Western advocates as secular, sanitary and environmentally sound, but it was also being advanced as the only safe and effective means for the disposal of the huge numbers of both human bodies and animal carcases left by war—the Franco-Prussian War in 1870–1871 and the Russo-Japanese war of 1904–1905 being cases where hygienic fears favoured burning the dead.[47] It was suggested (but promptly rejected) for the disposal of the dead of the Western Front in the First World War, Hindu and Sikh soldiers alone being honoured in this way.[48] Cremation was clearly high on the international mortuary agenda at the time when plague struck Bombay in 1896. Plague did not create the photographic trope of the Hindu cremation pyre but it certainly gave it fresh impetus and a wider circulation.

Yet in India and its photographic archive scientific objectivity and Orientalist sensationalism are not easily distinguished, just as cremation itself veered between the respectfully ceremonial and the abruptly functional. Hindu cremation had many negative—indeed stridently hostile—associations in the nineteenth century, as reflected in both visual imagery and the textual narration that accompanied or underpinned it. Cremation epitomised, to many Western eyes, the horror that was Hinduism. Until the 1830s Hindu cremation was little discussed or visually represented, except through sati and the funeral pyre of their husbands on which 'devoted', drugged or deluded widows were burned or on which they immolated themselves. Only with the abolition of sati in 1829 did the cremation pyre itself acquire a distinct iconographic existence, and even then Western commentators wilfully persisted in conflating the two, the imagined sati an ever-present spectre at the husband's funeral pyre.[49] There was, besides, a deep repugnance felt for cremation by many Western observers, especially missionaries, who believed cremation unchristian and ungodly, disrespectful to the dead, and offensive to the human senses: it was one of the constraints of the camera that it was only able to capture the sight of a Hindu cremation and not its impact on the other senses. Into the 1890s and beyond cremation in the conventional Indian manner remained one of the declared abominations of Hinduism, and as such it regularly featured in travel narratives, missionary tracts, and, not least, photography.

The sacred city of Benares (Varanasi) was pivotal here. Perceived as the very epicentre of the Hindu religion, Benares became a living tableau of all that was exotic and, to foreign eyes, uplifting or appalling about Hinduism. As the nineteenth century progressed, the identification of Benares with its famed Manikarnika cremation *ghat* on the banks of the Ganges grew ever more prominent and emblematic in Western accounts of the city.[50] Photography played a critical part in this. The first photographs of the *ghat* appeared in the 1850s, within twenty years of the introduction of photography to India.[51] In 1863 the British photographer Samuel Bourne visited Benares, declaring it to be 'well worth the attention of the photographer'.[52] He made several images of the *ghats*, in some of which burning bodies, or bodies awaiting cremation, are visible. However, Bourne's written account of the burning *ghat* is still more explicit in its message than his photographs. 'Five or six savage looking men were heaping wood on the blazing piles', he wrote, 'but I could discern through the flames, the roasting skull and feet of one of the bodies.' These belonged to a woman, 'whose husband stood by evidently regarding the horrid spectacle with the highest satisfaction'. Bourne added: 'On every hand [in Benares] you are reminded of the religious zeal of this deluded people. Their gods—hideous, shapeless monsters—are daubed on every wall, and on hundreds of little dirty so-called temples.'[53] Bourne has been hailed as a photographer of the Indian picturesque, but in Benares, in image and in text, the burning *ghat* stood for India's anti-picturesque, cremation construed as a sign of a pervasive pagan horror.[54]

Not every European (or American) was so shocked by what they saw of Indian cremation, whether at Benares or in Bombay and Calcutta. Alongside the discourse of loathing and the politics of repugnance, one can see the emergence of a very different response—one grounded in curiosity and voyeurism perhaps, but also appreciation and respect, a shift in sentiment that was reflected in the Western, as well as Indian, photography of the period. Several factors lay behind this. One was the rise of the cremation movement in the West from the early 1870s and hence Western curiosity about how cremations were performed in India, so much more visibly and in far greater numbers than in Europe or North America. Through a combination of Cook's tours, religious cults like Theosophy, the rise of photojournalism and the growth of both amateur and commercial photography, India began to witness the emergence of a kind of photographic necro-tourism, in which travellers visited and photographed not just Mughal mausolea—the Taj Mahal at Agra, Akbar's tomb at Secundra—or the memorial to the 1857 massacre of Europeans at Cawnpore (Kanpur), but also the cremation *ghat* at Benares, the burning ground at Sonapur, and the Parsi Towers of Silence on Bombay's Malabar Hill.[55] Postcards, travel narratives, illustrated guidebooks and even the text to accompany slideshows for armchair travellers—all dwelt on the allure of these exotic funeral sites and the strange practices they embodied.[56]

## Indian Photography and the Hindu Dead

Visiting, describing and imaging the dead and their cremation was not just a matter of Western fascination. It was also a subject for the Indian photographer. Although the Gatacre album is the most extensive and widely referenced source, it is evident that some of the photographic images of the Bombay plague were taken by Indians and conceivably they can be taken to represent their view of the plague rather than a mimicking of Europeans. As previously noted, the cremation scene depicted in Fig. 5.3 was also published in London's *The Graphic* weekly in September 1897 (few Indian newspapers had the technical capacity to reproduce photographs before the First World War). It appeared there alongside three other images of the Bombay plague—the others showing Hindus propitiating the goddess of plague, the use of a water hose to cleanse an infected building, and the house in Mandvi Bunder Road, close to the docks, where plague first arrived in the city. Only two of the four images appear in the Gatacre album: the cremation ground at Sonapur and the hosing down of buildings. Taken as a group these four photographs are very different in their quality and appearance from the rather over-exposed and fuzzy images taken by Moss for the Gatacre album and the two that appeared there are generally presumed to be Stewart's work. Certainly, they are more carefully staged and appear more empathetic to their Indian subjects—as suggested by the worship of the plague goddess and the observance of religious rites in the cremation scene. They suggest that the Indians portrayed had been willing participants in the photographing process and not merely the subjects of a more distant or censorious colonial gaze. Might they in fact have been the work of an Indian photographer?

At the bottom of the page of plague photographs, *The Graphic* carried the note in small type: 'From Photographs by Khemji Hirji Kayani.' Does this mean *taken* by Kayani or simply *supplied* by him? Kayani was a well-known individual in Bombay at the time, a leading figure in the city's Dasa Oswal Jain community, a cotton merchant and landowner originally from Kutch in western Gujarat. A frequent correspondent to Bombay's *Times of India* newspaper in the 1890s and 1900s, one of his main concerns was the urgent need for sanitary improvement in Mandvi, 'our unfortunate and voiceless ward' as he described it, and the initial locus of the city's plague epidemic.[57] When in April 1900 a small plague hospital was opened in Mandvi for Dasa Oswal Jains, Kayani was appointed its manager.[58] He could be critical of the authorities but his tone was moderate compared to the many, more vociferous critics of the government's anti-plague regime.[59]

Although not a Hindu (Jains, too, practised cremation), Kayani's images steer us away from the semi-official oeuvre of Moss and Steward and into the realm of Indian agency and the representation of plague as an Indian catastrophe. Indeed, the four photographs in *The Graphic* could be read as an Indian, rather than colonial, narrative sequence that contrasts with that of the Gatacre album, which, after all, opens with a portrait of Lord Sandhurst and the all-white Plague Committee. *Image 1*: plague arrives in Mandvi and results in Indian deaths 'before' (as the caption pointedly states), 'the municipality began to take official notice of it'. *Image 2*: the municipal authorities belatedly arrive on the streets to enforce anti-plague measures and disinfect buildings. *Image 3*: as plague continues, devout Hindus turn to propitiation of the plague goddess, seeking divine intervention to end the epidemic. *Image 4*: as deaths continue to mount, Indians respectfully dispose of their dead by the most appropriate religious means—cremation.[60]

Although the photographs and their captions might reflect Kayani's personal mindset, it is unlikely that he took the photographs himself. They may perhaps have been commissioned by him and reflect his impressions of plague in and around Mandvi, but I have found nothing to indicate that he practised photography himself. It is far more likely that the photographs were merely forwarded by Kayani to *The Graphic*. Perhaps they were taken by Stewart but it seems more likely that they were the work of an Indian professional photographer, Shivshanker Narayan, perhaps working on behalf of Stewart. Possibly Stewart, based in Poona and acting as the complier of the Gatacre album, simply incorporated some of Narayan's photographs along with those taken by Moss but did so without any attribution to his Indian colleague. Describing his studio in Bombay as offering (rather like that of Taurines) 'views of India and native characters', Narayan had earlier (in the 1870s) made a series of empathetic ethnographic images of Indian artisans and taken high-quality interior views of the Calcutta International Exhibition of 1883–1884.[61] More to the point, in June 1897 he exhibited a collection of sixteen photographs illustrating scenes from the plague epidemic. Presumably these were his own: there is no reference in the press report to the work of any other photographer. From the descriptions given, four of these directly correspond to the images published in *The Graphic* three months later. Moreover, according to *The Times of India*, this 'souvenir' series would have been incomplete if

Narayan had not included a 'funeral scene, which shows the cremation of plague victims at the Hindoo burning ground in Sonapore'.[62] This is undoubtedly the image discussed here as Fig. 5.3. We seem, therefore, to have images of the Bombay plague produced by both European and Indian photographers, and so the possibility of discerning differences in the perspectives they reveal—the Europeans more concerned with the state-directed sanitary campaign against plague, the Indians with the impact of plague on their own lives and on the (respectful) disposal of their dead.

One further point needs to be made about Indian photography in this period. Photo-iconography was becoming by the 1880s and 1890s central to Indian religious and public life: this includes photographic images of the illustrious Indian dead, taken post-mortem but prior to their cremation. Among these were Keshub Chandra Sen, leader of the reformist Brahmo Samaj, and Vidyasagar, a renowned Bengali educationalist and social reformer, whose cremation was halted at the burning *ghat* so that final photographs could be taken.[63] The image of Sen was taken by a photographer from the Calcutta firm of Bourne and Shepherd (of which Samuel Bourne had been a founder), that of Vidyasagar by a photographer from a Bengali studio. Professional photographers were often called upon to do this work. The task of photographing the Indian dead might transcend the racial divide, but it seems likely that Indians either took such photographs or gave specific directions as to how they were to be taken. In August 1886, when the Hindu sage Ramakrishna Paramahamsa died at Cossipore near Calcutta, he was photographed, his half-exposed body decked with flowers and surrounded by loyal disciples. This image was taken by a Bengali photographic studio at the behest of Dr Mahendralal Sarkar, Ramakrishna's former physician and interlocutor, and an influential figure in the promotion of Indian science and medicine in late nineteenth-century Calcutta. It was surely intended to memorialise a Hindu saint, to present an object for contemplation, and to show the dedicated cohort of his closest followers.[64]

By the plague years of the 1890s photography in India was no stranger to the dead. One can, of course, see this genre of post-mortem photography in India as following the Western practice of photographing the dead.[65] But this was more than mere iconographic mimicry. By the late nineteenth century there had emerged several forms of photographic representation of the dead and of cremation. One of the most conspicuous of these related to the Hindu idea of *darshan*, the spiritual boon attached in

India's 'visually oriented culture' to seeing an image of a god or the living presence of a sage, a saint or highly revered public figure.[66] Darshan is conventionally associated with the living and with representations of the divine, but it also applied to the memorialising of those whom Thomas Laqueur refers to in a very different cultural context as the 'special dead'.[67] Observing and imaging (in this case through photography) became a way of remembering, visualising and revering the exemplary dead. Perhaps the most celebrated illustration of this photographic darshan of the dead is that of Mohandas Gandhi, following his assassination in January 1948, where he is shown naked to the waist, his body strewn with flowers and an auspicious tilak on his forehead.[68] In an instance such as Gandhi's, the intimate and reverential darshan photograph of the body before cremation is supplemented by the image of the cremation pyre, the spectacle of the actual burning of the dead, a sight that attracted emotional crowds and produced scenes of intense collective grief. This had once been the spectacle of sati. Now it was a spectacle the modern technology of the camera was able to encapsulate and memorialise.

But there were other kinds of iconographic idioms of the dead, in which the element of darshan was far less evident. One of the most arresting images of cremation was a photograph taken by Lala Deen Dayal of the cremation of the Maharaja of Indore in 1886, early in his photographic career.[69] Where European photographers, such as Bourne, had often sought to sensationalise Hindu cremation and to present it as a 'horrid spectacle', Dayal's image suggests a spectacle of a quite different sensory order, calmer, reflective, even rather unremarkable. The remains of the ruler's body are barely distinguishable in the smoke of a dying fire. Indians stand around, looking relaxed, even bored. Some, in uniform, are from the maharaja's band; others are simply dressed in white. There are hardly any women to be seen, and certainly no satis. Some onlookers squat peacefully on the ground around the pyre and along the riverbank; others gaze down from a bridge. Many of the spectators have turned away from the smouldering pyre to look up at the camera, no grief evident on their faces. Like the later plague photographs, Dayal's image was reproduced in *The Graphic* newspaper, this time in the form of an engraving, less than a month after it was taken, one of several images of Indian cremation to appear in Britain's illustrated press in the late nineteenth century. *The Graphic* carried a brief commentary, written by Dayal himself. As the body was carried to the cremation site, he wrote, 'Every point of vantage was speedily filled by the large crowd, many of whom were visibly moved with grief' (this is *not* what his photograph actually shows since it was taken at a later stage of the proceedings). The late ruler's elder son lit the pyre 'amidst the deafening noise of drums, rifles, and cannons, the latter firing fifty-one rounds to indicate his age'. However, 'at the time the picture was taken', the commentary concluded, 'little remained of the whole'.[70] Dayal's photograph speaks to a different kind of audience from the earlier photographs and textual descriptions of Samuel Bourne: cremation has—at least for the moment—been stripped of its horrors. Those 'horrors' were to be iconographically revisited in the plague years soon after, but for the moment the camera captured a more subdued and contemplative funeral scene.[71]

## Conclusion

The photographic images of Bombay's plague dead principally discussed in this chapter (Figs. 5.3 and 5.4) present a wealth of interpretive possibilities. Like all of the contemporary plague photographs, they represent the unnamed dead, not the 'special dead' who were individually named and honoured. The scene revealed in Fig. 5.4 is an example of the mundane dead, the ordinary men and women whose bodies were burned unceremoniously in their thousands in city cremation grounds like Sonapur in Bombay: plague was above all else a destroyer of the Indian poor and this cremation scene, like many others, alludes to the nature and scale of that great mortality. But Fig. 5.3, while similar in some ways, including the 'modern' setting of an urban cremation ground, suggests a more respectful and religious disposal of the dead: the man is properly clothed and attended by priests. The image unlocks the possibility of an Indian photo-iconography that was more attuned to Indian experience and suffering than the more colonial administrative and sanitary worldview suggested by most of the images in the Gatacre album. Moreover, with respect to plague, photography and cremation, the 1890s represented a highly significant moment of conjunction in India. Plague struck at a time when photography had already become a widespread and versatile visual technology, a vehicle for amateurs and professionals alike. More than just an official medium or one confined to supposedly objective scientific, scenic and technical uses, it was also a means by which Europeans and Indians alike could record and communicate affect, a sense of belonging or of alienation, and a whole gamut of emotion from fear to delight, from disgust at burning corpses to ritual respect for the dead. Although a minority of the photographs in the archive of Indian plague images in the 1890s pertain directly to the body or the plague-stricken corpse, those that do open up a world of interpretive possibilities that are richly informative about the human responses to plague but also the ideological usages of photography and the social and political agendas of those who practised it in colonial India.

## Notes

1. Graham Twigg, *Bubonic Plague: A Much Misunderstood Disease* (Ascot: Derwent Press, 2013), 9, 69, following L. Fabian Hirst, *The Conquest of Plague: A Study in the Evolution of Epidemiology* (Oxford: Clarendon Press, 1953).
2. For plague in India, see: David Arnold, *Colonizing the Body: State Medicine and Epidemic Disease in Nineteenth-Century India* (Berkeley: University of California Press, 1993), chapter 5; Mark Harrison, *Public Health in British India: Anglo-Indian Preventive Medicine, 1859–1914* (Cambridge: Cambridge University Press, 1994), chapter 6; Raj Chandavarkar, *Imperial Power and Popular Politics: Class, Resistance and the State in India, c. 1850–1950* (Cambridge: Cambridge University Press, 1998), chapter 7.
3. On the need for a critical and contextual approach to imperial photography, see: James R. Ryan, *Picturing Empire: Photography and the Visualization of the British Empire* (London: Reaktion Books, 1997), 16–18.
4. For major contributions to the history of photography in India, see: John Falconer, 'Ethnographical Photography in India, 1850–1900', *Photographic Collector* 5, no. 1 (1984): 16–46; Christopher Pinney, *Camera Indica: The Social Life of Indian Photographs* (Chicago: University of Chicago Press, 1997); Maria Antonella Pelizzari (ed.), *Traces of India: Photography, Architecture, and the Politics of Representation, 1850–1900* (Montreal: Canadian Centre of Architecture, 2003); Zahid R. Chaudhary, *Afterimage of Empire: Photography in Nineteenth-Century India* (Minneapolis: University of Minnesota Press, 2012); Malavika Karlekar, *Visual Histories: Photography in the Popular Imagination* (New Delhi: Oxford University Press, 2013).
5. On the corpse as evidence, see Christopher Pinney, *The Coming of Photography in India* (London: British Library, 2008), pp. 18–19; for photography in crime detection: B. Sardar Singh, *A Manual of Medical Jurisprudence for Police Officers* (3rd ed., Moradabad: Lakshmi Narayan Press, 1916), pp. 165–68. By the time of the third plague pandemic the use of photography extended to micro-images of the plague bacillus itself.

6. See, for instance, Lukas Engelmann, '"A Source of Sickness": Photographic Mapping of the Plague in Honolulu in 1900'. In Lukas Engelmann, John Henderson and Christos Lynteris (eds.), *Plague and the City*, pp. 139–58 (London and New York: Routledge, 2019).
7. On famine photographs, see: J. Forbes Watson and John William Kaye, *The People of India* (8 vols, London: W. H. Allen, 1868–75), vol. 4 (1869), p. 206; John Falconer, 'Willoughby Wallace Hooper: A Craze about Photography', *Photographic Collector* 4, no. 3 (1983): 258–86; Chaudhary, *Afterimage*, chapter. 4.
8. Robert Flynn Johnson, 'Reverie and Reality'. In Robert Flynn Johnson (ed.), *Reverie and Reality: Nineteenth-Century Photographs of India from the Ehrenfeld Collection* (San Francisco: Fine Arts Museums of San Francisco, 2004), p. 4.
9. See, for example, the textual and photographic representation of the recent famine in George W. Clutterbuck, *In India, the Land of Famine and Plague* (London: Ideal Publishing Union, 1897), pp. 221–22.
10. This photographic absence is discussed in David Arnold, 'Death and the Modern Empire: The 1918–19 Influenza Epidemic in India', *Transactions of the Royal History Society* 29 (2019): 181–200.
11. Copies of this album are held in London by the British Library, the Wellcome Library and the National Army Museum, at the University of Cambridge's Visual Plague Photographic Database, and in Los Angeles by the Getty Research Institute. The number of images in these albums varies between 125 and 142. For a discussion of the contents, see Shivani Sud, 'Water, Air, Light: The Materialities of Plague Photography in Colonial Bombay, 1896–97', *Getty Research Journal* 12, 2020, 219–30; Shivani Sud, 'Bombay Plague Visitation, 1896–97', https://blogs.bl.uk/asian-and-african/2020/07/bombay-plague-visitation-1896-97.html. The Getty also holds an album of 147 photographs of 'Poona Plague Pictures, 1897–1898', complied by C. H. B. Adams-Wylie of the Indian Medical Service: http://hdl.handle.net/10020/cifa96r95
12. On this distinction, see the British Library catalogue note for photo collection 311/1.
13. As in the case of Bombay; see: *The Artful Pose: Early Studio Photography in Mumbai, c. 1855–1940* (Ahmedabad: Mapin Publishing, 2020). By 1913, Bombay had twenty-six photographers or photographic studios, Calcutta 17, and Madras 23: *The Madras Press Almanack and Directory of Madras and Southern India, 1913* (Madras: Lawrence Asylum Press, 1913), pp. 1287, 1311, 1337–38.
14. See advertisements for Taurines's studio in *Times of India*, 25 November 1884, p. 8, and ibid., 15 April 1885, p. 7.

15. As an army officer, Moss may have known Gatacre as the military commander of Bombay District in 1896–97. However, unlike many officers, Moss does not appear to have been involved in plague duties in Bombay.
16. See, for instance, the committee's unscheduled visit to the Muslim burial ground and Hindu cremation site on Queen's Road in Bombay on 14 April 1897, which may have been when several of Moss's photographs were taken: *Times of India*, 15 April 1897, p. 5.
17. As in the case of Thomas Biggs of the Bengal Artillery in the 1850s: John Falconer, 'A Passion for Documentation: Architecture and Ethnography'. In Vidya Dehejia (ed.), *India Through the Lens: Photography, 1840–1911* (Washington, D.C.: Smithsonian Institution, 2000), p. 71. See also the work of a Madras Artillery officer, Captain Allan N. Scott, *Sketches in India: Taken at Hyderabad and Secunderabad in the Madras Presidency* (London: Lovell Reeve, 1862).
18. *Times of India*, 10 June 1895, p. 5. For Devlin's photographs, see Sergeant-Major Devlin, *Views in Chitral Taken During the Advance of the 3rd Brigade of the Chitral Relief Force Under the Command of Brigadier-General W. F. Gatacre* (London: Maclure, 1895).
19. W. F. Gatacre, 'Preface', *Plague Visitation, Bombay, 1896–97*. For an assessment of Gatacre's plague work, see: Samuel K. Cohn, Jr., *Epidemics: Hate and Compassion from the Plague of Athens to AIDS* (Oxford: Oxford University Press, 2018), 320–31.
20. M. E. Couchman, *Account of Plague Administration in the Bombay Presidency from September 1896 till May 1897* (Bombay: Government Central Press, 1897), pp. 75–76. For suspicion of, and opposition to, the state's anti-plague measures, see: Ian Catanach, 'Plague and the Indian Village'. In Peter Robb (ed.), *Rural India: Land, Power and Society under British Rule*, pp. 216–243 (London: Curzon Press, 1983); David Arnold, 'Touching the Body: Perspectives on the Indian Plague, 1896–1900'. In Ranajit Guha (ed.), *Subaltern Studies V*, pp. 55–90 (Delhi: Oxford University Press, 1987).
21. For photography as either 'poison' or 'cure', see: Pinney, *Coming of Photography*. See, too, the photographs of famine relief measures in Hyderabad taken by Lala Deen Dayal and reproduced in Judith Mara Gutman, *Through Indian Eyes: 19th and Early 20th Century Photography from India* (New York: Oxford University Press, 1982).
22. For instance, R. W. Hornabrook, *Report on the Dharwar Plague Hospital, August 28th–December 18th 1898* (Dharwar: Dharwar Plague Hospital, 1899).
23. Robert Peckham, 'Hong Kong Junk: Plague and the Economy of Chinese Things', *Bulletin of the History of Medicine* 90, no. 1 (2014): 32–60; Robert Peckham, 'Plague Views: Epidemics, Photography, and the Ruined

City'. In Lukas Engelmann, John Henderson and Christos Lynteris (eds.), *Plague and the City*, pp. 91–115 (London and New York: Routledge, 2018).

24. Especially *Plague Visitation*, folios 15 and 21. Two photographs in the Wellcome Collection, unattributed but dated to c. 1897, show the dead bodies of a woman and a man, laid out respectively on a stretcher and a rattan bed.
25. Couchman, *Account*, p. 138.
26. *Graphic*, 30 January 1897, p. 120.
27. *Harper's Weekly*, 3 June 1899, p. 551.
28. These include 'Hindu Funeral Party En Route to Burning Ground'; 'Sonapore: General View of Hindu Burning Ground'; 'Sonapore: Corpse Ready for Cremation'; 'Sonapore: Corpse Burning'; 'Sonapore: Interior of Burning Ground with Corpse on Pyre', *Plague Visitation*. A separate image can be found in the Wellcome Collection, no. 2988, 'Burning Ghats, Bombay', dated c. 1900.
29. Couchman, *Account*, p. 138.
30. This page of photographs from *The Graphic* is reproduced in Sud, 'Bombay Plague Visitation', as fig. 2.
31. *Graphic*, 18 September 1897, p. 22.
32. The Wellcome Collection description of this image reads: 'A group of men lower the body of a dead man on to a pyre of logs prior to a Hindu cremation ceremony in Bombay during the time of the plague.'
33. For this and other aspects of modern Indian cremation, see: David Arnold, *Burning the Dead: Hindu Nationhood and the Global Construction of Indian Tradition India* (Oakland: University of California Press, 2021), chapter 3.
34. On the differential social impact of plague, see: Ira Klein, 'Plague, Policy and Popular Unrest in British India', *Modern Asian Studies* 22, no. 4 (1988): 723–55.
35. *Times of India*, 25 January 1897, p. 5.
36. S. M. Edwardes, *The Bombay City Police: A Historical Sketch, 1672–1916* (London: Oxford University Press, 1923), p. 49.
37. Clutterbuck, *In India*, pp. 35–36.
38. See Cambridge Visual Plague database, photos: https://doi.org/10.17863/CAM.29850 and https://doi.org/10.17863/CAM.29851
39. For the burning of bodies in the Pali plague, see: C. Renny, *Medical Report on the Mahamurree in Gurhwal in 1849–50* (Agra: Secundra Orphan Press, 1851), pp. 49–50.
40. 'Destruction versus Decomposition', *Indian Medical Gazette* 22 (1887): 175.

41. On the sanitary case for cremation, see: Edmund A. Parkes, *A Manual of Practical Hygiene* (London: John Churchill, 1864), pp. 457–58; Patrick Hehir, *Hygiene and Diseases of India: A Popular Handbook* (3rd ed., Madras: Higginbothams, 1913), pp. 407–12; A. J. H. Russell (ed.), *McNally's Sanitary Handbook for India* (6th ed., Madras: Superintendent, Government Press, 1923), pp. 247–50. For the burning of the possessions of the dead, see: *Amrita Bazar Patrika*, 21 April 1898, p. 5; ibid., 28 April 1898, p. 6; ibid., 12 May 1898, p. 5; and the use of fire to purge cities of plague, see: Christos Lynteris, 'A "Suitable Soil": Plague's Urban Breeding Grounds at the Dawn of the Third Pandemic', *Medical History* 61, no. 3 (2017): 354–56.
42. *Transactions of the Cremation Society of England* 9 (1896): 11; ibid., 13 (1900), p. 11.
43. On the rise of cremation in the West, see Thomas Laqueur, *The Work of the Dead: A Cultural History of Mortal Remains* (Princeton: Princeton University Press, 2015), part 4.
44. For cremation as Orientalist fantasy, see the depiction of the burning of the body of a Hindu merchant in Algiers in *L'Illustré du Petit Journal*, 1 January 1933, Getty Images online, no. 144848613.
45. *Plague Visitation*, folio 16.
46. For reference to Indian cremation, see: William Eassie, *Cremation of the Dead: Its History and Bearings upon Public Health* (London: Smith, Elder, 1875), pp. 90–98.
47. On the first of these, see: Edmund A. Parkes, *A Manual of Practical Hygiene* (4th ed., London: J. & A. Churchill, 1873), p. 442.
48. David Crane, *Empires of the Dead: How One Man's Vision Led to the Creation of WWI's War Graves* (London: William Collins, 2013), pp. 66–67.
49. As in E. H. Nolan, *The Illustrated History of the British Empire in India and the East* (2 vols, London: James S. Virtue, n.d. [c. 1859]), Vol. 2, pp. 37–38, 489–90.
50. Arthur Parker, *A Hand-Book of Benares* (Benares: E. J. Lazarus, 1895), pp. 44–45.
51. Photograph of Manikarnika Ghat in Benares by Felice Beato, c. 1858–60, in Partha Chatterjee, 'The Sacred Circulation of National Images', in Pelizzari, *Traces of India*, 281.
52. Samuel Bourne (ed. Hugh Ashley Rayner), *Photographic Journeys in the Himalayas* (Bath: Pagoda Tree Press, 2009), p. 5.
53. Ibid., p. 6.
54. Gary D. Sampson, 'Photographer of the Picturesque: Samuel Bourne', in Dehejia, *India Through the Lens*, pp. 163–75; Sophie Gordon, *The Imperial*

*Gaze: The Photographs of Samuel Bourne* (New York: Alkazi, 2000), p. 5. For the use of cremation photographs to represent heathen barbarity, see: Ryan, *Picturing Empire*, pp. 196–97.

55. Piers Brendon, *Thomas Cook: 150 Years of Popular Tourism* (London: Secker & Warburg, 1991), pp. 146–48. On the Cawnpore memorial, see: Chaudhary, *Afterimage*, pp. 12–13. For images of a Tower of Silence and Sonapore cremation, see *Views of Bombay: Old and New* (Bombay: 'Times of India' Office, n.d. [c. 1919]).
56. Advertisement for colour postcards by Thacker & Co., *Times of India*, 31 July 1901, p. 2; James Ricalton, *India Through the Stereoscope: A Journey Through Hindustan* (New York: Underwood & Underwood, 1907), pp. 30–36, 182–89. On plague postcards, see: https://scroll.in/magazine/891745/how-the-british-used-postcards-as-a-propaganda-tool-during-the-bombay-plague
57. *Times of India*, 14 July 1899, 4; ibid., 3 August 1900, p. 3.
58. *Report of the Municipal Commissioner on the Plague in Bombay for the Year Ending 31st May 1900* (Bombay: 'Times of India' Press, 1901), p. 257.
59. *Times of India*, 18 March 1898, p. 6; ibid., 19 April 1898, p. 5; ibid., 24 May 1899, p. 3; ibid., 6 July 1899, p. 6. In 1901 Kayani sent the government a letter of condolence on the death of Queen Victoria: ibid., 2 February 1901, p. 7, but in 1915 was listed as a delegate to the Indian National Congress: *Report of the Thirtieth Indian National Congress Held at Bombay on 27th, 28th and 29th December 1915* (Bombay: Commercial Press, 1916), p. 313.
60. This sequence is indicated by the order of the photographs described in 'A Souvenir of the Plague in Bombay', *Times of India*, 21 June 1897, p. 3.
61. E.g., 'Dyers at Work in Western India' (c. 1873) in British Library, Online Gallery, image 1000/52 (4909); advertisement, *Times of India*, 1 March 1910, p. 11; Peter H. Hoffenberg, 'Photography and Architecture at the Calcutta International Exhibition', in Pelizzari, *Traces of India*, pp. 174–93.
62. 'A Souvenir of the Plague in Bombay.' The report continues: 'This series will form a very useful addition to notes and reports on the epidemic, and will also be an interesting and expressive record of the plague in Bombay, 1896–97': *Times of India*, 21 June 1897, p. 3.
63. *Times of India*, 14 January 1884, 6; Subal Chandra Mitra, *Isvar Chandra Vidyasagar: Story of His Life and Work* (2nd ed., Calcutta: New Bengal Press, 1907), pp. 268–89.
64. Gwilym Beckerlegge, *the Ramakrishna Mission: The Making of a Modern Hindu Movement* (New Delhi: Oxford University Press, 2000), chapter 6;

Liam Buckley, 'Photography, Contemplation, and the Worship of Sri Ramakrishna', *Visual Anthropology Review* 35, no. 1 (2019): 50–58.

65. Jay Ruby, *Secure the Shadow: Death and Photography in America* (Cambridge, MA: MIT Press, 1995), chapter 2; Beth Ann Guynn, 'Post-mortem Photography', in John Hannavy (ed.), *Encyclopedia of Nineteenth-Century Photography* (New York: Routledge, 2008), pp. 1164–67.
66. Diana L. Eck, *Darsan: Seeing the Divine Image in India* (Chambersburg, PA: Anima Press, 1981), pp. 1–4.
67. Laqueur, *Work of the Dead*, 83.
68. As reproduced in Judith M. Brown, *Gandhi: Prisoner of Hope* (New Haven: Yale University Press, 1989), p. 382.
69. For this image and its significance, see Arnold, *Burning the Dead*, 139–41. For the contrary claim that Dayal did not substantially depart from the conventions of the picturesque and a Western photographic aesthetic, see Chaudhary, *Afterimage*, 122–31.
70. *Graphic*, 24 July 1886, cutting in Cremation Society of England Archive, University of Durham, CRE H9.
71. Photographs of the cremation of India's Hindu princes were not uncommon: see, for instance, 'Unique Pictures of the Cremation of a Raja', *Times of India*, illustrated supplement, 13 February 1927.

CHAPTER 6

# Reflexive Gaze and Constructed Meanings: Photographs of Plague Hospitals in Colonial Bombay

*Abhijit Sarkar*

## Introduction: Towards 'Visual Literacy' in Plague

'Plague is the most deadly of all known diseases. Obscure in its origin, persistent in its duration, terrible in its effects, it has baffled alike the investigations of science and the observations of the most perspicacious.'[1] Such was the observation in 1899 by Captain James Knighton Condon, a member of the Indian Staff Corps who, under the orders of the colonial Government of Bombay, compiled a detailed illustrated history of the

Research leading to this chapter was funded by a European Research Council (ERC) Starting Grant under the European Union's Seventh Framework Programme/ERC grant agreement no 336564 for the project 'Visual Representations of the Third Plague Pandemic'.

A. Sarkar (✉)
Oxford University, Oxford, UK
e-mail: abhijit.sarkar@history.ox.ac.uk



C. Lynteris (ed.), *Plague Image and Imagination from Medieval to Modern Times*, Medicine and Biomedical Sciences in Modern History, https://doi.org/10.1007/978-3-030-72304-0_6

progress of bubonic plague in Bombay (Mumbai) covering the period from September 1896 to June 1899. India bore the full brunt of this zoonotic disease. To give a clear idea of the colossal scale of the human disaster, the report of the Indian Plague Commission for the year 1898–1899 calculated that 'the mortality from plague in the 36 months from September 1896 to September 1899 amounts to over 376,000 in the Bombay Presidency, and to about 54,450 in the rest of India'.[2] Later, in his 1925 study of epidemiology, leading US public health expert Isidore Sydney Falk noted the continuance of the calamitous scale of plague mortality in India: 'In recent years it has claimed some six hundred thousand deaths per annum in India. Within the fortnight we have learned that there were reported 11,388 deaths from plague in India in the period November 11 to December 8, 1923.'[3] Ira Klein's research has calculated that from the time of plague's arrival in India in 1896 to the Second World War, the country suffered 'about ninety-five percent of the world's plague mortality'.[4] Periodic outbreaks of the disease killed approximately twelve million Indians.[5]

Though the gargantuan scale of the Bombay Plague has expectedly given rise to an enormous body of historical literature, visual representations of the epidemic have not received similar attention.[6] This is particularly lamentable as the outbreak was covered extensively by the photographic lens. These photographs have found no space in the historiography of Indian photography either, which is equally extensive.[7]

The technology of photography reached India in the 1840s, soon after its invention was revealed to the public in 1839 by Louis Daguerre in France and Henry Fox Talbot in England.[8] In the history of photography in India, the Bombay Plague of 1896–1897 was the first epidemic to be captured by the camera lens. The importance of studying plague photography partially lies in the fact that this epidemic photography cemented the practice of having suffering commoners and subalterns as photographic subjects in colonial India, a process that had started with the photographing of the Great Madras Famine of 1876–1878 by Willoughby Wallace Hooper, an English captain (later colonel) in the Madras Cavalry, A. T. W. Penn, a commercial photographer, and by Nicholas brothers, of whom little information survives.[9]

However, until plague struck in 1896, this precedence of photographing famine victims appears to have had little impact on the overall selection of photographic subjects in colonial India, as it did not engender any sustained interest in distressed subalterns as photographic subjects, which was distinct from the widespread interest in taking physiognomic photographs of various 'tribes' or people of different ethnicities and castes, or ethnographic photographs of exotic 'natives' such as snake charmers and religious ascetics. The subjects of the photographs taken in India in the two decades between the Great Madras Famine and the Great Bombay Plague continued to be primarily architectural marvels, picturesque landscapes, hunts, military manoeuvres, individual and family portraits of Indian and foreign elites and rulers, and durbar (court) ceremonies.[10] In nineteenth-century India, the majority of the photographers of these subjects were British and other Europeans, primarily military officers or army surgeons (or both) doubling as amateur photographs, plus a few civil servants and some commercial photographers (both ex-army men and complete civilians).[11] However, there were also numerous Indian photographers active in these genres, mostly civilian commercial photographers, in addition to a few rulers of Indian princely states as well as some Indian academics and authors who turned into avid amateur photographers.[12] A major distinction between these non-medical genres and medical photography was in fact that the latter was a monopoly of European photographers—mostly the British, plus a few members of the Russian, French, Austrian and German medical missions to India in 1897–1898. This is a distinction that will prove crucial in my conclusion of the photographic analysis in this chapter.

The onset of plague introduced such a paradigm shift in the understanding of what were 'photographable' subjects in India, that henceforth the suffering of the common people occupied a central place in photography in general. Plague's cementing effect on the selection of the agony of subaltern people as photographic subjects was such that soon after its outbreak in Bombay, even an elite photographic firm like the Raja Deen Dayal & Sons Studio, run by Dayal who was one of the official photographers to the Viceroy of India and to Queen Victoria, was commissioned in 1899 to take photographs of the famine relief works carried out by his patron Mahbub Ali Khan, the sixth Nizam of Hyderabad.[13] In post-plague India, subaltern victims have firmly held their place as a steady photographic subject, a fact that is established by the photographs of the Bengal Famine of 1943–1944, and of the religious genocide and unspeakable suffering of the refugees before, during, and after India's blood-soaked partition in 1947 that triggered arguably the largest and bloodiest mass migration in history.[14]

Christos Lynteris has convincingly demonstrated that plague photography that emerged in the late nineteenth century was actually 'the first and paradigmatic form of epidemic photography'.[15] However, despite the existence of a rich archive of photographic representations of plague not only in Bombay, but in every part of the globe that was afflicted by the third plague pandemic (1890–1959), the historiography of the pandemic lacks what Jeffrey Mifflin calls 'visual literacy' in the context of historical medical photographs.[16] So far, histories of plague globally have been reliant too heavily on textual sources. The lack of photographic sources in these histories can be explained by the methodological challenges that these sources bring with them. Unlike textual sources that readily dish out data, descriptions, opinions and arguments to historians, images do not 'speak out' on their own. Messages and narratives exist *wordlessly* in the visual of images as their existential constituents. Therefore, to attain the status of being sources of history, images compulsorily require to be deciphered by two-fold intervention by the historian—firstly intervention in the form of the cognisance of a message, meaning, and narrative mutely embedded in them, and secondly intervention in the form of the historian's inference or interpretation of that message, meaning, and narrative. Thus, transforming images into useable sources of history is a much more convoluted and elaborate undertaking than working with self-expressive textual sources. This explains the severe underuse of photographs in the histories of plague, or in history in general. Against this backdrop, this chapter aims to afford greater visibility to photographs as sources in the writing of histories of plague, and thereby also put the history of plague in conversation with the history of photography.

Among all the non-textual sources, in the case of Bombay at least, photographs dominated the visual regime of plague; they outnumbered by far other forms of visual representation of the outbreak such as paintings, etchings, and sketches. Consequently, in any endeavour to analyse and understand the visual depiction of the Bombay Plague, photographs must be the starting point. Globally, photographs of anti-plague measures, clinical manifestations of the disease, and plague hospitals have sporadically made their way to scholarly articles and monographs on plague as evidence of, or visual attestation to, the themes discussed on the basis of textual sources. However, in the case of plague in Bombay, even such evidential use of plague photographs in scholarly histories is hard to come by, though a small number of contemporaneous official histories, for instance, Captain Condon's 1899 history of the Bombay Plague used some plague-related photographs, such as photographs of inoculation in Bombay and research laboratories, as mere illustrations of the contents of the written words.[17] Another rare illustrative use of plague photographs is found in the published lecture *The Cause and Prevention of the Spread of Plague in India* by army doctor Captain William Glen Liston, a member of the Plague Research Commission, on December 11, 1907 before the Bombay Sanitary Association, which included photographs of unsanitary chawls and gullies (narrow alleys) in residential neighbourhoods gravely infected by plague to demonstrate the speaker's point about the enormous rat infestation of Indian houses.[18]

Nevertheless, it cannot be stressed enough that 'visual literacy' in plague photography does not mean mere ocular acquaintance with the photographic archive and its use as an apparatus to attest to, or visually demonstrate or cement a narrative or point presented in textual-source-based histories of plague. Instead, fine-grained and critical reading of the photographs themselves as subjects as well as independent sources of the history of plague is an indispensable constituent of 'visual literacy' in plague.

## What Can Hospital Photographs Offer?

This study will demonstrate, through a granular examination of some representative photographs of plague hospitals in Bombay, that in the context of these hospitals, photographs offer a fascinating history of the colonial medico-political concerns and agendas. While proffering his theses on the 'photography of history', Eduardo Cadava asserts that 'the event of photography is necessarily anterior to any history of photography—photography does not belong to history; it offers history.'[19] In the case of plague hospitals in Bombay, a critical dissection of the subjects' positioning, their poses, and camera's angle reveals unseen but embedded histories of colonial medico-politics. In defying the limits of being a mere captor of sights, here photography does indeed offer histories that are beyond the visuality of captured sights. In this sense, as John Tagg points out, 'Photographs are never "evidence" of history; they are themselves the historical.'[20] It is precisely this understanding in which lies the purpose of my study of the photographs of plague hospitals as an independent subject of history.

Lukas Engelmann has revisited the foundational quandary: 'what are medical photographs of plague?', and subsequently argued for extending the definition or scope of medical or clinical photographs to the extent of incorporating photographs of built environments that do not apparently exhibit any connections with medical subjects or issues.[21] Engelman underscores that 'clinical photographs—capturing symptoms of people with plague—constitute just a fraction of plague's archive'.[22] His research has revealed that instead of plague-racked human bodies, 'the focus of doctors, epidemiologists and government officials was fixed largely on the locales, the environment and the buildings in which plague cases appeared or threatened to arrive. The photographic focus of plague images was emphatically the epidemic's ecology.'[23] What was the point of such a focus on plague's ecology? What was to be achieved by it? Christos Lynteris argues:

> Its aim was to generate an archive of the terrain of plague. A total record of what was imagined to be the environment or milieu where plague manifests itself, so that, with the development of further knowledge on the disease, future scientists could study this archive and draw vital conclusions and methods for eradicating plague. It was this colonial apparatus, rather than the microscope, that promised to capture the breeding grounds of plague; in other words, plague as not-yet-manifested, plague that-will-be manifested, and plague as what is always and already here.[24]

In furtherance of these observations about the focus of the lens of plague photography, my research reveals that in the particular case of Bombay at least, in addition to locales actually or potentially affected by plague, hospitals that sought to provide a relieving space from the malady also became a prime and recurrent subject of photography, fuelled by the fact that in plague-ravaged Bombay, hospitals still had novelty-value as a subject. In fact, it was the cataclysmic effect of the plague outbreak that triggered the actual growth of modern Western hospitals in the city. Hospitals were fundamental to the efforts at fighting plague. The chapter on hospitals in the 1897 *Report on the Bubonic Plague in Bombay*, authored by British Brigadier-General Sir William Forbes Gatacre, Chairman of the Bombay Plague Committee, opened with the acknowledgement: 'An important basis of the operations for the eradication of plague in the City of Bombay was the establishment of temporary hospitals to which plague cases could be sent.'[25] These precipitously initiated but rapidly developing hospitals were an integral part of the built environment of plague in the city. That is why the Bombay hospitals received so much photographic coverage. They received photographic attention to such an extent that, borrowing from Georges Didi-Huberman's description of photographic iconography of the Salpêtrière hospital in Paris, we can safely call the Bombay hospitals the 'image factory' of plague in India.[26] Photographs of these hospitals essentially captured the built environment that not only served as the physical site of a major part of the plague pandemonium, but also contributed to its causes. All these make the study of hospital photographs exceedingly important and an ideal entry point to understanding the overall visual representation of the built environment of plague in Bombay, and in turn of the pandemic as a whole.

If we read photographs of plague hospitals (and the captions that are existentially contingent on them) as self-standing sources of history, they emerge as quite handy for piecing together a narrative of the physical and social construction of the newly emerging treating sites in colonial Bombay. Given the already discussed centrality of hospitals in the medical responses to plague in Bombay, critical analysis of a set of their representative photographs in this study demonstrates how the novel visual medium of photography captured, and thereby also revealed, social realities at play on the site of hospitals. In addition, it shows how miscellaneous meanings were constructed, and messages conveyed, through the careful composition of the photographs.

Christopher Pinney has asserted: 'Pre-photographic representations always depended on the trustworthiness of the author/artist, and many early volumes of lithographs included assurances of the closeness of fit between the image and the reality.'[27] Against this pre-photographic past, photography was perceived as a 'cure' or 'solution' to the flaws in the truth claims in earlier technologies of representation.[28] Because, as Pinney puts it: 'Photography required no additional autoptic testimony for eyewitnessing was the ontological condition of the very existence of the photograph.'[29] It was as if a 'prosthetic eye' or the 'scientist's "true retina."'[30] Consequently, photographs were widely considered to proffer a 'stern fidelity' to reality, as Reverend Joseph Mullens, missionary pastor at the Native Church at Bhowanipore in Calcutta (Kolkata) and avid amateur photographer, argued before the Photographic Society of Bengal on October 29, 1856.[31] Because of this seemingly 'stern fidelity of photography', it seemed like the ideal medium to Mullens to document 'all the minute varieties of oriental life; of oriental scenery, oriental nations and oriental manners'.[32]

However, in stark contrast to Mullens's self-deceiving faith in the 'stern fidelity' of photographs to reality, the present study will demonstrate that instead of using photography to capture and thereby freeze unmediated sights for present and posterity with the possibility of reproducing each of those frozen sights in one or more numbers, varying 'realities' and corresponding meanings were actually being *constructed* through the careful composition of the photographs of plague hospitals. These realities were not real so to speak; they were realities staged for the camera pre-photography and finalised and fixed by the camera in their post-photography avatars. In revealing this process of manipulating reality, I follow the lead of Laikwan Pang's apt observations made in the context of visual modernity in China:

> If the imprecision of lithography indicates some unbridgeable distance between representation and reality, photography conflates the two. The new lithographic culture strove to stabilize meanings and objects through detailed visual representation, while photography has the opposite tendency of destabilizing reality, because the desire to manipulate reality becomes a new point of departure.[33]

Notably, with the advantage of hindsight, after living in an age of ubiquitous videography, we may note that manipulation or outright construction of 'reality' by photography had and still has an advantage, that it could be more memorable, and therefore its realities could be longer lasting than the realities constructed by moving images. Susan Sontag has highlighted: 'Photographs may be more memorable than moving images, because they are a neat slice of time, not a flow.'[34] It is this efficacy of still photography as a visual medium in engendering in the audience an intense and long-lasting impression that in turn enables it to construct effective realities or meanings.

## The Photographs and Their Questionable Authorship

The photographs that I examine in this study are all held at the Wellcome Library (London), mounted in the original hammered brown half-leather album with gilt-hatched border designs prepared in circa 1897, with the title 'Plague Visitation, Bombay, 1896–1897' printed in letterpress on the front cover in gold lettering.[35] A formal portrait photograph of Lord William Mansfield Sandhurst, the governor of Bombay presidency between 1895 and 1900, was mounted on the first page, presumably to afford the album an aura of state endorsement coming from the highest British authority in the Bombay presidency. However, the British royal coat of arms stamped in gold above the title made the imperial character of the album already conspicuous.

Notably, the album was not compiled as a mere matter-of-factly visual record of plague in Bombay, instead, in addition, its creators hoped that it would be enjoyed by the viewers as a curio or collector's item made with 'compliments', as indicated by the letterpress label pasted on the front fixed endpaper which presented the album 'With the *Compliments* of the Bombay Plague Committee'.[36] The front pastedown bears a one-page printed preface authored by Gatacre, which elaborates the contents of the album: 'The ensuing pages contain a series of Photographs taken during the prevalence of Plague in Bombay and during the period that the preventive measures for its extinction had been placed by the Bombay Government in the hands of the following Committee' [the Bombay Plague Committee].[37]

The album contains thirty-one pages with 143 mounted photographs. Eight of the pages contain one large-format albumen print each. In addition, there are four pages that contain respectively nine, five, four, and three photographs each. Most importantly, there are nineteen pages containing six smaller gelatine silver prints each, mostly arranged around certain themes. The dominant theme is hospitals (built environment as well as patients, nurses, doctors et al.) which are the subject of forty-six photographs. All hospital photographs analysed in the present study are among these smaller tipped-in prints. The main subjects of the remaining photographs are cremation and burial of plague corpses (see Arnold, this volume), search parties and house-to-house visitations in search of hidden plague patients and dead bodies, plague infected tenements and deserted houses, disinfection of chawls and streets, segregation camps, and medical inspection of passengers at train stations and boat jetties.

Gatacre's preface acknowledged a British army officer, Captain C. Moss of the Gloucestershire Regiment, as the photographer of most, *but not all*, of the photographs.[38] The preface also declared: 'Mr. F.B. Stewart, Photo Artist, Poona, compiled the Album for the Committee, and holds the copy-right.'[39] In 1903, *A Manual of Plague*, authored by Major William Ernest Jennings, the Chief Medical Officer for Plague Operations in the Bombay presidency, contained reproductions of six photographs from this album, and expressed gratitude to 'photo-artist' Stewart for permission to reproduce the photographs, without mentioning Moss at all.[40] The reproduced photographs showed the exterior and interior of a temporary plague hospital, a plague cot with a patient accompanied by an Indian woman, two plague infected dwellings opened up for desiccation, and a scene of medical examination of passengers about to embark at a port in Bombay.[41]

Original 1897 copies of the album are also at the British Library (henceforth BL) and the National Army Museum, in London, and at the Getty Research Institute (Los Angeles)—which makes it clear that planned dissemination of the album's visual messages was an integral component of this photographic project and accordingly the album was made in multiple copies.[42] The BL copy was gifted by D. B. G. Line (née Dimmock), daughter of Surgeon-Major Henry Peers Dimmock of the Indian Medical Service, who was a member of the same Bombay Plague Committee chaired by Gatacre.[43] The BL catalogue divides the photographs into two categories: 'In attributing the prints it has been assumed the small quantity of large format albumen prints of professional quality are the work of Stewart, while the majority, smaller prints of lower standard, are Moss's contribution.'[44] Two of the large photographs of better quality (along with two other plague photographs from unknown source[s] showing respectively Hindus offering *yagnaya* or prayers to the Goddess of Plague in front of a sacred fire, and the house where the first plague case was found) were reproduced in September 1897 in a one-page photo-essay titled 'The Plague in India: Fighting the Epidemic in Bombay' in the illustrated weekly newspaper *The Graphic*, published from London.[45]

Intriguingly, the weekly mentioned neither Moss nor Stewart as the photographer. Instead the partially faded tiny printed letters at the bottom of the photo-essay can be deciphered to read 'From Photographs by Khimji Hirji Kayani, Bombay'.[46] To make things more problematic, approximately two dozen photographs from the 'Plague Visitation, Bombay' album, mostly smaller prints attributed to Moss, are also found in another contemporaneous album titled 'Poona Plague Pictures', which is in the collection of the Getty Research Institute and contains 148 photographs.[47] The Institute's catalogue mentions: 'Many of the 148 photographs (including the two-part group portrait panorama) in the album can be attributed to Pune-based British photographer F. B. Stewart ... Other photographs of a more amateur nature were likely taken by Adams-Wylie or his close associates.'[48] This Charles Adams-Wylie was a junior British doctor at the General Plague Hospital in Poona in 1897–1898. Notably, there is again no mention of Captain Moss here as the photographer of any photograph whatsoever.

Out of the eight photographs to be analysed in this study, four (Figs. 6.4, 6.6, 6.7, and 6.8) appear in both albums. This naturally makes their attribution to Moss questionable. Further, a fine-grained scrutiny of the two albums reveals a significant disparity in the degree of visual agency accorded to Indian patients in the overall composition of their photographs. In most of the photographs in the 'Plague Visitation, Bombay' album, the patients lack visual prominence, they are often relegated to the position of indistinct bodies in the background and margins of the photographs. In contrast, several photographs in the 'Poona Plague Pictures' album afford greater visual agency to Indian patients (males as well as females) as they occupy much more space within the photographic frames and are placed prominently in the foreground, often offering a full-frontal view. This variance in the composition style confirms that the large majority of the photographs in each of these albums are the works of two different photographers with different ideas and styles. In addition, it acts as a caveat against rushing to the conclusion that in Western medical photography in colonial India indigenous people were *systematically* subjected to invisibilisation.

## Negotiating Gender and Age

Jeffrey Mifflin in his powerful study of historical medical photographs highlights: 'Photographs are created by the convergence of photographer, subject, camera, and other variables, such as who is or isn't present, and the authority or influence they may have.'[49] Following in this vein, I shall demonstrate how the composition of photographic frames vis-à-vis plague hospitals in Bombay, and the question of *who is or isn't present* in those frames, can be read productively to decipher the intended or desired effect or meaning of the photographs.

As Figs. 6.1 and 6.2 visually convey, some hospitals had separate wards for women and children. The establishment of women's ward may well be explained by the inflexibility of gendered segregation prevailing in the Bombay society, and the consequent need to provide women patients seclusion from the gaze and touch of males, other male patients as well as male medical practitioners. In this connection, it must be noted that gender segregation and women's seclusion were so important for Indian societies that their violation by military search parties who entered Indian houses in search of hiding plague patients to transfer them to hospitals (see Cohn, this volume), was one of the reasons why two extremist Hindu

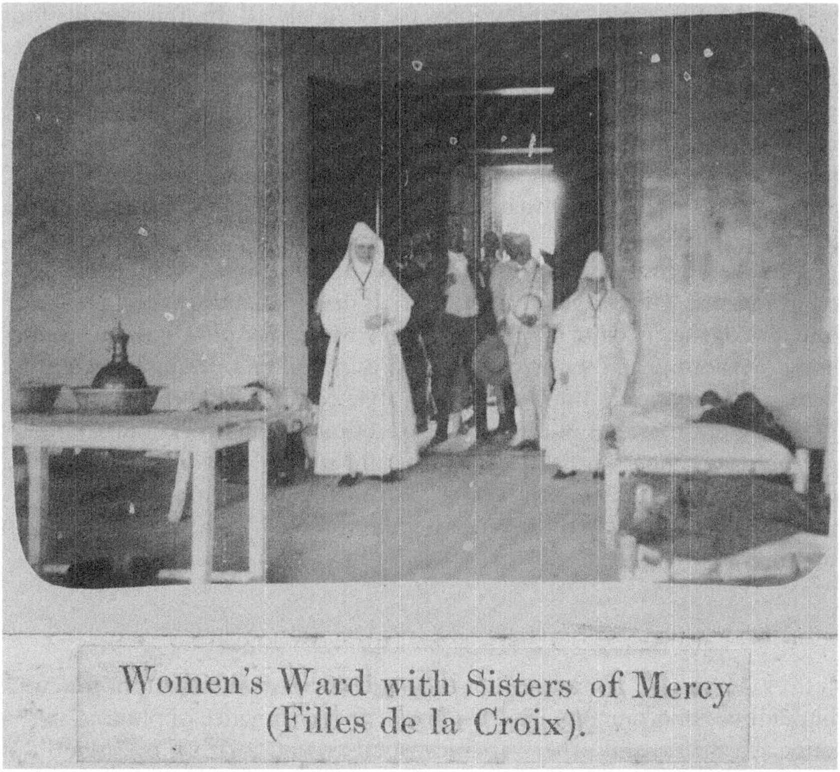


**Fig. 6.1** 'Women's Ward with Sisters of Mercy (Filles de la Croix)', *Plague Visitation, Bombay, 1896–1897*. (Courtesy of Wellcome Collection)

brothers from the 'high'-caste Chitpavan Brahmin community, Damodar Hari Chapekar and Balkrishna Hari Chapekar, assassinated Walter Charles Rand, the British Special Plague Officer in Poona (Pune), and Lieutenant Charles Egerton Ayerst, Rand's military escort, on the evening of Queen Victoria's Diamond Jubilee celebrations in Poona, on June 22, 1897.[50]

Antipathy to medical examination of women's body by male doctors was equally fierce among the Muslim communities. Notably, in the1890s, there was no qualified woman doctor in Bombay, neither Indian nor foreign. After Rand's murder, a four-volume report on plague in India was compiled in 1898 under the orders of the Home Department of the

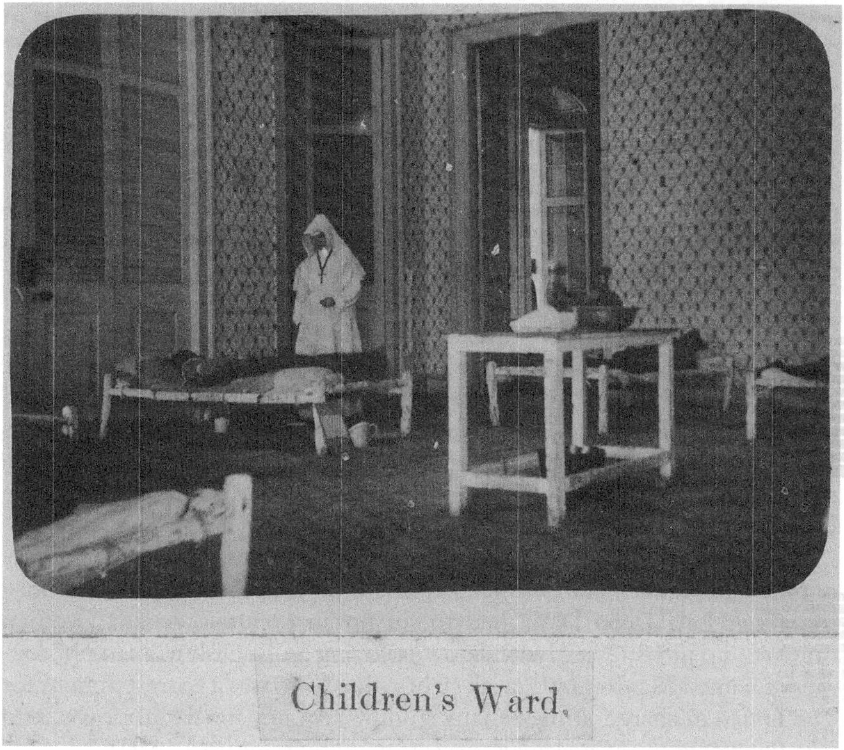


**Fig. 6.2** 'Children's Ward', *Plague Visitation, Bombay, 1896–1897*. (Courtesy of Wellcome Collection)

Government of India, by Robert Nathan, a civil servant and later a well-known British intelligence officer. The report emphasised that 'both Hindus and Muhammadans view with the greatest dislike any intrusion into their homes, and especially any possible interference with the privacy of their women'.[51] The same report quoted a Kazi from the Konkani Sunni Muslim community whose inflammatory street harangue (transcribed and translated into English) of March 13, 1897 opposing hospitalisation of plague patients asked the following: 'How could a husband be expected to tolerate the sight of his wife's hand being in the hand of another man?'[52]

The idea of allowing women to access medical treatment outside home was abhorred so much by most of the Indian families that on several occasions women were left to die slow and painful deaths at home in rooms locked from outside. The above report quoted from another report by Surgeon-Lieutenant-Colonel James Sutherland Wilkins, the doctor-cum-officer entrusted with the administration of anti-plague operations in the seaport town of Mandvi: 'In our visitation to the houses many cases were found either dead or dying, and all feeling of humanity amongst the people seemed to be blunted, as on several occasions we unlocked doors closed from the outside and found cases left to die, *chiefly* women.'[53] In this context of the paramount importance of gender segregation in plague-ravaged Bombay, David Arnold's study of the Bombay plague explains: 'Because most of the doctors were male as well as white, their touch was considered either polluting or tantamount to sexual molestation, especially when it involved the examination of women's necks, armpits and thighs.'[54] Notably, to make things worse for both Brown female patients and White male doctors, it was precisely these parts of a woman's body that needed to be medically examined during the pandemic, as plague buboes typically occur in neck, armpits, and groins.

The practice of gender segregation and purdah was standing in the way of photographing Indian women in such a way that the noted Indian photographer Lala Deen Dayal had to set up an exclusive zenana (women only) studio in 1892 in Hyderabad employing an English woman photographer named Mrs Kenny-Levick (whose husband was a correspondent for *The Times*) in charge of the zenana studio.[55] An unnamed author wrote in *The Journal of the Photographic Society of India* in January 1892:

> As this studio is for photographing native ladies only, special arrangements had to be made to protect them from the gaze of the profane and the stern. So the place is surrounded by high walls, and all day long within this charmed enclosure Mrs Kenny-Levick, aided by native female assistants, takes the photographs of the high-born native ladies of the Deccan.[56]

Previously, woman photographers Mrs E. Mayor, Mrs D. Garrick, and Mrs Allan had started zenana studios in Calcutta respectively in 1863, 1877, and 1885, followed by a similar zenana studio established in the same city in 1897 by the photographic firm Mitter & Co.[57] Sarojini Ghosh was perhaps the first professional Indian woman photographer who opened her own studio named The Mahila [Women] Art Studio and Photographic Store, in 1899 at 32 Cornwallis Street in Calcutta. An article in the Indian English newspaper *Amritabazar Patrika* on February

16, 1899 felt the need to highlight the seclusion provided to women at the Mahila Art Studio, praising it as: 'A thoroughly secluded studio for ladies under Sreemuttee Sarojini Ghosh.'[58]

Given such stringent gender segregation standing in the way of photography in India in the nineteenth century, it is indeed striking that the photograph of a women's ward (Fig. 6.1) does comprise a group of six White Western males, presumably either medical practitioners or administrators, clad in immaculate Western attire and standing cramped in the narrow space within the frame of the open door to pose for the camera lens. However, noticeably, they stood at the threshold of the women's ward, without entering it. Contrary to this exclusionary physical positioning of male subjects in the photograph, the two White female subjects, Sisters of Mercy, in their distinctly nun clothing, were photographed standing a few steps inside the women's ward. In the overall composition of this photograph, the women patients are almost invisible; they merely form indistinct bodies lying in wooden disposable beds on both sides. Thus, evidently, bodily manifestation of their disease itself was not a matter of interest or use in the message that the photographer sought to convey. The inclusion of Western male subjects, albeit outside the threshold, gives a clue to the purpose of composing such a photograph, which appears to be peddling an impression to the local patriarchal society that *mahilancha sammaan* (the honour of women) was still being maintained in the hospitals by keeping them out-of-bounds from male touch, while at the same time keeping them under the supervising 'guardianship' of males, albeit foreign. Thus, the photograph seems to have been aimed at dispelling particularly gender-based fears of, and aversion to, plague hospitals, and thereby serving as a publicity tool vis-à-vis the hospitals in the attempt to secure a social license for them. Obtaining such social acceptance of the hospitals was essential because, as pointed out by James Alfred Lowson, the Plague Commissioner of Bombay, in his 1897 report on plague: 'The mortality amongst Asiatic races is usually from 90 to 100 per cent outside hospital, whilst careful treatment in hospital often brings the death-rate down by 20 per cent or even more.'[59]

In this respect, Fig. 6.1 is a politico-medical and medico-administrative photo, rather than a medical photo per se in the conventional sense. Photographs of this genre are fundamentally different from the innumerable close-ups of the bodies of plague victims capturing the most dreadful buboes and gruesome lesions caused by the lethal disease.[60] Photographs of the latter genre clearly had a different audience and different purpose;

they were primarily aimed at professional medical researchers and practitioners to aid them in advancing scientific enquiry into the disease, and secondarily at government authorities to convey to them the degree of ghastliness and painfulness of the disease. Thus, these were images that were produced to contribute to the generation and advancement of scientific knowledge about plague, as well as to prompt swift and wholehearted actions from relevant authorities against the scourge by triggering a sense of alarm.

Some medical texts in the late nineteenth and early twentieth centuries reproduced these grisly photographs to demonstrate the clinical features of plague. An instance of such clinical use of photographs was the influential 1905 *Treatise on Plague* authored by William J. Simpson, a distinguished doctor of tropical medicine and formerly the first full-time government Health Officer in Calcutta from May 1896 to September 1897. The treatise contained eleven photographs of buboes, necroses and carbuncles on the bodies of plague patients, although without mentioning the names of the photographed patients and without acknowledging the source(s) or photographer(s) of the images.[61]

In this connection, it is important to note that close-up photographs of disfigured plague corpses and moribund plague patients in the late 1890s were not used to prompt humanitarian aid or to raise relief funds. In this respect, plague photography was fundamentally different in its purpose from that of famine photography in India in the nineteenth century. In her study of 'atrocity photography' and its link to humanitarian movements in the late nineteenth century, Christina Twomey has convincingly demonstrated that photographing the Indian famine of 1876–1878 'did introduce the practice of displaying shocking images of bodily suffering and deprivation as truth claims in order to prompt humanitarian action'.[62] Along the same line, Zahid Chaudhary's study of photography in nineteenth-century India reads famine photographs as pleas for sympathy and attempts at reproducing affect.[63] This was not the case for plague photography, because the disaster of the pandemic was caused not by the want of any material items like food, but by the abundance of some living beings—rats and fleas—and the expeditious community spread of the disease's agent. As for hospitals in India, they struggled not for funding, but for social acceptance in the first place. Therefore, humanitarian aid in the form of material help was thought to be unfit for fighting the crisis. In addition, in comparison to feeding famine victims, humanitarian aid in the form of amateur volunteers attending and nursing plague patients was naturally thought to be a much more hazardous task that risked getting the volunteers readily infected by the excruciatingly agonising and lethal disease. Therefore, we do not come across any deliberate effort to use plague photographs in the 1890s to engender sympathy for the plague victims in the hearts of the audience.

The politico-medical and medico-administrative photographs of hospitals were inextricable constituents of the whole process of making sense of plague by the British in colonial Bombay. Plague photography reflected the different paths in which the colonial gaze travelled in the city, ranging from the avenues where the gaze was fixated on the close-ups of the physical manifestation of the disease on indigenous bodies for purely clinical purposes, to ways in which the gaze was peering at the scars caused by the disease on the authority of the state over the Indian social body. As I demonstrated earlier, the plague effect was often experienced in the form of mounting suspicion of and resistance to government intervention in public health matters. Against this backdrop, photographs of socially acceptable plague hospitals where local practices of gender and caste segregation were maintained, and in turn convalescence cases, were consciously put into the service of procuring indigenous trust in colonial medical inventions like the establishment of modern Western hospitals, and thereby healing the wounds that the epidemic caused onto the hegemony of the colonial state. In this sense, photographic representation of plague lent support to the process of 'colonial accommodation of indigenous values' through orientalising Western medical establishments to make them socially and culturally acceptable on Indian soil.[64]

In addition to patients' sex, their age too led to the construction of separate wards, as is evident from Fig. 6.2, captioned 'Children's Ward'. The photograph shows a built environment similar to the one in Fig. 6.1. Both photographs were taken inside wards housed in buildings made of concrete and bricks with high walls in the Indo-Gothic style of architecture favoured by the colonial state in big Indian cities in the latter half of the nineteenth century. The printed wallpapers in Fig. 6.2 clearly convey a distinctly Western style of interior décor, whereas in both photographs the tall doors with built-in wooden blinds represent Indian adaptation for air ventilation in hot Bombay weather. In the composition of Fig. 6.2, the only living subject that is clearly visible is that of a Christian nun. She, like the two Sisters in Fig. 6.1, wears necklace with big cross, which adds a potent religious element to the overall effect of the photograph. In Fig. 6.2, the child patients again form almost invisible subjects lying in four wooden disposable beds in three directions of the room. Their presence is pronounced in the text captioning the visual, but in the composition of the photograph itself, they have been relegated to an almost invisible presence. Thus, presence of children in the visual of the children's ward is 'suggested' as opposed to 'shown'. Borrowing Shawn Michelle Smith's expression in the context of photographic visibility, we can say that the children and women in these photographs remained 'at the edge of sight'.[65] Unlike in the photographs of women and child patients in the 'Poona Plague Pictures' album, their visibility in the 'Plague Visitation, Bombay, 1896–1897' album is low; their presence is not independently established in the photographs here. Because of this low visibility accorded to their photographic presence, it was only through the combined effect of the photographed visual and the accompanying texts in the caption that the photographs could become what they claimed to be, that is, photographs of women's and children's wards respectively. The visuals alone cannot and do not produce the aspired understanding or meaning of the photographs, captions are integral to the generation of the intended understanding or meaning, and in turn subsequent circulation.

Photographs of children's wards alert us to the issue of children's fate in the Bombay Plague, a theme which stands remarkably neglected in the existing corpus of literature. The existence of children's wards indicates that in addition to accepting the necessity of accommodating the local culture of gender segregation (that provided the rationale behind separate women's wards), the authorities in the Western-style hospitals were also drawing a connection between the bodily effect of plague on the one hand, and different stages of physical maturity of its victims on the other. The establishment of an epistemological connection between age-specific bodily conditions and the degree of the effect or danger of plague seems to have been the rationale behind separate wards for plague-affected children, as otherwise they could have been easily accommodated with adult patients (often their parents) in the same rooms where several adult patients were accommodated in physically distanced beds, which would have also reduced the anxiety of separation for the child patients and their families. Instead, the setting up of separate children's ward indicates the doctors' conscious decision to separately study the impact of plague on younger lymph nodes, lymphatic vessels, and lymphoid tissues in children.

In fact, the existence of children's wards warns us against rushing to the conclusion that gender segregation in the local society was the sole factor behind the establishment of women-only wards. A supplementary explanation becomes possible—the medical practitioners in Western-style hospitals were perhaps drawing connections, right or wrong, between the specificities of female body and the impact of plague on women. Thus, photographs of women's and children's wards in plague hospitals force us to re-think the medical epistemologies at play in the decision to have separate wards for separates sexes and different ages.

Coming back to the photographs, the location of the camera itself in both Figs. 6.1 and 6.2 merits careful thinking. In both cases the camera was placed many meters inside the ward facing the exit. The physical location of the photographer behind the scene was well exposed to feared-plague contagion. This raises the unanswered but intriguingly important question—what went on during the photographing process in those hospital rooms when the cameraperson was placed in the same room with patients carrying a highly infectious and fatal disease? Was he provided any personal protective equipment whatsoever? This novel experience of photographing patients of such a supposedly contagious and deadly disease at close quarters was bound to have been different for him from the usual safe experience of photographing architectural marvels or taking portraits in palaces and studios. In fact, the first European who caught plague in Bombay was a plague photographer.[66] On February 18, 1899, doctor C. H. Freeman Underwood, a private practitioner in Bombay, told the Indian Plague Commission in his oral evidence, 'There was the case of the very first European that got plague, Mr. Orford Sherman. He had been photographing the cases at Mandvi, and subsequent to that he developed plague, and died of it.'[67] Thus, photographing plague cases was a task of proven peril. Notably, some doctors, such as Dr Underwood himself, took photographs of their patients. Underwood informed the Plague Commission that, in late July/early August 1896, he took three photographs of three patients, two 'natives' and one European, out of medical curiosity as they were suffering from a disease that appeared to him of 'a peculiar type', as no one had seen a plague patient in Bombay before that year.[68] In addition to this clinical use of photography, the report of the Indian Plague Commission for 1898–1899 chaired by T. R. Fraser pointed out the use of plague photographs in the praxis of medical governance of local societies—they were used by the state for surveillance of passengers and crew disembarking from vessels in Bombay.[69] The Commission explained the usefulness of photography in medical governance:

> Great difficulty was experienced in finding the house to which the new-comer had gone, and the inspecting officer found it impossible to satisfy himself that the people he examined in the house were the people who had passed out of the camp under surveillance. It was, therefore, found necessary to bring people under surveillance to the observation camp for examination, and an attempt was made to prevent the danger of personation by photographing groups of persons admitted to surveillance.[70]

The Commission was satisfied with the efficacy of photography as an apparatus of state-surveillance as it noted: 'Many of the difficulties in the matter of surveillance can be, and have been, removed by the use of photography.'[71] However, these photographs are naturally incapable of telling us anything about the photographers' gear during the *event* of photography that preceded the existence of the photographs. Did the photographer in the Bombay plague hospitals interfere with the posture or positioning of the patients in the ward according to his plan of composing his photographic frames? Did he invade their personal space? Was the act of photographing affected in any way by the photographer's personal feelings and emotions in the grim ambience of the room? Alternatively, did he ignore the agony and depression in the room and the imagined possibility of contagion, and executed the photographs in the patients' natural posture and position with business like aloofness? Of course, unlike the 'behind the scene' or 'behind the shoot' photographs of the present day that allow us visual access to the event of picture-taking, the event of photography itself was not photographed in the Bombay hospitals at the time. My research so far has encountered not even a stray photograph of a photographer in any Bombay hospital. Therefore, the above questions about the event of photography remain unanswered, making any visual route to write the history of the *event* of photography in Bombay hospitals an impossible one to pursue.

## From Ball Room to Slaughterhouse

Figures 6.3 and 6.4 contain the sights of respectively a ballroom and a slaughterhouse which were converted into plague hospitals. Before examining these photographs in detail, it is analytically useful to see them beyond the parochial context of Bombay and situate them within the larger national context of private plague hospitals, particularly of their built environment. This allows us to trace the general pattern of exigency

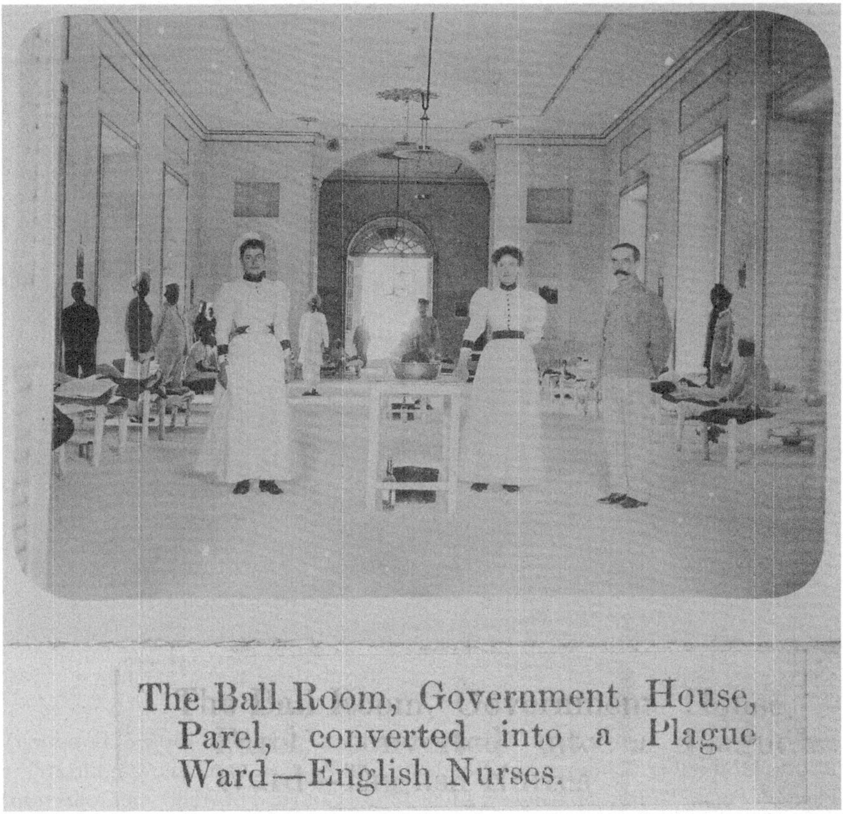


**Fig. 6.3** 'The Ball Room, Government House, Parel, converted into a Plague Ward—English Nurses', *Plague Visitation, Bombay, 1896–1897*. (Courtesy of Wellcome Collection)

and contestation in the development of hospitals. Away from Bombay, the development of private plague hospitals around 1898 in Calcutta, then the capital city of India, provides the most succinct summary of the debates about private hospitals. Among the big cosmopolitan cities of India, Calcutta was only second to Bombay in terms of the intensity of the onslaught of plague. In 1903, Frank Clemow, a British doctor who spent approximately fifteen months in India and who was also the British Delegate to Istanbul's Conseil Supérieur de Santé, published his

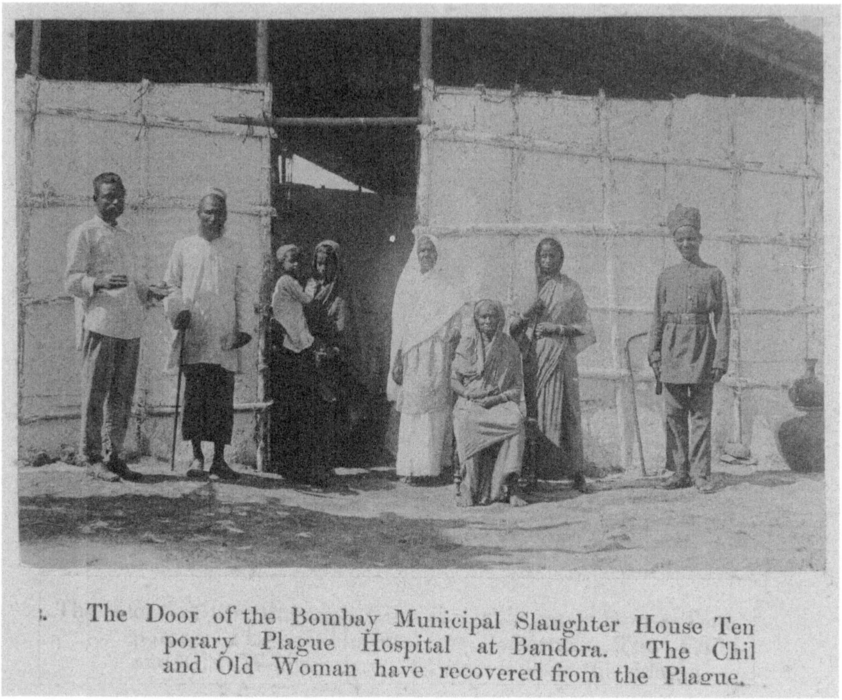


**Fig. 6.4** 'The Door of the Bombay Municipal Slaughter House Temporary Plague Hospital at Bandora. The Child and Old Woman have recovered from the Plague', *Plague Visitation, Bombay, 1896–1897.* (Courtesy of Wellcome Collection)

pioneering study of the geographical distribution and diffusion of all known diseases, where he clearly noted about plague that in some periods 'a mortality comparable to that in Bombay, prevailed in Calcutta'.[72] Parallels between hospitals in Bombay and Calcutta endow us with a good sense of the general trend in their development, as the two cities were comparable in both area and population.

*The Calcutta Journal of Medicine* edited by Mahendra Lal Sircar, an MD in allopathy turned champion of homeopathy and physician of the influential Hindu guru and mystic Ramakrishna Paramahamsa, wrote in January 1898 in a report titled 'Disallowance of Home Segregation in Cases of Plague—A Mistake and A Danger' that 'our belief is that physicians are the most efficient disseminators of infection'.[73] The journal disputed the perception of hospitals as being effective treating spaces away from individual homes, and subsequently proposed the introduction of hospitals-at-home:

> Plague hospitals, without the aid of disinfection, would have been intolerable nuisances, and would have proved the most fertile breeding ground of the disease. And if the nurses and the physicians are saved from catching the disease it is partly because, as we have said, from their getting slowly immunised, but chiefly because its contagium is efficiently destroyed by disinfectants. Our contention, therefore, is that if hospitals can be so treated and managed as to be rendered safe for the inmates, why cannot the homes of patients be similarly treated and managed?[74]

Though the journal conceded that in most cases the location of private houses in congested and unsanitary quarters and their small size and often complete lack of ventilation 'forbid their conversion into hospitals suitable for patients', it continued to argue that

> there are houses with big compounds of good length and breadth, and there are houses with high and extensive terraces. In the compounds or on the terraces of these houses sheds can be erected for the accommodation of patients, and kept well-disinfected and isolated from the rest of the buildings.[75]

Notably, even before such a major medical journal endorsed the establishment of hospitals-at-home on sanitation grounds, the Plague Notification of November 19, 1897, after much deliberations, had already allowed the construction of private hospitals in a bid to appease the feelings of those who resisted the transfer of plague patients to hospitals run by the colonial state, rigidly insisting on their 'obligation' to maintain untouchability along caste separation, as well as gender segregation and seclusion for the women in their families. On January 6, 1898, at the invitation of the British Indian Association, a conference of the influential members of the 'native community' and representatives of various public bodies as well as of the 'Native press' was held in Calcutta to review the revised plague regulations. At this conference, the patriarchal culture of maintaining gender segregation even when medically dealing with a pandemic, was endorsed. The chair of the conference, Maharaja Bahadur Jotindra Mohun Tagore, proudly reported to the secretary of the Municipal Department that the conference acknowledged that the practice of purdah was in fact more rigid in Calcutta than in Bombay, and that in Bombay 'the native dwelling-houses are rarely found divided into two separate parts—the outer and the inner—as is the case with the great majority of the native dwelling houses in Calcutta, where the *purdah* system prevails in its utmost rigour, and separate accommodation has to be provided for the female inmates'.[76] The conference insisted that this already existing division of the house of any 'native gentleman' into an outer quarter (*bahirmahal*) and an inner quarter (*andarmahal/antahpur*) along gender lines was conducive to home segregation/quarantine of plague patients along medical lines as well.[77]

However, though *The Calcutta Journal of Medicine* was in favour of hospitals-at-home, it took a slightly differing stance on the matter as it did not fail to note the formidable financial obstacles that stood in the way of converting *bhadralok* (gentlemanly) houses into private hospitals:

> It is true that a concession has been made in allowing private hospitals, but how many will be able to avail themselves of such hospitals? These hospitals, as we said in a previous article, "cannot be numerous enough or capacious enough for the numerous respectable poor, who, while they have good ancestral houses, have not the means to contribute towards the buildings of such hospitals; and for them to 'expect the richer members to assist them in preserving their social usages intact by providing special hospitals and camps for them to take refuge in,' would most likely turn out to be depending upon a vain expectation."[78]

However, if the number of private hospitals in Calcutta is anything to go by, there was considerable progress in the building of private plague hospitals. *The History and Proceedings of the Bengal Plague Commission* (1899) informs us that the power of granting licenses for plague hospitals and camps in and around Calcutta was entrusted to the Plague Commission, and the commission received 205 applications of which '95 for hospitals in garden houses situated outside town limits'.[79] Notably, 'large numbers of applications for family plague hospitals began to flow in'.[80]

Cut to Bombay, the *Report of the Bombay Plague Committee* for the period from July 1, 1897 to April 30, 1898 first noted: 'In the middle of June 1897, with the rapid decline of the Plague, all the temporary hospitals were closed owing to the monsoon.'[81] However, plague soon recrudesced in the following month.[82] Regarding private hospitals during this 'Second Epidemic', the report wrote:

> The system of Private Hospitals inaugurated last year by the Committee has been adhered to and considerably extended during the second Epidemic. In spite of much adverse criticism, it has been found that Private Hospitals, open to all members of any particular caste or sect, have as a whole been a distinct success. The many intricate questions relating to caste habits, food, religious scruples and other similar difficulties, have in this way been overcome and the various apprehensions and fears of the people have been set at rest. The result has been a moral and practical support to the measures adopted by the Committee that has proved invaluable.[83]

The aforementioned history of plague, compiled by Captain Condon in 1899, too observed: 'The sanctioning of caste camps and hospitals did away with objections on such grounds as caste and rank.'[84] However, the report of the Bombay Plague Committee noted in 1898: 'The choice of a building satisfactory in every sanitary detail, together with convenience of position for the special caste for whom it was intended, has often proved most difficult. Undoubtedly there has been much left to be desired in this respect as regards certain private hospitals.'[85]

Within the larger context of hospitals-at-home and hospitals in garden houses, Figs. 6.3 and 6.4 demonstrate the ad hoc nature of the growth of state medical infrastructure on formerly non-medical sites, triggered by the urgency of dealing with plague's onslaught. In the absence of adequate infrastructure, sites that were most unlikely to have been envisaged as treating spaces for patients under usual circumstances, were hurriedly converted into state hospitals. Figure 6.3 shows the interior of the ballroom of the Government House at Parel in Bombay which was converted into a plague ward, and Fig. 6.4 captures the exterior of the Bombay Municipal Slaughter House at Bandora (Bandra) which was also converted into a temporary pesthouse. The rushed conversion of some of the available non-medical spaces into treating sites initiated a process what Estela Duque has termed 'medicalisation of space' in the context of her study of medical architecture in Philippines.[86]

Figures 6.3 and 6.4 underscore the enormous degree of difference between sundry built environments, from the stately ball room of the Government House to the rudimentary hut of a slaughterhouse, that were deployed in the war against the plague onslaught in Bombay. Analogous to the contrast in the built environments, the human subjects of these two photographs exhibit striking contrast. In Fig. 6.3, the composition of the frame is dominated by three white persons, two English female nurses and a male (presumably a doctor or a health administrator) positioned in the foreground. Notably, unlike the Sisters in Figs. 6.1 and 6.2, the nurses here posed in standard Western female nursing uniforms without displaying any recognisable religious symbol like crosses. This conveys the existence of nonreligious professional nursing by European nurses in plague-ravaged Bombay, many of whom had been trained in Britain. In addition to numerous nuns who doubled as nurses, there was a growing corps of professional European nurses who found employment in plague hospitals.[87]

On three sides of the photograph, several Indian ward boys/attendants have been placed in unmistakably synchronised poses. They were all in formal Indian servants' clothing complete with turbans on their heads which are fairly common features of servant costumes in the nineteenth- and early twentieth-century photographs of Indian servants and attendants of the British. As the patients are again indistinct subjects lying and sitting on *charpais* (knitted coir bed with wooden frame and legs) in the distance in the background of the photograph, it is difficult to decipher the class and cultural background of the patients who received medical care in such a stately ballroom. In this sense, the patients as social entities are again visually buried in the photograph. Their physical presence is marginally registered in it, whereas their socio-cultural background is conspicuous only by absence. Thus photography, which was popularly perceived as a medium to 'show', in effect acted as a dual medium which hid the visual agency of some subjects in photographs like this, while concurrently raising the visibility of others. While the patients in Fig. 6.3 have been subjected to a process of invisibilisation, the English nurses and the man have been accorded a conspicuous visual dominance in the photographic frame.

As for Fig. 6.4, which captured the rudimentary hut of a slaughterhouse-turned-hospital, it is indeed striking that even a slaughterhouse, which was perhaps the most unhygienic candidate to be converted into a treating site, was put into the service of battling plague. Reviewing French photographer Eli Lotar's celebrated photographs of Parisian La Villette's slaughterhouses, Georges Bataille once observed: 'In our time … the slaughterhouse is cursed and quarantined like a plague-ridden ship.'[88] Such was the public image of slaughterhouses that they could effortlessly attract the deployment of plague imageries. Yet, a slaughterhouse in Bombay was made part of the war against plague. However, as the clothing of the photographed subjects indicates, it was the people from the lowest rungs of society who were condemned to be treated there. Even during the social cataclysm of plague, ballrooms, even in their hospital avatar, remained out-of-bounds for them. Though the low-class population, who overlapped almost entirely with 'low'-caste population as well as with most Muslims, remained outside the built environment of better treating sites in Bombay, it was paradoxically the same urban poor at whom the state's anti-plague measures were primarily directed. Prashant Kidambi's study of localist etiological doctrines in government plague policies in Bombay has demonstrated that the state's 'anti-plague offensive' was primarily directed at 'the neighbourhoods as well as the bodies of the urban poor'.[89] Yet, the state often had to condemn the bodies of the same poor to treating sites of poor quality in order to maintain quotidian social distance between the worlds of the 'high'-caste Hindu patients on the one hand, and of the 'low'-caste and Muslim poor patients on the other, which unambiguously manifested the intersectional link between class, caste, religion, and medicine.

Intriguingly, the fact that it was originally a slaughterhouse, the rudimentary hospital in Bandora had attracted stricter 'European supervision' of its sanitary arrangements. This fact, coupled with the social exclusion of Muslim butchers from the Bombay society, initially proved to be a boon in warding off plague around the slaughterhouse. Major J. Mills, a veterinary surgeon as well as First-Class Magistrate who was put in charge of treatment at the slaughterhouse-hospital, wrote in his 1897 report on plague:

> Though plague had existed in Bombay in a severe form since September [1896] and at Bandora since the beginning of November 1896, strange to say no cases took place amongst the inhabitants of the slaughter-house compound until the 14th February 1897 ...This may be due to the fact that the sanitary arrangements of the place are under strict European supervision, and that there was not much intercommunication between the employees and people living in the infected areas.[90]

Notably, Fig. 6.4 is markedly at odds with the other photographs examined so far, as the gaze of the lens in this photograph is firmly focused on the patients who, unlike in the already discussed photographs, are the prime subject in the composition of the photographic space and dominate almost the entire frame. Thus, this photograph assigns full and clear visual agency to patients. This shift in emphasis followed the shift in the intention of the plague photographers to convey messages through their photographs. The caption of the photo states: 'The Child and Old Woman have recovered from the Plague.' Thus, it is a convalescence photograph aimed at publicising the success of plague hospitals, and thereby dispelling popular fears and misgivings about them. Notably, for the purpose of conveying the message of convalescence, a photo of recuperated urban poor usually living in 'filthy' chawls or slums, was more effective than a photo showing better-off patients from better living conditions, because if the poor living in crammed unsanitary conditions could survive the plague, it conveyed greater chances of survival for the better-off people living in relatively hygienic environments. In the photo, the convalesced child and the old woman convey a happy ending to the suffering brought about by the pestilence; they are united with their family in recovered health and, presumably, ready to go home. Thus, Fig. 6.4 highlights the 'rupturing' effect that plague had on the continuance of usual human life, while concurrently suggesting that the 'rupture' could be ended by accepting medical care at hospitals. It marks a shift from the endeavours at producing a spectacle of affliction in the overwhelming majority of plague photographs to an attempt at producing a spectacle of convalescence.

In addition to the obvious fact that the patients in the photograph had no better alternative to accepting treatment at a former slaughterhouse, their modest clothing and unpretentious style of wearing them make their low locus in the class hierarchy conspicuous. Remarkably, none of the women is veiled in this photograph, and male and female subjects were captured standing next to each other within a single frame. Thus, it challenges our understanding of the overall social phenomenon of plague in Bombay as being unvaryingly affected by an impermeable custom of gender segregation and women's seclusion. This photograph of women with uncovered faces looking straight into the lens clearly conveys that among the lowest rungs of working-class Muslims such as butchers (as well as among the 'low'-caste communities), who were at the margins of the local society, gender segregation and purdah were unaffordable luxuries as the women of their families had to labour side by side with men to keep the families fed. Therefore, violation of women's seclusion was not a concern for them, neither when accepting state medicine at hospitals nor when facing the gaze of a camera, operated mostly by men in nineteenth-century India. In contrast, the dismantling effect that the colonial state's initial anti-plague crusade was having on the stringently guarded custom of women's seclusion in better-off Muslim and Hindu 'high'-caste/Savarna families, considerably hardened their resistance to hospitals as well as to the photographing of their women. The way gender played out in the plague pandemic as well as in the photographing of it, is a clear testimony to the comparative liberality among the low-class and 'low'-caste communities vis-à-vis the mixing of genders, in comparison to Indians of higher classes and 'upper' castes.

## Publicised Convalescence and the Photographable Other

'Convalescence itself is a space in-between, a hazy yet paradoxically crystal-clear state between sickness and health. Introduced as a third term in the rhetoric of sickness and health, convalescence becomes the vehicle for a series of in-between states'—this is the observation of Barbara Spackman while analysing scenes of convalescence in select literary texts.[91] In the context of Bombay in the late 1890s, publicising this in-between state between plague's bodily lesion and full recuperation was essential to render a raison d'être to the existence of plague hospitals. Therefore, publicising convalescence became a foremost concern of plague photography. Akin to Fig. 6.4, Fig. 6.5 is also from the genre of convalescence photographs. This is one of the very few plague photographs from India that captured a child in full visibility. In comparison to Fig. 6.4, this one portrays a better built environment—a spacious room with walls made of tin sheets and a roof built of similar sheets and bamboos along with sturdy wooden beams. The child is seated on a *charpai* with decoratively curved wooden legs, looking straight into the camera. She or he is flanked on both sides by two English female nurses. Through the aperture between the two nurses, the silhouette of an Indian woman, presumably an Indian nurse or a relative of the convalescing girl, is visible in the background. In the composition of the photograph, the two English nurses have been assigned high visibility, whereas the Indian woman has been relegated to a silhouette in the background. Through this visual highlighting of the English nurses, the credit for causing convalescence has been appropriated for them, while the Indian woman has been relegated to playing a second fiddle in the visual narrative of publicising convalescence in a Western hospital. This relegation of the Indian agency for convalescence to a dark background is noteworthy, given the Indian woman's courage in

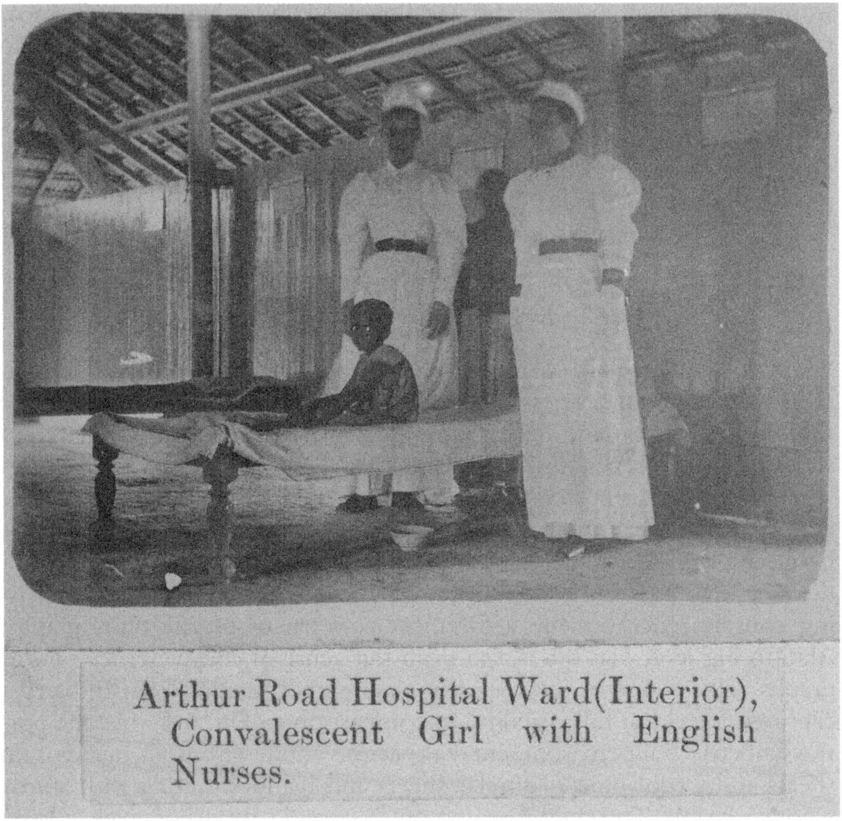


**Fig. 6.5** 'Arthur Road Hospital Ward (Interior), Convalescent Girl with English Nurses', *Plague Visitation, Bombay, 1896–1897.* (Courtesy of Wellcome Collection)

overcoming social interdiction and personal dread of catching the disease to nurse a plague patient in a hospital.

However, in the vein of convalescence photographs, Fig. 6.6 represents a different strand. Here both the convalesced patient and the nurse are adult Brown women who are dominant subjects in the composition of the frame with clear frontal visibility assigned to them. This is a photograph where an Indian subject (the nurse) has been visually allowed full share of the credit for causing convalescence.

Notably, in both Fig. 6.4 (as discussed above) and Fig. 6.6, adult Indian women looked straight into the camera offering full frontal visibility of

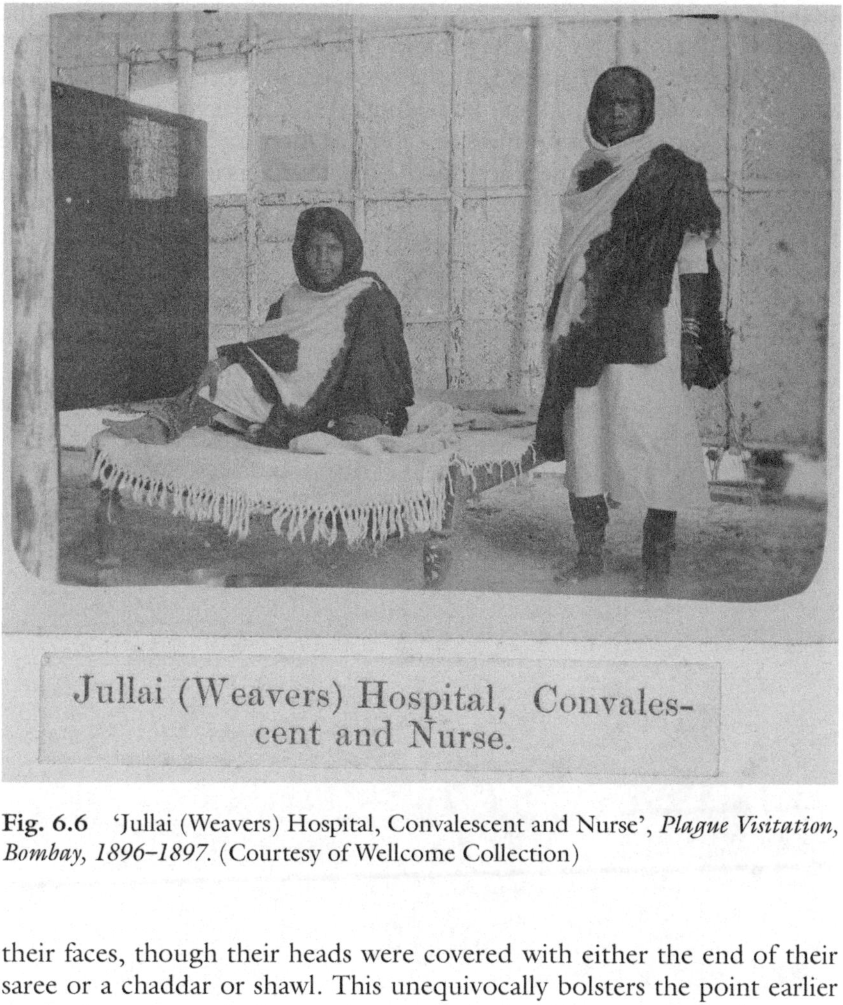


**Fig. 6.6** 'Jullai (Weavers) Hospital, Convalescent and Nurse', *Plague Visitation, Bombay, 1896–1897*. (Courtesy of Wellcome Collection)

their faces, though their heads were covered with either the end of their saree or a chaddar or shawl. This unequivocally bolsters the point earlier made that among the lower classes of Muslim female patients there was not much aversion to the photographic gaze of a camera lens. The exception of the 'very lowest' classes of Indian population in allowing unveiled women to appear in front of men outside their families had been recognised even by Rand, the Special Plague Officer of Poona. After Rand's murder, the abovementioned 1898 report compiled by Nathan quoted from late Rand's report on plague: 'The females of the better class of Muhammadan families were all purdah women. Among all classes of the population, *except perhaps the very lowest*, there existed an aversion to the invasion of their privacy which is unknown in Western Europe.'[92] The caption of Fig. 6.6 clearly states that the photograph was taken at a hospital set up exclusively for patients belonging to the Jullai (Muslim weavers) community, a community indispensable for Bombay's booming textile mills. The centrality of the labour of the Jullai weavers in Bombay's cotton textile industry prompted the local state to attempt minimising the pandemic's effects on this community, resulting in special hospitals built exclusively for them.[93]

The 'Plague Visitation, Bombay' album contains an intriguing photograph captioned 'Nariel Wadi Hospital. Women at Breakfast' (Fig. 6.7).

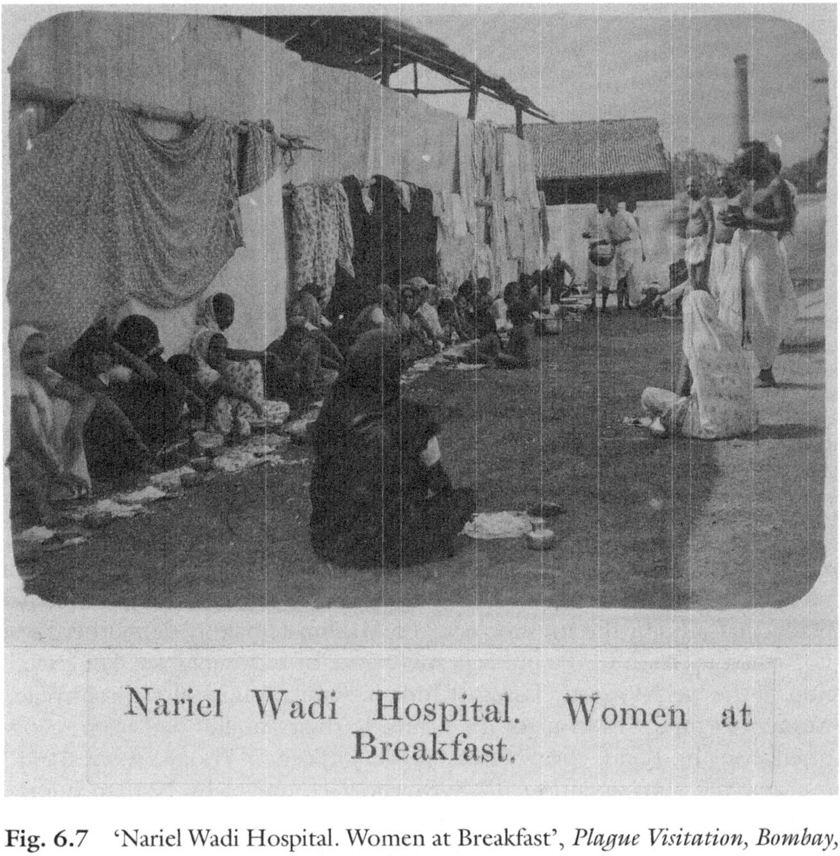


**Fig. 6.7** 'Nariel Wadi Hospital. Women at Breakfast', *Plague Visitation, Bombay, 1896–1897*. (Courtesy of Wellcome Collection)

Within the genre of hospital photography, this photograph is strikingly atypical as it does not include any subject that instantaneously gives away its identity of being a hospital or medical photograph. Instead, it captured a scene from quotidian life; it merely shows about fifteen humbly clothed Indian women together with almost the same number of children sitting on the ground in an enclosed compound next to the corrugated tin wall of a large shed and having food in either banana leaves or disposable leaf plates placed on the ground. The visual content of the photograph itself does not offer any clue to it being a photograph of a hospital. Only after perusing the caption the viewer is informed that the photograph is of women having breakfast at Nariel Wadi Hospital. Thus, it is the accompanying caption that divulges the identity of the subjects and furnishes a context or setting to the captured visual. Without the concomitant text, the captured visual alone is unable to generate any sense of it being a hospital photograph. Like Figs. 6.1 and 6.2, it is again the coalesced effects of an anterior sighting of visual and posterior reading of text (or the two actions in reverse order) that complement each other to complete the meaning of the photograph.

The raison d'être of photographing everyday life at a hospital merits some reflection. A photograph of a mundane event such as a modest breakfast arrangement for women and children within such minimalistic setting could be a product of neither the intention to publicise the usefulness of hospitals away from homes, nor any desire to advance scientific and medical enquiry into the physical suffering of the patients. This, perforce, leads us to the conundrum: why was the breakfast scene considered 'photographable' in the first place, when taking even a single photograph was an expensive and onerous task involving bulky instruments? In Pierre Bourdieu's observations about photography, we find a tentative answer which deserves to be quoted in some length:

> [T]he range of that which suggests itself as really photographable for a given social class (that is, the range of 'takeable' photographs or photographs 'to be taken', as opposed to the universe of realities which are objectively photographable given the technical possibilities of the camera) is defined by implicit models which may be understood via photographic practice and its product, because they objectively determine the meaning which a group confers upon the photographic act as the ontological choice of an object which is perceived as worthy of being photographed, which is captured, stored, communicated, shown and admired.[94]

The ostensibly striking use of 'object' rather than 'subject' in Bourdieu's exposition becomes less striking if we consult *Camera Lucida*, where Roland Barthes is certain, 'Photography transformed subject into object, and even, one might say, into a museum object.'[95] Therefore, if we employ Bourdieu's framework, the elevation of the sight of women's breakfast to the status of 'photographable' appears to have taken place due to the meaning/worth conferred upon the scene by either the photographer or the commissioning medical authority or both. But what precisely is this meaning/worth? I argue that the photograph was a product of the Western gaze that strove to record 'native' everyday life for visual circulation and consumption within Western circles both in India and abroad, much in line with Reverend Mullens's abovementioned urge to use photography to visually document 'all the minute varieties of oriental life; of oriental scenery, oriental nations and oriental manners'.[96] The 'otherness' of 'native' lifestyle as seen and captured by the camera lens on the site of hospitals was merely an extension of the general Western gaze that dominated the novel medium of photography not only in British India, but in every colony of the Western powers.[97] After all, as observed by Elizabeth Edwards, it is incontrovertible that 'The cultural circumscriptions included under the modern interpretative blanket 'Western perception of the "Other"' are central to the creation and consumption of photography in the second half of the nineteenth and the first half of the twentieth centuries'.[98] Photographing quotidian events in the everyday being of others, such as their breakfasts, provided visual longevity and reproducibility to the already stark everyday differences between 'us' and 'them'. While examining colonial photography in relation to the international exhibitions or 'world fairs' in big Western cities in the nineteenth century, Anne Maxwell has aptly explained the production and consumption of photographs delineating otherness: 'Ideological positioning was not just a function of consumers' interpretation of images, but was also determined by the maker's own relationship to dominant culture.'[99] As the makers of the plague photographs in India in the late 1890s were all Europeans (predominantly British), confident of the 'superiority' of European habits and everyday living cultures, they were riveted by an irresistible curiosity in the otherness of the quotidian life of Indians, which resulted in the elevation of 'native' everyday life to the status of photographable.

In this connection, it is imperative to bear in mind that the attraction to otherness was a two-way phenomenon—whenever Indians, particularly the non-Westernised commoners and subalterns like the women in the 'Women at Breakfast' photograph (Fig. 6.7), came in contact with the Europeans in India, they were of course fascinated by the otherness of European lifestyle. But they lacked the power of the new technology of photography in nineteenth-century India where it was still a monopoly of the elite Indian and European photographers. The audience of the photographs taken by Indian elite photographers such as the Indian Maharaja Sawai Ram Singh II of Jaipur, or Bengal renaissance elite Raja Rajendralal Mitra, or court photographer Raja Deen Dayal or by elite commercial studios, was naturally confined to Indian and European elites and royalties, and at the lowest to the higher echelons of Indian bourgeoisie (anglicised as well as traditional) in cosmopolitan cities.[100] As even the photographs by Indians, let alone the photographs by European photographers, were thus inaccessible by the Indian masses, the photographs of Europeans and European lifestyle were unable to convey the otherness of the European everyday living cultures to the wider Indian public, unlike the photographs of Indian quotidian lives taken by European upper and middle-class military officers and commercial photographers which were quickly reproduced in newspapers in Europe, and in commercially produced popular photo-illustrated accounts of life in 'exotic' India, and were thus consumed by the wider Western public. For instance, as discussed before, four photographs related to the Bombay Plague were reproduced in September 1897 (i.e. immediately after they were taken in Bombay) in the illustrated weekly newspaper *The Graphic,* published from London and priced at nine pence per copy.[101] Thus, having the power of mass circulation of photographs enabled the Europeans to firmly delineate the otherness of Indians, whereas the lack of the same (emanating partly from the disinterest of Indian elite photographers to 'plebeianise' the elite art of photography) prevented the Indian public from visually highlighting and widely relaying the otherness of European foreigners.

Importantly, similar to Figs. 6.4 and 6.6, barring three women (including the two sitting with their backs to the camera), the faces of the rest in Fig. 6.7 were exposed to the camera lens. Though their heads are covered with the end of their saree, their faces are exposed and some of them are even staring at the lens wearing a look of confused curiosity. The photo is akin to Fig. 6.4 in one more respect—it also contains approximately a dozen Indian male subjects towards the right end of the frame; some eating, some serving food, and some just standing and seemingly supervising the breakfast arrangements. The presence of males alongside unveiled females and the occurrence of inter-sex dining, and the success of the photographer in being allowed to capture such a mixed-sex visual within a single frame, is a testimony to the permeability of the defence of gender segregation on the site of plague hospitals. Again, a caveat applies: this relaxation of the rules of gender segregation was confined to patients who were low class and 'low' caste, two statuses that in most cases overlap in India.

It is worth noting that barring the 'Women at Breakfast' photograph, none of the photographs analysed in this study captured the human subjects 'in-action', they did not photograph anyone in the middle of doing anything. This in fact is the case in the overwhelming majority of plague photographs from India. In them, all the dominant subjects posed looking straight into the camera. Thus, staginess was intrinsic to these photographs. The act of posing for the camera 'created' the photographs rather than simply allowing the camera to 'capture' the human subjects in their lived reality during the pandemic. In this connection, we may recall Barthes's reflections on photography:

> Now, once I feel myself observed by the lens, everything changes: I constitute myself in the process of 'posing,' I instantaneously make another body for myself, I transform myself in advance into an image. This transformation is an active one: I feel that the Photograph creates my body or mortifies it, according to its caprice.[102]

Such a transformation through the process of posing was ingrained in the photographs of plague hospitals as the patients as well as the medical personnel were required to pose for the 'prosthetic eye' of the camera. The very fact that in most cases human subjects were made to pose for the camera, instead of being caught in action, dents the truth claim of hospital photography in late nineteenth-century Bombay. The alteration in the instinctive posture and motion of the human subjects prior to and during the process of photographing in the hospitals falls neatly in line with Barthes's thesis: 'the Photograph is the advent of myself as other.'[103]

The final photograph discussed in this study, Fig. 6.8, is the epitome of posed photographs. In it, seven Bandora Convent Sisters who nursed plague patients in Parel and Mahim in Bombay posed in the most formal postures and the most pronounced missionary nun clothing completed

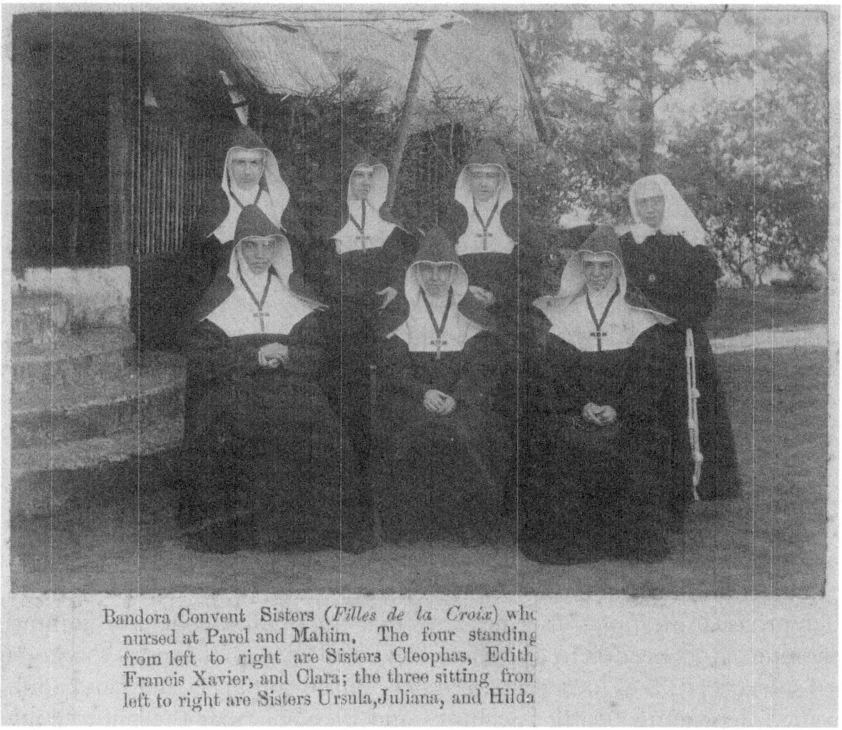


**Fig. 6.8** 'Bandora Convent Sisters (*Filles de la Croix*) who nursed at Parel and Mahim. The four standing from left to right are Sisters Cleophas, Edith, Francis Xavier and Clara; the three sitting from left to right are Sisters Ursula, Juliana, and Hilda', *Plague Visitation, Bombay, 1896–1897*. (Courtesy of Wellcome Collection)

with big cross pendants. Unlike the other photographs analysed in this chapter, the caption of this photograph identified the subjects by individual names.

The service of nuns-cum-nurses was significant in the Bombay hospitals, particularly during the initial phase of plague in the city. In January 1898, *The Indian Lancet* put together a report titled 'The Sisters During the Plague', which pointed out: 'Before the arrival of the Nurses sent for by the Plague Committee from England, almost all the nursing in the plague hospitals was undertaken by the Sisters of the different religious bodies … Their services were entirely gratuitous.'[104] The genre of group photographs of these nuns-cum-nurses was unmistakably aimed at visually propagating a correlation between Christian missionary ethos and medical care in colonial India, publicising care for the sick as something innate in the missionary project. Thus, photographs of nuns-cum-nurses were handy publicity tools in advancing Christian missionary programmes. Group photographs of foreign nurses in the most pronounced version of Christian clothing, as well as the positioning of nun nurses by the side of patients in the composition of photographs of hospital wards (e.g. in Fig. 6.1), injected a strong Christian element in these photographs. They served to visually claim the due credit for serving the religious 'other' in a colonised society in the time of the most crippling medical emergency, and thereby to sustain the claim of 'intrinsic' spirit of service in missionary activism in colonial India.

## Conclusion

The analysis of the photographs in this study demonstrates how epidemic photography during the Bombay Plague became an apparatus to construct multiple meanings and diverse messages according to the demands for different visual meanings and messages that the photographers or commissioning actors needed to circulate. The visual agency of sundry categories of subjects (such as local and foreigner, female and male, child and adult) varied depending on the meanings and messages that the photographs were required to construct and convey. In the context of the power of photographs in the praxis of governance, Elizabeth Edwards's study of photograph collecting practices in the Colonial Office in London in the mid-nineteenth century has pointed out: 'The historiographical and methodological challenge therefore is to link the production of the photographic record with the effective *realpolitik* that connects the representational power of photographs with the praxis of government.'[105] My study has addressed this historiographical and methodological challenge by dovetailing hospital photography with their deployment in the praxis of medical governance as well as in missionary activism in colonial Bombay. It has demonstrated that every hospital photograph was chaperoned by the demands for specific meaning and visual message that they were to engender, meaning and message that would align with the praxis of medical governance. It argues that hospital photographs were always aligned with the tilt of the colonial state's governmentality—when the sights of the plague hospitals were needed to be used to dispel gender-based aversion to public hospitals and thereby secure an indigenous social license for them (particularly from the rich and 'high'-caste Indians), then male subjects in the photographs were thoughtfully positioned at a distance from the female subjects. In the same vein, when advertising the prospect of recuperation through accepting treatment at hospitals was necessary to persuade patients to visit hospitals, then sights of convalesced patients were made a prominent theme in plague photography and were given visual dominance in the photographic composition. Likewise, where the visuals of nuns-cum-nurses could be used to aid Christian missionary activism in India by publicising medical care for the indigenous people as a foundational ethos of missionary works, they were assigned high visual agency in the photographs. However, by contrast to convalescence photographs, in these medico-missionary photographs, where the intended visual message required Indian patients to play nothing more than a second fiddle in the narrative of missionary service, they were relegated to the level of indistinct bodies in the background and margins of the photographs.

The reflexive gaze that was engendered in the Western circles in Bombay by the otherness of Indian societies, was instrumental in kindling a keen photographic interest in the quotidian life and habits of Indians, particularly during the pandemic which many British doctors (such as Captain Liston, the abovementioned army doctor and member of the Plague Research Commission) attributed to the 'filthy' lifestyle of the 'natives' which harboured rats and plague fleas in Indian houses.[106] This photographic interest in the everyday living cultures of Indians during the pandemic is exemplified by the photograph of the women's breakfast at Nariel Wadi Hospital. Notably, Western reflexive gaze was instrumental not only in the photographic documentation of the everyday lives of Indians, but in all branches of medical photography in general. As I pointed out before, unlike other genres of photography in nineteenth-century India, such as architectural and portrait photography in which numerous Indian photographers were active, medical photography was a monopoly of European photographers—mostly the British, plus a few members of the Russian, French, Austrian and German medical missions to India in 1897–1898. To the Indian photographers, medical sights lacked aesthetic appeal, they were grim and therefore not worthy of photographing. As the British were by far the most prolific producers of plague photographs in Bombay, it was their reflexive gaze that dominated plague photography, and in turn built the visual regime of the plague pandemonium for both foreign and local audiences, lay as well as medical. Thus, the visual perceptions and imaginations of the Bombay Plague by Indian and foreign contemporaries as well as subsequent generations were and have been shaped by the reflexive gaze that was inherent in British epidemic photography in India in the 1890s.

## Notes

1. J. K. Condon, *The Bombay Plague, Being A History of the Progress of Plague in the Bombay Presidency from September 1896 to June 1899* (Bombay: Education Society's Steam Press, 1900): p. ix.
2. House of Commons Parliamentary Papers (henceforth HCPP), Cd. 810, Indian Plague Commission, 1898–1899, *Report of the Indian Plague Commission with Appendices and Summary*, Vol. V, p. 50.
3. I. S. Falk, 'Some Riddles in Epidemiology', *The Scientific Monthly* XX, no. 4 (1925): 383–404, p. 393.
4. Ira Klein, 'Plague, Policy and Popular Unrest in British India', *Modern Asian Studies* 22, no. 4 (1988): 723–755, p. 725.
5. Ibid., p. 724. See also: India, Public Health Commissioner, *Annual Report of the Public Health Commissioner with the Government of India, 1929* (Calcutta: Superintendent of Government Printing, 1932), p. 69.
6. Sandhya L. Polu, *Infectious Disease in India, 1892–1940: Policy-making and the Perception of Risk* (Basingstoke: Palgrave Macmillan, 2012); Mridula Ramanna, *Health Care in Bombay Presidency, 1896–1930* (Delhi: Primus Books, 2012); Myron Echenberg, *Plague Ports: The Global Urban Impact of Bubonic Plague, 1894–1901* (New York: New York University Press, 2007); Prashant Kidambi, *The Making of an Indian Metropolis: Colonial Governance and Public Culture in Bombay, 1890–1920* (Aldershot: Ashgate, 2007); Prashant Kidambi, "An Infection of Locality": Plague, Pythogenesis and the Poor in Bombay, c. 1896–1905', *Urban History* 31, no. 2 (2004): 249–267. I. J. Catanach, 'South Asian Muslims and the Plague, 1896–c. 1914', *South Asia: Journal of South Asian Studies* (hereafter *SAJSAS*) XXII, no. sl (1999): 87–107. I. J. Catanach, '"Who are your leaders?" Plague, the Raj and the "communities" in Bombay, 1896–1901'. In Peter Robb (ed.), *Society and Ideology: Essays in South Asian History Presented to Professor K. A. Ballhatchet*, pp. 196–221 (Delhi: Oxford University Press, 1993); I. Catanach, 'Plague and the Tensions of Empire: India, 1896–1918'. In David Arnold (ed.), *Imperial Medicine and Indigenous Societies*, pp. 149–171 (Manchester: Manchester University Press, 1988); Rajnarayan Chandavarkar, *Imperial Power and Popular Politics: Class, Resistance and the State in India, c. 1850–1950* (Cambridge: Cambridge University Press, 1998), chapter 7; David Arnold, *Colonizing the Body:*

*State Medicine and Epidemic Disease in Nineteenth-Century India* (Berkeley: University of California Press, 1993); David Arnold, 'Touching the Body: Perspectives on the Indian Plague, 1896–1900'. In Ranajit Guha (ed.), *Subaltern Studies* V, pp. 55–90 (New Delhi: Oxford University Press, 1987).

7. Susan Hapgood, *Early Bombay Photography* (Ahmadabad: Mapin Publishing, 2015); Malavika Karlekar, *Visual Histories: Photography in the Popular Imagination* (New Delhi: Oxford University Press, 2013); Malavika Karlekar, *Re-Visioning the Past: Early Photography in Bengal, 1875–1915* (New Delhi: Oxford University Press, 2005); Stéphanie Roy Bharath, 'The Alkazi Collection of Photography: Visual Heritage from South Asia', *Photo Researcher* 13 (2010): 42–53; Christopher Pinney, *The Coming of Photography in India* (London: The British Library, 2008); Christopher Pinney, *Camera Indica: The Social Life of Indian Photographs* (London: Reaktion Books, 1997); Sophie Gordon, 'Uncovering India: Studies of Nineteenth-Century Indian Photography', *History of Photography* 28, no. 2 (2004): 180–190; Sophie Gordon '"A Silent Eloquence": Photography in 19th-Century Lucknow', in Rosie Llewellyn-Jones (ed.), *Lucknow: Then and Now* (Mumbai: Marg Publications, 2003): 134–145; Maria Antonella Pelizzari (ed.), *Traces of India: Photography, Architecture, and the Politics of Representation, 1850–1900* (New Haven: Yale University Press, 2003); John Falconer, *India: Pioneering Photographers, 1850–1900* (London: British Library and Howard and Jane Ricketts Collection, 2002); Vidya Dehejia (ed.), *India Through the Lens: Photography 1840–1911* (Washington: Freer Gallery of Art and Arthur M. Sackler Gallery at the Smithsonian Institute, 2000).
8. Louis Jacques Mandé Daguerre, *An Historical and Descriptive Account of the Various Processes of the Daguerréotype and the Diorama* (London: McLean, 1839). H. Fox Talbot, *The Pencil of Nature* (London: Longman, Brown, Green and Longmans, 1844).
9. Willoughby Wallace Hooper, Album: *Secunderabad* ['Scenes of the Madras Famine'], J. Paul Getty Museum, Los Angeles, Object Number: 84.XO.940.7. Nineteen photographs of the famine taken by Hooper are in the Picture Library, Royal Geographical Society, London, Images: S0001994–S0002012. For twelve more photographs of the famine, see Hooper, Album: *Hungersnot in Indien 1876–1878*, Photography Collection, Museum Ludwig, Cologne, Inv.-Nr. ML/F/SL 0788/01-12. For discussions on photographs by Penn and Nicholas brothers, see: Christopher Penn, *In Pursuit of the Past: The Discovery of the Life and Work of A. T. W. Penn, Pioneering Photographer of South India* (Worplesdon: C. F. Penn, 2010); Christopher Penn, *The Nicholas Brothers*

*& A. T. W. Penn: Photographers of South India 1855–1885* (London: Bernard Quaritch, 2014).

10. Rosie Llewellyn-Jones (ed.), *The Alkazi Collection of Photography: The Uprising of 1857* (Ahmedabad: Mapin Publishing, 2017); Rosie Llewellyn-Jones (ed.), *Portraits in Princely India, 1700–1947* (Mumbai: Marg Publications, 2008); Sean Willcock, 'Aesthetic Bodies: Posing on Sites of Violence in India, 1857–1900', *History of Photography* 39, no. 2 (2015): 142–159; Hugh Rayner (ed.), *Photographic Journeys in the Himalayas, 1863–1866: Samuel Bourne,* 4th corrected edition (Bath: Pagoda Tree Press, 2014); Hugh Rayner (ed.), *Early Photographs of Ladakh* (Bath: Pagoda Tree Press, 2013). Julie F. Codell (ed.), *Power and Resistance: The Delhi Coronation Durbars, 1877, 1903, 1911* (Ahmedabad: Mapin Publishing, 2012); Sophie Gordon, 'A City of Mourning: The Representation of Lucknow, India in Nineteenth-Century Photography', *History of Photography* 30, no. 1 (2006): 80–91; Sophie Gordon, *Monumental Visions: Architectural Photography in India, 1840–1901*, PhD thesis, School of Oriental and African Studies, University of London, 2011; Robert Flynn Johnson, et al., *Reverie and Reality: Nineteenth-Century Photographs of India from the Ehrenfeld Collection* (San Francisco: Fine Arts Museum of San Francisco, 2003); Ray Desmond, *Victorian India in Focus: A Selection of Early Photographs from the Collection in the India Office Library and Records* (London: Her Majesty's Stationary Office, 1982); G. Thomas, *History of Photography, India, 1840–1980* (Hyderabad: Andhra Pradesh State Akademi of Photography, 1981).
11. The most notable among the nineteenth-century army officers and military surgeons (or both) who doubled as photographers were Captain Linnaeus Tripe, Colonel Thomas Biggs, Major Robert Christopher Tytler (his wife Harriet Christina Tytler was also an accomplished photographer in her own right), Colonel Alexander Greenlaw, Captain Melville Clarke, Dr John Murray, Dr John McCosh, and Dr William Henry Pigou. For works by these photographers, see: Linnaeus Tripe, *Photographic Views of Ryakotta and Other Places in the Salem District with Descriptive Notes by J. A. C. Boswell (s. n., 1858)*, British Library (hereafter BL), London; Thomas Biggs, Theodore Hope, and James Fergusson, *Architecture at Ahmedabad, the Capital of Goozerat, Photographed by Colonel Biggs* (London: John Murray Press, 1866). For a collection of photographs taken by Harriet Tytler and Robert Tytler in 1858 see Gale Primary Sources (hereafter GPS), Nineteenth Century Collections Online (hereafter NCCO), Collection: Photographs from the India Collection at the British Library (hereafter PICBL). M. S. Nagaraja Rao (ed.), *Vijayanagara through the Eyes of Alexander J. Greenlaw, 1856, John Gollings, 1983* (Mysore: Government of Karnataka, Directorate of Archaeology &

Museums, 1988); Melville Clarke, *From Simla through Ladac and Cashmere, 1861* (Calcutta: Savielle and Cranenburgh, 1862). For a collection of seventy-one photographs taken by John Murray between 1850 and 1858, see GPS, NCCO, Collection: PICBL. John Murray, *Photographic Views in Agra and Its Vicinity* (London: J. Hogarth, 1858), Royal Collection Trust, London. John Murray and J. Boileau, *Picturesque Views in the North-Western Provinces of India, Photographed by John Murray with descriptive letter-press by J. T. Boileau* (London: J. Hogarth, 1859). John McCosh, *Advice to Officers in India* (London: W. H. Allen, 1856); Philip Henry Egerton, *Journal of a Tour Through Spiti to the Frontier of Chinese Thibet, with Photographic Illustrations* (London: Cundall, Downes & Co., 1864). Among the European commercial photographers who made their names in photography in nineteenth-century India were Captain Edmund David Lyon, Samuel Bourne, Felice Beato, John Burke, and John Edward Saché. For works by these commercial photographers, see: Edmund David Lyon, *Notes to Accompany a Series of Photographs Designed to Illustrate the Ancient Architecture of Southern India: Taken for Government, and Described by Capt. Lyon, edited by James Fergusson* (London: Marion & Co., 1870); Samuel Bourne, 'Samuel Bourne's India', original photographs, The Anne S. K. Brown Military Collection (hereafter ASKBMC), Brown University Library (hereafter BUL); For 155 original photographs by Bourne and Charles Shepherd, see: 'Photographs of India', c. 1862–c. 1872, Royal Commonwealth Society Library at Cambridge University Library; Felice Beato, 'Photographic Views of Lucknow Taken after the Indian Mutiny', original photographs, ASKBMC, *Beato*, BUL; John Burke, original photographs, ASKBMC, *John Burke*, BUL; Omar Khan, *From Kashmir to Kabul: The Photographs of John Burke and William Baker, 1860–1900* (Munich: Prestel, in association with Ahmedabad: Mapin, 2002); Hugh Rayner, *John Saché: A Photographer in British India, 1864–1882* (Bath: Pagoda Tree Press, 2012).

12. Among the Indian commercial photographers, Lala Deen Dayal, Darogah Haji Abbas Ali, Shapoor N. Bhedwar, and Ahmad Ali Khan achieved fame. For Deen Dayal's photographs, see: The Alkazi Collection of Photography (hereafter ACP), New Delhi. Darogah Haji Abbas Ali, *An Illustrated Historical Album of the Rajas and Taaluqdars of Oudh* (Allahabad: North-Western Provinces and Oudh Government Press, 1880); Darogah Haji Abbas Ali, *The Lucknow Album: Containing a Series of Fifty Photographic Views of Lucknow and Its Environs Together with a Large Sized Plan of the City* (Calcutta: G. H. Grouse, Baptist Mission Press, 1874); Ahmad Ali Khan, *Lucknow Album*, BL, Vol. I and Vol. II, References: Photo 269/1 and Photo 269/2 (Archives and

Manuscripts) Rahaab Allana, 'Performance for Camera: Shapoor N. Bhedwar and the Dimensions of Studio Photography in Bombay'. In Rahaab Allana et al., *The Artful Pose: Early Studio Photography in Mumbai, c. 1855–1940* (Ahmadabad: Mapin Publishing, 2010): 56–72. Maharaja Sawai Ram Singh II of Jaipur and Maharaja Birchandra Manikya of Tripura were the most notable nineteenth-century Indian kings-cum-photographers; Laura Weinstein, 'Exposing the Zenana: Maharaja Sawai Ram Singh II's Photographs of Women in Purdah', *History of Photography* 34, no. 1 (2010): 2–16; Vidya Dehejia, 'Maharajas as Photographers'. In Dehejia (ed.), *India Through the Lens*: 226–229; Karlekar, *Re-Visioning the Past*. Bengali antiquarian and archaeologist Raja Rajendralal Mitra and Bengali children's author Upendrakishore Raychaudhuri are the most notable examples of nineteenth-century Indian academics and authors who turned photographers; G. Thomas, 'Babu Rajendra Lall Mitra', *History of Photography* 8, no. 3 (1984): 223–226; Malavika Karlekar, 'Eye on the Empire—Rajendralal Mitra Took Photography Beyond the Studio', *The Telegraph*, November 11, 2007. For Upendrakishore Raychaudhuri's research into photography's technological aspects, see: Emily Kristin Bloch, *Making Sense of Nonsense: A Contextual Study of the Art of Sukumar Ray*, PhD thesis, The University of Chicago, 2013, pp. 15–19.

13. Deborah Hutton, 'Raja Deen Dayal and Sons: Photographing Hyderabad's Famine Relief Efforts', *History of Photography* 31, no. 3 (2007): 260–275.
14. Sunil Janah, *Photographing India* (New Delhi: Oxford University Press, 2013). For more famine photographs by Janah, see: 'The March of Death', *People's War*, 'Special Bengal Number' (November 7, 1943). For photos of the Bengal Famine by William Vandivert, see the *LIFE* Picture Collection via Getty Images, object names: 595687–595720, and 18997554, https://www.gettyimages.co.uk/photos/william-vandivert-famine?mediatype=photography&phrase=William%20vandivert%20famine&sort=best# (accessed August 6, 2020). Google Arts & Culture, 'William Vandivert, Aug 16, 1912–Dec 1, 1989', selected year: 1943–1944, most photographs with a generic title 'Calcutta and Bengal Famine', https://artsandculture.google.com/entity/g122t1kxf (accessed August 6, 2020). For eighty photos of the famine taken by Vandivert, see the website of *Old Indian Photos: Historical Photographs of Indian Subcontinent*, 'Bengal Famine of 1943—A Photographic History', parts 1–4, https://www.oldindianphotos.in/2009/12/bengal-famine-of-1943-part-1.html (accessed December 21, 2020). 'Margaret Bourke-White's The Great Migration: Five Million Indians Flee for Their Lives', *LIFE* Magazine (November 3, 1947); Margaret Bourke-White, *Halfway*

*to Freedom: A Report on the New India in the Words and Photographs of Margaret Bourke-White* (New York: Simon and Schuster, 1949); Margaret Bourke-White, *Interview with India, in the Words and Pictures of Margaret Bourke-White* (London: Travel Book Club, 1951).

15. Christos Lynteris, 'The Prophetic Faculty of Epidemic Photography: Chinese Wet Markets and the Imagination of the Next Pandemic', *Visual Anthropology* 29, no. 2 (2016): 118–132, p. 125.
16. Jeffrey Mifflin, 'Visual Archives in Perspective: Enlarging on Historical Medical Photographs', *The American Archivist* 70, no. 1 (2007): 32–69, p. 33.
17. Condon, *The Bombay Plague*, Plates I–X.
18. W. Glen Liston, *The Cause and Prevention of the Spread of Plague in India: A Lecture Delivered before the Bombay Sanitary Association on 11th December 1907* (Bombay: The Times Press, 1908).
19. Eduardo Cadava, *Words of Light: Theses on the Photography of History* (New Jersey: Princeton University Press, 1997), p. 128.
20. John Tagg, *The Burden of Representation: Essays on Photographies and Histories* (Basingstoke: Macmillan, 1988), p. 65.
21. Lukas Engelman, 'What are Medical Photographs of Plague?', *Remedia* (January 31, 2017) https://remedianetwork.net/2017/01/31/what-are-medical-photographs-of-plague/ (accessed April 9, 2018).
22. Ibid.
23. Ibid.
24. Lynteris, 'The Prophetic Faculty of Epidemic Photography', p. 126.
25. W. F. Gatacre, *Report on the Bubonic Plague in Bombay by Brigadier-General W.F. Gatacre, C.B., D.S.O., Chairman, Plague Committee, 1896–1897, with Plans* (Bombay: Times of India Steam Press, 1897), Chapter II 'Hospitals', p. 19.
26. Georges Didi-Huberman, *Invention of Hysteria: Charcot and the Photographic Iconography of the Salpêtrière*, trans. Alisa Hartz (Cambridge: MIT Press, 2003), p. 44.
27. Christopher Pinney, 'The Prosthetic Eye: Photography as Cure and Poison', *Journal of the Royal Anthropological Institute* 14, no. S.1 (2008): S33–S46, p. S34.
28. Ibid., p. S34.
29. Ibid., p. S35.
30. Ibid., p. S33. Georges Didi-Huberman has discussed how 'in the nineteenth century, photography became the paradigm of the 'scientist's "true retina"'; Didi-Huberman, *Invention of Hysteria*, pp. 32–33.
31. Reverend Joseph Mullens, 'On the Applications of Photography in India', *The Journal of the Photographic Society of Bengal*, no. 2 (January 21, 1857): 33–38, p. 33.

32. Ibid.
33. Laikwan Pang, *The Distorting Mirror: Visual Modernity in China* (Honolulu: University of Hawai'i Press, 2007), p. 70.
34. Susan Sontag, *On Photography* (London: Penguin, reprint 2002 [1977]), p. 17.
35. *Plague Visitation, Bombay, 1896–1897*, Album Creation Date: c. 1897, Wellcome Library, London, catalogued as 'The Bombay plague epidemic of 1896–1897: work of the Bombay Plague Committee. Photographs attributed to Capt. C. Moss, 1897'.
36. Ibid., front fixed endpaper. Emphasis mine.
37. W. F. Gatacre, 'Preface', *Plague Visitation, Bombay*, front pastedown.
38. Ibid.
39. Ibid.
40. William Ernest Jennings, *A Manual of Plague* (London: Rebman Limited, 1903): p. xi.
41. Ibid., plates X–XV.
42. BL, Reference: Photo 311/1 (Archives and Manuscripts catalogue). National Army Museum, London, Accession No: NAM. 1992-08-74. GRI Library, Special Collections, Accession no.: 96.R.81.
43. 'Immediate Source of Acquisition' in the Library's catalogue.
44. See the Library's catalogue entry.
45. 'The Plague in India: Fighting the Epidemic in Bombay', *The Graphic: An Illustrated Weekly Newspaper*, London, September 18, 1897, p. 394.
46. Ibid.
47. Charles Henry Benjamin Adams-Wylie (compiler), *Poona Plague Pictures*, Album Creation Date: 1897–1908, GRI Library, Special Collections, Accession no.: 96.R.95.
48. Ibid. See the catalogue entry.
49. Mifflin, 'Visual Archives in Perspective', p. 34.
50. For Damodar Chapekar's extremist 'Hindutva' thoughts, see excerpts from the English translation of his *Autobiography* in Marathi that he wrote in jail before being executed, in Sedition Committee, 1918, *Report* (Calcutta: Superintendent Government Printing, 1918), pp. 2–4, 7, 13, Dhananjayrao Gadgil Library, Gokhale Institute of Politics and Economics, Pune. For the most extensive discussion so far of the Chapekars' extremist Hindutva ideas and actions, see: John McLane, *Indian Nationalism and the Early Congress* (Princeton: Princeton University Press, 1977), pp. 332–357. For an incisive discussion on the 'Hindu nationalist resistance to the plague emergency', see: Echenberg, *Plague Ports*, particularly sub-chapter 'The Pune Revolt'. For gender dimensions of the Maharashtrian societies' responses to the colonial government's anti-plague actions, and the politics around the murders of

Rand and Ayerst, see: I. J. Catanach, 'Poona Politicians and the Plague', *SAJSAS* 7, no. 2 (1984): 1–18; I. J. Catanach, "The Gendered Terrain of Disaster'? India and the Plague, c. 1896–1918', *SAJSAS* xxx, no. 2 (2007): 241–267; Arnold, 'Touching the Body', particularly pp. 59, 64, 77–79, 84–87.

51. R. Nathan, *The Plague in India, 1896, 1897*, Vol. I (Simla: Government Central Printing Office, 1898), p. 147.
52. As quoted in ibid., p. 148.
53. As quoted in ibid., p. 217. Emphasis mine.
54. Arnold, 'Touching the Body', p. 64.
55. Karlekar, *Re-Visioning the Past*, p. 63. See also: Thomas, *History of Photography*, p. 41.
56. Anonymous, 'Mr. Lala Deen Dayal', *Journal of the Photographic Society of India* 5 (January 1892): p. 10, as quoted in R. Desmond, '19th Century Indian Photographers in India', *History of Photography* 1, no. 4 (1977): 313–317, p. 317.
57. Thomas, *History of Photography*, p. 41; Karlekar, *Re-Visioning the Past*, p. 125. See also: Siddhartha Ghosh, *Chobi Tola: Bangalir Photography-Chorcha* [Taking Pictures: The Practice of Photography by Bengalis] (Calcutta: Ananda Publishers, 1988), excerpt from Bengali translated by Debjani Sengupta, 'Zenana Studio: Early Women Photographers of Bengal', *Trans Asia Photography Review* 4, no. 2 (2014); Siddhartha Ghosh, 'Early Photography in Calcutta'. In Pratapaditya Pal (ed.), *Changing Visions, Lasting Images: Calcutta through 300 Years*, 143–158 (Bombay: Marg Publications, 1990).
58. As quoted in Sengupta, 'Zenana Studio'.
59. James A. Lowson, *Report on the Epidemic of Plague from 22nd February to 16th July 1897* (Bombay: Government of Bombay, 1897), p. 4.
60. Many of these disturbing photographs showing gruesome wounds on the bodies of dying plague victims are in Visual Representations of the Third Plague Pandemic Photographic Database, University of Cambridge. Close-ups of plague patients in Bombay in 1899–1900 are also in the collection of the Photothèque of the Pasteur Institute in Paris.
61. W. J. Simpson, *A Treatise on Plague Dealing with the Historical, Epidemiological, Clinical, Therapeutic and Preventive Aspects of the Disease* (Cambridge: Cambridge University Press, 1905), Chapter XIV.
62. Christina Twomey, 'Framing Atrocity: Photography and Humanitarianism', *History of Photography* 36, no. 3 (2012): 255–264, p. 258.
63. Zahid R. Chaudhary, *Afterimage of Empire: Photography in Nineteenth-Century India* (Minneapolis MN: University of Minnesota Press, 2012), Chapter 'Famine and the Reproduction of Affect'.

64. I have borrowed the expression 'colonial accommodation of indigenous values' from Arnold, *Colonizing the Body*, p. 251.
65. Shawn Michelle Smith, *At the Edge of Sight: Photography and the Unseen* (Durham NC: Duke University Press, 2013).
66. Transcription of the oral evidence of Dr C. H. Freeman Underwood, given in person to the Indian Plague Commission at the Bombay Secretariat on February 18, 1899, in: HCPP, Cd. 141, Indian Plague Commission, 1898–1899, *Minutes of Evidence Taken by the Indian Plague Commission, with Appendices. Vol. III. Evidence Taken from 11th February 1899 to 20th May 1899*, p. 84.
67. Ibid.
68. Ibid., p. 81.
69. HCPP, Cd. 810, pp. 387–388.
70. Ibid., p. 387.
71. Ibid., p. 388.
72. Frank G. Clemow, *The Geography of Disease* (Cambridge: University Press, 1903), p. 332.
73. 'Disallowance of Home Segregation in Cases of Plague—A Mistake and A Danger', *The Calcutta Journal of Medicine* xvii, no. 1 (January 1898): 1–12, p. 7.
74. Ibid., p. 8.
75. Ibid.
76. Official letter from Maharaja Bahadur Sir Jotindra Mohun Tagore to H. H. Risley, Secretary to the Government of Bengal, Municipal Department, as reproduced in: 'Plague Prevention in Calcutta: Native Opinion on the Rules', *The Indian Lancet* 11 (January–June 1898): 40–41, p. 40.
77. Ibid.
78. 'Disallowance of Home Segregation', p. 10.
79. Bengal Plague Commission, *History and Proceedings of the Plague Commission, Bengal, 1896 to 1898* (Calcutta: Bengal Secretariat Press, 1899), p. 10.
80. Ibid.
81. *Report of the Bombay Plague Committee Appointed by Government Resolution No. 1204/720P on the Plague in Bombay, For the Period Extending from the 1st July 1897 to the 30th April 1898* (Bombay: Times of India Steam Press, 1898): p. 110.
82. Ibid., p. 111.
83. Ibid., p. 113.
84. Condon, *The Bombay Plague*, p. 27.
85. *Report of the Bombay Plague Committee*, p. 113.
86. Estela Duque, 'Modern Tropical Architecture: Medicalisation of Space in Early Twentieth-Century Philippines', *Architectural Research Quarterly* 13, nos. 3–4 (2009): 261–271.

87. 'Plague Prevention in Calcutta', p. 39.
88. As cited in Yve-Alain Bois & Rosalind E. Krauss, *Formless: A User's Guide* (New York: Zone Books, 1997), p. 46.
89. Kidambi, "An Infection of Localit", p. 252.
90. 'Report by Veterinary-Major J. Mills, A. V. D., 1st Class Magistrate in Plague Charge, Bandora Slaughter-House', in Gatacre, *Report on the Bubonic Plague*, p. 193.
91. Barbara Spackman, *Decadent Genealogies: The Rhetoric of Sickness from Baudelaire to D'Annunzio* (Ithaca: Cornell University Press, 1989), p. 42.
92. As quoted in Nathan, *The Plague in India*, pp. 203–204. Emphasis mine.
93. Aditya Sarkar briefly touches on the relative bargaining power that the Jullai (or Julaha) community wielded over the cotton mill owners in the late 1890s; Aditya Sarkar, 'The Tie That Snapped: Bubonic Plague and Mill Labour in Bombay, 1896–1898', *International Review of Social History* 59, no. 2 (2014): 181–214, pp. 197–198.
94. Pierre Bourdieu et al., *Photography: A Middle-Brow Art*, translated by Shaun Whiteside (Cambridge: Polity Press, 1990, reprint 1998), p. 6.
95. Roland Barthes, *Camera Lucida: Reflections on Photography*, translated by Richard Howard (New York: Hill and Wang, 1981), p. 13.
96. Mullens, 'On the Applications of Photography', p. 33.
97. For an excellent exposition of this issue, see: Anne Maxwell, *Colonial Photography and Exhibitions: Representations of the 'Native' People and the Making of European Identities* (London and New York: Leicester University Press, 1999).
98. Elizabeth Edwards (ed.), *Anthropology and Photography, 1860–1920* (New Haven: Yale University Press, 1992), p. 5.
99. Maxwell, *Colonial Photography and Exhibitions*, p. 9.
100. Weinstein, 'Exposing the Zenana'; Dehejia, 'Maharajas as Photographers'; Karlekar, *Re-Visioning the Past*; Thomas, 'Babu Rajendra Lall Mitra'; Karlekar, 'Eye on the Empire'.
101. 'Plague in India', *The Graphic*, September 18, 1897, p. 394.
102. Barthes, *Camera Lucida*, p. 12.
103. Ibid., pp. 10–11.
104. 'The Sisters during the Plague', *The Indian Lancet* 11 (January–June 1898): 39–40, p. 39.
105. Elizabeth Edwards, 'Photographic Uncertainties: Between Evidence and Reassurance', *History and Anthropology* 25, no. 2 (2014): 171–188, p. 173. Emphasis in original.
106. Liston, *The Cause and Prevention of the Spread of Plague*.

CHAPTER 7

# Plague in India: Contagion, Quarantine, and the Transmission of Scientific Knowledge

*Samuel Cohn Jr.*

I begin with the famous debate on contagionism versus anticontagionism, initiated by Erwin Ackerknecht's Fielding Garrison Lecture of 1948, which thirty years later Margaret Pelling challenged in her seminal work on cholera in Britain. In less than thirty pages, Ackerknecht reached back to Hippocrates and ran through quarantine regulations tied to plague during the late Middle Ages and early modern period before unravelling aspects of nineteenth-century medicine that focused on a myriad of physicians and reformers in the US, France, Germany, Russia, and Britain. His spectrum of diseases was also wide-ranging. He concentrated on plague, yellow fever, and cholera, but typhus, typhoid, smallpox, tuberculosis, and endemic childhood diseases also entered his frame of analysis.[1]

In short, Ackerknecht concluded that anticontagionism experienced an Indian summer from 1821 to 1850, after which it declined steadily with the biological breakthroughs that ushered in the 'laboratory revolution'.

S. Cohn Jr. (✉)
University of Glasgow, Glasgow, UK
e-mail: Samuel.Cohn@glasgow.ac.uk





C. Lynteris (ed.), *Plague Image and Imagination from Medieval to Modern Times*, Medicine and Biomedical Sciences in Modern History, https://doi.org/10.1007/978-3-030-72304-0_7

Taking his cue from Rudolf Virchow (1821–1902), the radical reformer and founder of cell biology, Ackerknecht placed this Indian summer within a larger economic and political context.[2] Those supporting anticontagionism, he maintained, were ideologically intertwined with liberal commercial interests and hence were antagonistic to quarantine regulations because of the deleterious impact on trade. By contrast, 'despotic' regimes were more willing to develop and impose quarantine.

In her careful analysis of the writings of John Snow, Edwin Chadwick, Southwood Smith, William Budd, and other English reformers, medical practitioners and pioneers of epidemiology and microbiology, Margaret Pelling challenged Ackerknecht's division of scientists and reformers into strict camps of contagionists and anticontagionists. Moreover, against Ackerknecht's earlier position, she maintained that the victory of the contagionists came seventeen years later than mid-century in 1867.[3]

A re-reading of Ackerknecht shows that his contrast between the two camps was not as stark as Pelling suggested. First, as with his hero, Virchow, who slides from one side to the other, Ackerknecht held that mixtures of miasma and contagion continued to influence views within both groups through the nineteenth century. Moreover, different epidemic diseases produced different notions about contagion. Not even the most hardened anticontagionists, he argued, denied that smallpox was a contagious disease. Furthermore, Pelling never questioned whether more autocratic regimes such as Russia tended to support notions of contagion or enforced quarantine more readily than liberal governments such as Britain and Holland, the first countries to reject their quarantine acts.[4]

But, despite Ackerknecht's comprehensive investigation across time, diseases, and nationalities, and his intellectual inheritance from Virchow, I contend that this debate has remained principally a study in the history of ideas. A social history of contagion and its primary battlefield—quarantine and isolation camps and resistance to them—have yet to be written, especially from the perspective of the victims of quarantine and isolation and not from the ideas of reformers and physicians or the policies of states.[5] This chapter will begin to investigate this question by examining one pandemic disease—plague—at the beginning of 'the third pandemic' and in one subcontinent alone—India—where around 94% of human plague deaths have been estimated since the disease reached Hong Kong in 1893.[6]

First, although cholera may have sparked more social violence over time than any disease in world history, neither notions of contagion nor of quarantine were usually the root causes of this violence.[7] Instead, across political regimes from liberal Manchester to Czarist Russia, similar forms of protest flared: marginal populations—recent Irish Catholic immigrants in English cities or in Edinburgh, Glasgow, or New York; Asiatic Sarts in Tashkent; impoverished women and children in Glasgow; or fig-growers and fishermen in Puglia—believed physicians and other health workers were allied with the state to cull populations of the poor.[8] True, hospitals were major targets of the rioters' ire, but these perceived chambers of death were not quarantine centres as were segregation plague camps in India or smallpox pesthouses in America. Moreover, unlike smallpox violence, those destroying cholera hospitals generally appeared fearless of cholera's potential infection or contagion: they mobbed cholera hospitals to 'liberate' their neighbours and loved ones, whom they 'triumphantly' carried on shoulders back to their homes.[9]

The diseases to provoke social violence centred on fears of contagion or hatred of quarantine and isolation were smallpox, yellow fever, and plague, and with all three, violence came late in the day. With smallpox, it was the pandemic of 1880–1882; with yellow fever, it occurred only by the end of the nineteenth century and especially with yellow fever's epidemic finale in the US in 1905, after the mysteries of that disease's transmission had been dispelled. With plague, it came with its spread to China in 1893 and in India from 1896 to 1902. However, in terms of social violence, smallpox and especially yellow fever pale by comparison with plague in the Indian subcontinent. Smallpox violence concentrated largely in the US and could be vicious as with the burning of pesthouses with the intimates occasionally cremated alive, but rarely did it explode into mobs in the thousands as it did in Chicago in June 1894, Milwaukee, two months later, and Montreal for three months in 1885.[10] With yellow fever, despite mass panic and migration, its social violence and cruelty was far less than smallpox's and was limited almost entirely to newly arrived Sicilian sugarcane workers on Louisiana plantations in 1905. By contrast, I have discovered sixty-three incidents of collective protest triggered by plague in the Indian subcontinent alone. Moreover, several of these included numerous, unspecified villages, and crowds could mount in the thousands as at Calcutta in July 1897 with 5000, at the much smaller town of Seringapatam in November 1898 with 10,000, and, most significantly, in Mumbai on 9 June 1898, with Julai textile workers supported by shop closures and industrial strikes.[11]

Papers such as *The New York Times* reported these riots as though they were akin to European cholera riots, riveted by similar fantasies of purposeful poisoning by physicians and agents of the state.[12] Instead, with plague riots, such fantasies were rare.[13] In addition, to being political protests against specific policies tied to quarantine, decrying military searches, segregation camps, and destruction of property, especially of the poor, plague's collective action differed from cholera's in other ways. First, instead of tearing societies asunder as was the upshot of cholera riots, plague protest often united them across caste, class, and religious boundaries. As Indian newspapers in English and Indian languages indicate, elites might begin with caste and class prejudices against those who took their grievances to the streets, but by mid-1897, newspaper editors and other intellectuals began supporting those who risked their lives to change plague policies. Secondly, unlike the bulk of cholera riots, Indian plague protests often began with peaceful demonstrations, letters to editors and colonial officials and drafted resolutions in town halls to alter colonial plague policies. Finally, of India's plague riots and mass demonstrations, fifty-four of the sixty-three I have found, 86% of them clustered sharply into less than five years, from mid-1897 to 1901. By comparison, in parts of Europe, cholera riots stretched from the 1830s to cholera's last major European wave in 1910–1912.[14] Over these years, despite developments in understanding cholera's transmission, improvements in its treatment and mechanisms of prevention, much the same fantasies as in 1831 continued to rouse marginal groups to murderous violence. And in places such as Venezuela, these resurfaced during the seventh cholera wave into the 1990s and beyond.[15]

* * *

Why did plague riots in the Indian subcontinent abruptly soar in number and crowd sizes and then vanish so quickly? I argue that a certain clash of cultures sparked the rise towards the end of 1897. This clash was not, however, the usual one conjured up by this cliché or as the international press and occasionally native papers presented it at the end of the nineteenth century to berate the supposed 'superstition' and ignorance of indigenous Indian labourers[16] in contradistinction to supposedly educated, scientifically-minded Europeans. Senior officers in colonial governments often went further with their class and racial stereotypes as with India's sanitary commissioner, Robert Harvey, who in 1897 branded Indians across classes as 'suspicious of innovation, extremely conservative, very ignorant, full of prejudices and superstitions and of amazing credulity'.[17]

Instead, from the entry of plague into Mumbai in the autumn of 1896 to the Julai riots of March 1898, native newspapers did not lash out against Western science or colonial plague policies but initially only against the 'tyrannical' implementation of their policies. Western bureaucrats and native critics agreed: their bubonic plague fit the same patterns seen with historic plague since the Black Death; it was a highly contagious disease, transmitted person to person, and to contain it, stringent quarantine measures were necessary. Indian intellectuals wholeheartedly endorsed the principles behind quarantine, including segregation camps, even military searches, if administrated correctly without humiliating indigenous populations and especially the poor. As the *Indian Spectator* stated, 'Segregation is good, but not without reasonable provision … proper treatment and diet.'[18] Fear of humiliation and hardship caused by cruel and corrupt execution of sound plague operations encouraged Indians to conceal cases and flee, thereby fanning the contagion.[19] The native press could even criticise colonial quarantine measures for being too lax as in June 1897, when Bombay's Plague Commissioner, General W. F. Gatacre, reduced quarantine measures. In response, native papers pleaded for plague searches to be reinstated, at least for the monsoon season.[20]

However, shortly after the Julai riots, criticism from Indian newspapers and intellectuals shifted. No longer were activists such as Bal Gangadhar Tilak (journalist, newspaper owner, and future independence fighter) convinced that the science behind colonial plague prevention was sound. As early as April 1898, native papers began to criticise key concepts behind plague prevention. Their doubts did not derive from Eastern folklore, superstition, or fatalism; instead, the latest Western science combined with their own local statistics observed over several plague seasons had changed their minds. Tilak's paper, *Mahrátta*, sounded a new note: 'The surest and the best method of arresting the progress of plague, recommended by both Western and Eastern writers of scientific fame, is the temporary abandonment of the infected area.'[21] Clearly, neither Tilak nor other intellectuals of the Presidency of Bombay had been at the vanguard of these changes in attitudes and science. Already, as Tilak's paper indicated, practices in the less urbanised districts of Belgaum, Sátára, Solápur and Nagar had succeeded in lowering mortalities by recognising the rat as the plague's carrier and endorsing evacuation over quarantine.[22]

The native intellectual voice that previously defended the principles of quarantine while condemning its implementation began to crack. In criticising current plague policy of removing sufferers against their will,

Karachi's *Phoenix* no longer based its arguments on poor conditions of plague camps or the violence of military searches. Now, the paper questioned the theoretical foundations of colonial plague policies, and it was based on Western science as well as their figures charted over six plague seasons since 1896: 'Four English doctors say that segregation does no good.'[23] A week later, the paper was more emphatic, calling notions of segregation 'absurd'. Their argument rested on recent historical observations: 'In our city, notwithstanding disinfection, quarantine, segregation, hospitals, public and private, and other various means … the disease is on the increase.' It then analysed plague mortality figures in Bombay after the Julai riots, when 'people began nursing their relatives in their homes'. Against notions of strict segregation, 'the plague declined'.[24]

Government opinions soon followed, even if their practices lagged behind the science of their scientific advisors. In July, the Supreme Government of Bombay ordered the municipality to abolish all quarantines and sanitary cordons, declaring them 'of no use in stamping out the plague or diminishing its intensity'.[25] Yet the city continued its rigorous cordons and examinations at railway stations well into 1901, and at other places, these impositions persisted through 1902, despite repeated complaints and petitions from citizens for their repeal. By 1901, criticism against the old approaches and theories of plague control, many reaching back to the Middle Ages, were now nearly universal in newspapers across the Presidency of Bombay.[26] Pune's Kál condemned the theory and practice of segregation and burning clothes: 'Nobody requires any more of these measures … The People are disgusted with these troublesome and futile rules that … gave rise to riots in Bombay, Calcutta, and Cawnpur, and now are doing the same in the Punjab.'[27]

Instead of 'fatalism' or a clash between Western science and supposed indigenous superstition, native papers called for new scientific solutions. Reminiscent of British reformers of the mid-nineteenth century, native papers vehemently publicised the need for sanitary improvement, castigating municipal and colonial governments for their apathy and failure to cleanse filth from cities and villages or provide modern drains, latrines, fresh water, and proper burials for animals and humans. And these calls came not only from newspapers in major metropolitan regions.[28]

In addition, native criticism of plague policies and practices demanded new scientific experimentation and data collection.[29] The *Indian Spectator* berated Bombay's plague commission, declaring that it was 'strange in the midst of all scientific and medical activity, no systematic commissions … to investigate the causes and methods of treatment of plague' had been organised.[30] Lucknow's *Advocate* similarly pleaded for empirical analysis to decipher what factors caused some communities to be devastated by plague, while others escaped or were only slightly scathed. It proposed scientific studies into the conditions of life, 'modes of living', food consumption, and preventive measures to explain the divergences, and advocated committees to be appointed to compile surveys. Only through 'constant enquiries' and 'science', it maintained, could plague's present mysteries be unravelled.[31] These Indian calls for systematic scientific analysis of plague statistics anticipated the British plague commission reports that began in 1906 and lasted into the second decade of the twentieth century, filling thousands of pages in the Cambridge *Journal of Hygiene*. These included hundreds of temperature and humidity charts, their correlations with plague outbreaks mapped across vast regions of the subcontinent, with cases analysed by caste, religion, race, social class, occupation, and housing, down to types of masonry.[32]

Yet, despite new evidence of rats spreading plague and hospital reports since 1897, repeatedly stating that plague patients were not infecting their doctors or nurses, colonial bureaucrats as late as 1902 refused to dismantle quarantine stations or end forced confinement in segregation camps. In April 1900, 'overzealous' enforcement of quarantine regulations sparked a plague riot of thousands at Kanpur (Uttar Pradesh) supported by 3000 textile workers, who went on strike, protesting plague measures.[33] The rioters attacked constables, killed their captain, and threw five police alive (by some reports) and six (by others) into the camp's raging fires. It was one of the most widely reported riots connected with any disease of the late nineteenth and twentieth centuries, reaching papers from Mexico to Australia.[34] It remained in the press for the next eight months with court cases and appeals that reached the Viceroy of India and the houses of Parliament at Westminster against twenty-five rebels condemned to death.[35] The increased frequency of plague protest from late 1897 to 1901 derived not just from abusive and humiliating enforcement of colonial plague policies. Instead, increasingly, Indians across social classes came to realise that these quarantine measures, grounded in colonial conceptions of 'contagionism' and traditions of plague prevention that harkened back to Middle Ages, were utterly useless with the bubonic plague of the third pandemic.

* * *

In conclusion, plague protest in India differed radically from cholera riots that continued flaring in Europe for eighty years after their first appearances in 1831. The sharp clustering of plague riots from late 1897 to 1901, well before plague mortalities in the subcontinent reached their peak in 1907 nationally and later in other regions such as the Punjab, reflected a clash of cultures. Yet it was one that failed to follow the expected script. Against new knowledge about plague transmission and contagion that first came to understand rats and not people as the principal carriers and later discovered that fleas were the major vectors, British officials continued to impose useless and humiliating forms of quarantine and isolation until 1902. The stalwarts against modern Western science were not indigenous workers or villagers. In fact, they may have even been at the vanguard of the new scientific awareness of plague's mysterious paths of dissemination, ahead of Western or Eastern Science and of intellectuals such as Tilak, who began to question the underlying premises of colonial plague prevention based on person-to-person transmission only by June 1898. Earlier, Indian villagers had commented on the sudden strange appearance of dead rats falling from their rafters just before human plague appeared, ringing alarm bells to abandon their huts for a two-month vacation in neighbouring woods.[36] Evacuation, not quarantine, had early on become villagers' approach to plague prevention. By contrast, this switch from quarantine to evacuation does not appear as official colonial plague policy in Mumbai or other major cities until June 1901.[37] And scientists such as Major E Wilkinson in his meticulous medical reports on plague in the Punjab, published in 1904, continued to argue that the rat was not the plague's essential carrier. The web of influences across regions, social classes, and professions that finally convinced officials to change plague policies in India remains to be studied.

* * *

Back to the anticontagionist debate: seen from the perspective of social history as opposed to medical ideas, this debate assumed a new shape in the last years of the nineteenth century.[38] First, with two of the three big epidemic diseases of the nineteenth century—plague and yellow fever—a species of anticontagionism, not contagionism, eventually won the scientific debate. With these diseases dependent on insect vectors, and with plague, also rodent carriers, quarantine measures were not the answer. Moreover, as Indians argued in their newspapers, often such measures, in fact, hindered effective care of the afflicted. Second, with these diseases the debate had not ended in 1850 or 1867; rather the period of transition was later than either Ackerknecht or Pelling imagined: with plague in India, it was 1901 or 1902, and with yellow fever, it was in 1905. Third, a liberal-democratic nation, Britain, which from the early nineteenth century had led the march against quarantine, now, at the end of the nineteenth century had switched sides, becoming the bulwark of contagionism, quarantine, and isolation. In the Indian subcontinent, British colonial governments reacted with greater ferocity and exacted more lethal repression on protesters who turned against British medical doctrines than seen even with the Czarist crackdowns on cholera rioters into the 1890s. Not only did Britain silence native newspapers in 1897 after the assassinations at Pune, but its repression of rioters at Calcutta in 1897 resulted in killing over 600 protesters, the largest number tallied connected with any crackdown on disease protest anywhere in the world during the nineteenth or early twentieth century.[39] Finally, the history of plague policy in India shows that reformers and scientists were not the only actors to influence attitudes and policies on contagionism, anticontagionism or quarantine and isolation. In India, workers and villagers were instrumental in changing attitudes towards plague science that overturned outdated contagionist policies governing diseases transmitted from animals to humans via insect vectors. Only after the spate of plague riots in 1897 and 1898 did municipal and colonial governments begin to recognise that their anti-plague policies had been useless and that in addition to destroying the property of the poor and humiliating indigenous peoples, these policies had succeeded only in spreading distrust and violence.

## Notes

1. Erwin H. Ackerknecht, 'Anticontagionism between 1821 and 1867: The Fielding H. Garrison Lecture', *Bulletin of the History of Medicine* 22 (1948): 562–593. On Ackerknecht's conclusions, see among other places, J. N. Hays, *The Burdens of Disease: Epidemics and Human Response in Western History* (New Brunswick, N.J.: Rutgers University Press, 1998), p. 137; Peter Baldwin, *Contagion and the State in Europe, 1830–1930* (Cambridge: Cambridge University Press, 1999), pp. 28–35, 488–489, 491–492; Martina King and Thomas Rütten, 'Introduction'. In Rütten and King (eds.), *Contagion and Contagious Diseases: Medicine and Literature 1880–1933*, pp. 1–16 (Berlin: De Gruyter, 2013); Andrew

R. Aisenberg, *Contagion: Disease, Government, and the 'Social Question' in Nineteenth-Century France* (Sanford, California: Stanford University Press, 1999), pp. 5, 186; Roger Cotter, 'Anticontagionism and History's Medical Record'. In P. Wright and A. Treacher (eds.), *The Problem of Medical Knowledge: Examining the Social Construction of Medicine*, pp. 87–108 (Edinburgh: Edinburgh University Press, 1982); Mark Harrison, *Medicine in an Age of Commerce and Empire: Britain and its Tropical Colonies 1660–1830* (Oxford: Oxford University Press, 2010), pp. 1, 253. For a critique and substantial bibliography on this debate, see David S. Barnes, 'Cargo, "Infection," and the Logic of Quarantine in the Nineteenth-Century', *Bulletin of the History of Medicine* 88 (2014): 75–101, especially pp. 75–80.

2. Virchow argued: 'Medicine is a social science … anthropology in its widest sense, whose greatest task is to build up on a physiological foundation … Politics is nothing but medicine on a large scale. The 'ultimate aim' of medicine, 'is eminently social'; cited in George Rosen, *A History of Public Health* (New York: MD Publications, 1958), p. 13.
3. Margaret Pelling, *Cholera, Fever and English Medicine 1825–1865* (Oxford: Oxford University Press, 1978), pp. 1–18.
4. Also, see Pelling's return to aspects of the contagionist-anticontagionist debate: Margaret Pelling, 'The Meaning of Contagion: Reproduction, Medicine and Metaphor'. In Alison Bashford and Claire Hooker (eds.), *Contagion: Historical and Cultural Studies*, pp. 15–38 (London: Routledge, 2001). Here, Pelling maintains that 'the polarities of contagion and miasm, contagionist and anticontagionist … continues to prove irresistible', pointing to Baldwin (*Contagion and the State*). Yet for venereal diseases in the nineteenth century, Baldwin was at odds with Ackerknecht, claiming that his dichotomy 'liberal versus Conservative equals voluntary versus regulatory do not pair up (pp. 487–488). According to Barnes ('Cargo'), Baldwin is 'one of several historians … to consign Ackerknecht's thesis to the dustbin of historiography (p. 78).
5. For Baldwin's exhaustive analysis, *Contagion*, the actors were states and not social movements. Collective protest and its effects on change are scarcely mentioned.
6. Graham Twigg, *Bubonic Plague: A Much Misunderstood Disease* (Ascot: Derwent Press, 2013), pp. 9 and 69, based on L. Fabian Hirst's figures calculated in 1938 (*The Conquest of Plague: A Study of the Evolution of Epidemiology*, Oxford university Press: Oxford, 1953), puts it at 95%, as does Ira Klein ('Urban Development and Death: Bombay City, 1870–1914' *Modern Asian Studies* 20 (1986), p. 744) and others (Samuel Cohn, Jr., *The Black Death Transformed: Disease and Culture in Early Renaissance Europe*, Oxford: Oxford University Press, 2002, p. 13). Since the 1930s,

however, the proportion of plague deaths in the Indian subcontinent has fallen significantly. Using data from *Plague Manual: Epidemiology, Distribution, Surveillance and Control* (Geneva: WHO, 1999), table 1, pp. 18–25, I have recalculated these proportions. However, given the overwhelming preponderance of plague deaths of the 'third pandemic' that occurred before 1938, the overall decline in the India subcontinent's proportion is only around 1%. For the arrival of plague in Hong Kong, I use 1893, when the first cases were reported, rather than the canonical date of 1894, when Alexandre Yersin arrived in Hong Kong.

7. This goes against the arguments of Sir Richard J. Evans ('Epidemics and Revolutions: Cholera in Nineteenth-Century Europe'. In Terence Ranger and Paul Slack (eds.) *Epidemics and Ideas: Essays on the Historical Perception of Pestilence*, pp. 149–173, Cambridge University Press: Cambridge, 1992), and of Baldwin (*Contagion*), who follows Evans: 'By imposing measures that were unrelenting in their insistence that contagion should be fought first and foremost with cordons, quarantines and sequestration … Popular unrest, riot and rebellion were the Fruit.' But I know of no cholera mobs to have attacked cordons and only a few, principally in Italy after 1866, to have attacked quarantines; see: Samuel Cohn, Jr., *Epidemics: Hate and Compassion from the Plague of Athens to AIDS* (Oxford: Oxford University Press, 2018), chapter 9.
8. For this history, see: Cohn, *Epidemics*, chapters 7–10. For the historiography on cholera riots, see: ibid., pp. 180–182.
9. Such incidents in cholera riots not only characterised the first European wave of the disease in 1831–1837, but endured into the late nineteenth century in Russia as at Saratov in 1892 (*Morning Call* [San Francisco] 15 July 1892, front page; and *La Presse* [Paris], 20 July 1892) and even into the twentieth century in Italy, as at Gioia del Colle and Massafra (both in Puglia) in 1911 (*Times Dispatch* [Richmond, Virginia], 10 September 1911, front page; and Nicola Simonetti, and Mimma Sangiorgi, *Il colera in Puglia dal 1831 ai giorni nostri* (Fasano, Brindisi: Schena, 2003), p. 189.
10. Cohn, *Epidemics*, pp. 293–295.
11. Ibid., chapters 14–16; and for the Julai riot in Bombay: Prashant Kidambi, *The Making of an Indian Metropolis: Colonial Governance and Public Culture in Bombay, 1890–1920* (Aldershot: Ashgate, 2007), chapter 5.
12. *New York Times*, 28 May 1900; Ira Klein, 'Plague, Policy and Popular Unrest in British India', *Modern Asian Studies* 22 (1988): 723–755. 'The majority of the people believed that doctors were the chief men who spread the plague, and this some believed they did for their own living and maintenance'; ibid., p. 749. Also, see: Ian Catanach, 'Fatalism? Indian Responses to Plague and Other Crises', *Asian Profile* 12 (1984): 183–192, suggests

that rumours of well poisoning with Indian plague was similar to blaming the Jews during the Black Death (p. 190).

13. As with Mumbai mill workers attacking an Isolation hospital plague on Arthur Road on October 10, 1896, but even here the protest became one against quarantine and isolation; Cohn, *Epidemics*, p. 327.
14. Cohn, *Epidemics*, pp. 207–222, 227–230, 254–261.
15. In places, such riots continued into the 1920s in the Soviet Union, and during the seventh cholera wave in Naples, Torre del Greco, and Bari in 1973, Peru and Venezuela in the 1990s, and Zimbabwe, and Haiti in the twenty-first century; ibid., pp. 208 and 228–229.
16. See, for instance: Kalakankar's *Hindustan*, British Library, India Office Records and Private Papers (hereafter BL), Selections from the Vernacular Newspapers published in the North-Western Provinces and Oudh: L/R/5/77, week ending 1 May 1900, no. 12, which explained the violence at Kanpur as sparked by the 'most foolish and mischievous rumours', circulated 'among the ignorant masses', concluding that 'India is a land of superstitions and false beliefs'; 'the fault of the riots ... was king Mob [which] is impervious to reason.'
17. Cited in David Arnold, *Colonizing the Body: State Medicine and Epidemic Disease in Nineteenth Century India* (Berkeley: University of California Press, 1993), p. 232; and BL, Report of Native Papers Published by the Bombay Presidency, L/R/5/152, week ending 3 July 1897, no. 29, *Dnyán Prakásh.*
18. L/R/5/152, week ending 13 March 1897, no. 16, *Indian Spectator.*
19. Cohn, *Epidemics*, p. 338.
20. BL, Bombay Presidency, L/R/5/151, week ending 17 June 1897, no. 17; and Cohn, *Epidemics*, p. 331.
21. L/R/5/152, week ending 16 April 1898, no. 23, *Mahrátta.*
22. Ibid. By the time the scientific community had been convinced that the flea was the essential vector in plague transmission with Glen Liston's experiments published in 1910, the riots and plague protests had disappeared in the subcontinent.
23. L/R/5/152, week ending 14 May 1898, no. 24, *Phoenix.*
24. Ibid., week ending 21 May 1898, no. 54, *Phoenix.*
25. Ibid., week ending 16 July 1898, no. 12, *Bombay Samáchár.*
26. See: Cohn, *Epidemics*, pp. 339–340.
27. L/R/5/156, week ending 1 June 1901, no. 32, Kál; Wai's Moda Vritta claimed: 'Hundreds of instances can be cited showing plague regulations, however severe, are absolutely inefficient against the disease and are only a fruitful source of trouble both to Government and the people'; ibid., no. 35, *Moda Vritta.*

28. Below are just a few examples from the Punjab, the North-Western Provinces, and Bengal; many more could be cited: L/R/5/188, week ending, 8 July, no. 13, Tribune (Lahore); L/R/5/77, week ending 3 April 1900, no. 9, Hindustani; week ending 17 April 1900, no. 25, Tihfa-i-Hind (Bijnor); L/R/5/79, week ending 18 April 1903, no. 19, Hindosthan (Kalakankar). Also, see: Cohn, *Epidemics*, pp. 339–340.
29. An editorial in the *Indian Spectator* on October 20, 1901, anticipated the international plague reports and massive experimentation and accumulation of data that began five years later. On these later voluminous plague-commission reports, see: Cohn, *The Black Death Transformed*, pp. 26–33.
30. L/R/5/156, week ending 26 October 1901, *Indian Spectator.*
31. BL, North-Western Provinces, L/R/5/80, week ending 16 May 1902, no. 11.
32. On these reports, see: Cohn, *The Black Death Transformed*, pp. 26–33.
33. L/R/5/77, week ending 17 April 1900, no. 24, *Indian Daily Telegraph.*
34. See: *Mexican Herald*, 23 August 1900, p. 2; *South Australian Register* (Adelaide), p. 25 August 1900, p. 7; and *Bunbury Herald* (Western Australia), 25 August 1900, p. 3.
35. Cohn, *Epidemics*, pp. 334–337.
36. Charles Creighton, *History of Epidemics in Britain*, 2nd edition, ed. by D. E. C. Eversely, E. A. Underwood and L. Ovenall, I: A.D. 664–1666 (Cambridge: Cambridge University Press, 1894; London, 1965), pp. 168.
37. L/R/5/156 1901, no. 26. Sudharak, 17 June.
38. See, for instance: William Coleman, *Yellow Fever in the North: The Methods of Early Epidemiology* Wisconsin Publications in the History of Science and Medicine, number 6. (Madison: University of Wisconsin Press, 1987): 'The very notion of disease transmission by means of a nonvertebrate vector did not enter the mainstream of medical and epidemiological thought until the 1890s, primarily through Theobald Smith's starling demonstration that Texas cattle fever was caused by a tick-borne protozoan' (p. 12).
39. See: Cohn, *Epidemics*, p. 372 (Section 505 of the Criminal Procedure).

CHAPTER 8

# Bamboo Dwellers: Plague, Photography, and the House in Colonial Java

*Maurits Bastiaan Meerwijk*

'At the time when the number of plague cases was highest, no house rats could be produced by the inhabitants.'[1] This observation by Dr Willem de Vogel at the Far Eastern Association of Tropical Medicine (FEATM) meeting in Hong Kong, in January 1912, revealed a key scientific anxiety that had haunted Dutch colonial health officials struggling to contain an outbreak of plague in East Java. As cautious adherents of the

Research leading to this chapter was funded by a European Research Council (ERC) Starting Grant (European Union's Seventh Framework Programme ERC grant agreement no. 336564) for the project *Visual Representations of the Third Plague Pandemic* held at CRASSH (University of Cambridge) and the Department of Social Anthropology (University of St Andrews). I am grateful to Christos Lynteris and Susie Protschky for reading through earlier versions of this chapter. My deepest thanks to Johan van Langen, Liesbeth Ouwehand, Ingeborg Eggink, and the staff at the Arsip Nasional Republik Indonesia (ANRI) for helping to secure access to key archival materials.

M. B. Meerwijk (✉)
The University of Hong Kong, Pok Fu Lam, Hong Kong
e-mail: meerwijk@connect.hku.hk





C. Lynteris (ed.), *Plague Image and Imagination from Medieval to Modern Times*, Medicine and Biomedical Sciences in Modern History, https://doi.org/10.1007/978-3-030-72304-0_8

'rat-flea-man' transmission theory, the *sous-chef* of the Civil Medical Service and his companions were unnerved by this unexpected scarcity of rats from the plague-stricken *kampongs* (villages). 'We lived in uncertainty', confided another physician in his government report, 'and could not prove that in Malang human and rat plague coincided.'[2] Caught in the tenacious epistemic uncertainties surrounding the origins and transmission of plague, Dutch investigators resorted to a 'methodical search' of the houses of plague patients that at last resolved their conundrum.[3] Concealed beneath the picturesque *atap* (palm leaf) roofs, in between double walls made of woven mats, and inside the posts and beams of Java's principal building material—bamboo—they uncovered myriad shelters containing rat cadavers and rat nests. In the following months, Dutch health officials re-articulated longstanding associations between plague and the house. Photographs helped integrate the structure's materiality and design into plague's transmission theory, prompting the signature health intervention of the Dutch late-colonial period: home improvement.

To appreciate the impact of these developments, we must recall how the fast-growing *Bambuseae* had enmeshed themselves into the cultural fabric of Southeast Asia.[4] This organic 'skyscraper with hollow stems' served to construct dams, bridges, irrigation works, dwellings, and sailing craft, or was processed into food, utensils, and decorations.[5] 'The Bamboo is one of the most wonderful and most beautiful productions of the tropics', exulted the famed naturalist Alfred Russell Wallace in *The Malay Archipelago* (1869), and 'one of nature's most valuable gifts to uncivilized man.'[6] More dynamically, we might appreciate the relationship between humans and bamboo as an 'alliance' struck against the region's volatile typhoons and earthquakes, its heat and monsoon rains.[7] The bamboo house was light and airy, easily destroyed, quickly rebuilt, and highly economical. Praising the material for 'combining all the advantages of strength and elasticity', a Swiss traveller in Java described its use in house construction in 1829:

> To build their *pandoks* (homes or huts) the Javanese employ the bamboo, which is strong and easy to manipulate. For a home of four to six rooms one requires about seven or eight hundred pieces of bamboo. One piece from 20 to 50 feet costs 3 or 4 *duiten* or 2 cents. The tools are similarly made from bamboo. Four good craftsmen are able, within a week, to make a firm and strong home.[8]

With wild and cultivated groves proliferating across Java, the practical and economic value of bamboo—layered upon its spiritual and medical significances—was difficult to overstate.[9] Nevertheless, bamboo had by 1900 acquired an ambiguous quality in the eyes of European colonials in Asia.[10] The heterogeneous bamboo dwellings of Southeast Asia were alternatively praised as above, or rejected as dark, dirty, permeable, flammable, and impermanent.[11] Now, in Java, the bamboo house became pathogenic.

Dutch plague anxieties might as well have come to rest on a very different material (such as rice) or on the Javanese plague patient instead.[12] Their focus on bamboo and the house, however, fed into wider concerns of plague's relation to the built environment. Historically, plague was a 'house disease' with a distinct urban preference.[13] The disease had become pandemic following an outbreak in Hong Kong in 1894, and consequently urged an interrogation of the urban environment of 'plague ports' as diverse as Bombay, Sydney, Honolulu, and Hanoi.[14] In the Dutch colonial context, by contrast, plague assumed a rural character that posed new questions about its nature and transmission. Ultimately, persistent references to the 'plague house' no longer referred to the house as a mere 'site', 'space', or 'breeding ground' of the infection. Rather, the discovery of rats and nests hidden within the hollow bamboo frame rendered the house physically complicit in its spread.[15] This chapter asks: how did this transformed materiality of the plague house come about? It is a key question, for the reframing of the traditional bamboo dwelling underpinned a dramatic intervention in the built and natural environments of Java: home improvement. Between 1911 and 1942, over 1.6 million houses were reconstructed in an attempt to build out the rat.[16]

The 'methodical' search of the house was extensively documented on camera. In fact, epidemic photography of plague in Java was fixated on the house: its construction, its inhabitation, and the various stages of its reconstruction.[17] Photography was crucial to the initial re-imagination of the role of the house in plague transmission. Photographs documented the presence of the rat within 'infected' dwellings and captured the proximity of nesting spaces to human occupants. They betrayed the physiological gaze of the physicians who commissioned them, zooming in on those features of the house seen to aid human/rat interaction and thereby facilitate plague transmission. As I will demonstrate, they conspired with other scientific imagery to document a tubular network dissecting the house that was inhabited by the rat, positioning the structure as a missing link in a nascent and distinctively Javanese plague ecology grounded both in its design and in its materiality. In other words, photographs helped to insert a bamboo link in plague's putative rat-flea-man transmission scheme. In decades to come, Dutch physicians and officials relied on these photographs to defend this scheme—and the interventions it inspired—to Dutch lay and foreign expert audiences alike. Read alongside tables, diagrams, and maps also produced by Dutch colonial physicians, this quasi-anatomic photography was instrumental in placing the bamboo house—and not the soil, the air, the native body, the rats, the fleas, or even the plague bacillus itself—at the heart of Dutch plague anxieties.[18]

## Plague in Java

The history of plague in Java has remained remarkably understudied while the confusion surrounding its first appearance was vital to the subsequent re-articulation of Dutch plague anxieties.[19] Over the first months of 1911, physicians in East Java were alerted to individuals suffering from strange and unfamiliar febrile symptoms. Dead rats were found in unusual numbers at the warehouses of Soerabaja and by the station master at Sidoardjo. Reports of unexplained mortality in the interior reached officials of the Civil Service and indeed the press.[20] At Weltevreden, the administrative capital of the Dutch East Indies, the government medical laboratory received samples for investigation from apprehensive physicians who asked: could this be plague?[21] Several plague scares notwithstanding, the archipelago had thus far escaped inclusion in the ongoing pandemic: unlike neighbouring colonies and in spite of scant preventative measures.[22] The bacteriological investigations yielded negative results, however, leaving curious cases of malaria and typhus to be isolated without further consequence. But on 27 March, Dr J. de Haan, the director of the laboratory, confirmed the presence of plague-like bacilli in the blood sample of a Javanese woman sent to him by a private physician from the district Malang, in the *residentie* (district) Pasoeroean.

The next day, de Vogel departed for East Java in all haste: followed by de Haan, a retinue of assistants, physicians, and a regiment of soldiers.[23] Within days, they confirmed the presence of plague in the district to Governor-General Alexander Idenburg, who informed the Dutch government by telegram: '46 cases of plague at Malang, strict measures have been taken.'[24] Haunted by the millions of lives bubonic plague had claimed in India since 1896 as well as a devastating outbreak of pneumonic plague in Manchuria only months earlier, Idenburg promised extensive financial and administrative support to contain the disease.[25] Personally, he recommended the burning of entire kampongs but instead de Vogel sought to implement the directives of the Second Indian Plague Commission: isolating victims, separating their families, and evacuating affected villages.[26] Some kampongs went up in flames after all, but primarily the Dutch oversaw the burning, fumigation, and unroofing of individual 'infected houses' in the hope that fire, sulphur, and sunlight might eliminate rogue plague bacilli resting dormant in their earthen floors or hiding among a proliferation of 'things'.[27] Malang was placed under a military quarantine, nearby Soerabaja declared an infected port, and 'rat extermination' campaigns were launched across Java.[28]

The diagnosis of plague in the rural interior of East Java (rather than in a major port city) took the recently organised Civil Medical Service by surprise.[29] In a telegram from 26 May, Idenburg emphatically defended this oversight to the Dutch government: 'No physician had previously seen plague, medical literature paints different image from that seen here, symptoms not very severe, course decidedly atypical.'[30] Newspapers and officials had levelled an attack on Soemowidigdo, a local *doctor djawa* (Javanese physician) who had hesitantly reported an outbreak of glandular fever earlier in March, but de Vogel publicly and privately defended him stating that 'none of us knew the disease from experience'.[31] As a result, plague had already spread across East Java. Acutely aware of this deficient expertise, the Minister of Colonies asked the Amsterdam bacteriologist Dr J. J. van Loghem to travel to Malang and aid in the control efforts barely two days after being informed of the outbreak.[32] As the former director of the pathological laboratory at Medan, van Loghem had called attention to the looming threat of plague to the archipelago.[33] Still, his primary experience with the disease appears to have derived from a brief study/conference tour in India in 1909.[34] In a striking move that we might think of as 'maintaining viral sovereignty' the Dutch chose *not* to request the assistance of foreign plague experts at work closer by in Asia.[35]

In principle, both de Vogel and van Loghem were converts to the notion that plague was a rat disease transmitted to humans secondarily by the rat flea—a theory first proposed by the Pasteurian scientist Paul-Louis Simond in 1898.[36] The specifics of such a transmission scheme in Java were unclear, however. What rats were involved? Which fleas? Might other 'common parasites' shared by humans and rats play a role?[37] How were humans themselves implicated in the spread of plague? While the blanket quarantine imposed Malang was justified on the basis that 'we know almost nothing of the rats within the affected districts', the range of control measures deployed in the district in the meantime indicated that competing notions of plague transmission had not been eclipsed.[38]

The rat-flea-man transmission model was understood to require 'a very close contact between the sick rat and man'.[39] If a plague rat died too far away from humans, infected fleas could not hope to make the species jump. Given earlier observations in Sumatra that humans and rats lived in close proximity to one another, Dutch suspicions settled quickly on the normally ubiquitous house or *dessa* (village) rat at as the principal host of plague in Java.[40] The Malangese were encouraged by a small fee to catch as many rats as possible and deliver them to de Haan's improvised laboratory for examination.[41] Of the thousands of rats collected, however, the vast majority were field or *sawah* (rice paddy) rats. Skull measurements determined that both types belonged to the species *Mus rattus*, but they displayed critical biological and behavioural differences that brought the former into much closer contact with humans—only, where were they?[42]

When prompted, wrote de Vogel, the residents of Malang insisted all house rats had been caught.[43] Dr A. Deutmann, overseeing plague control in the sub-district Karanglo, likewise observed that 'not a single house rat was found, however much we searched for them'.[44] This phenomenon had previously been observed elsewhere and cast doubt on the notion that rats constituted the sole '*Infectionsquelle*' (well of infection) of plague.[45] At the laboratory, meanwhile, the mechanical processing of rats by de Haan—a miniature of the 'almost industrial' rat examination in India—elicited but few diagnoses of rat plague: thus failing to confirm a 'rodent link'.[46] When van Loghem finally arrived in Malang on 16 May, he observed that 'on superficial examination, there appears to be no connection between human plague and rat plague' in Java.[47]

By their sheer number, field rats had demanded the attention of investigators. If in meetings a *resident* (local governor) expressed the view that the hunt for the elusive house rat appeared 'a lot more important', de Vogel would reply that the role of all rats remained 'an open question'.[48] By June, however, significant objections had risen against the role of the field rat in plague transmission. They were healthy, lived apart from humans, and carried but few fleas.[49] Furthermore, plague continued its 'metastatic' (as opposed to contiguous) spread through Malang irrespective of the war waged against the rat—the oncological metaphor likening plague to a tumorous growth gaining traction in subsequent years.[50] In a private letter, de Vogel confessed that, if the number of cases had dropped slightly, he dared not say 'whether it is because of our interventions ... or because we removed people from their infected houses just in time'.[51]

In the absence of the house rat—or any type of infected rat—Dutch anxieties had lingered on the house: the space, as mentioned earlier, historically associated with plague transmission. But what exactly was 'infectious' about the houses of Javanese plague sufferers? Their earthen floors, cluttered objects, lack of ventilation, 'dark spaces', and of course the hygienic practices of their occupants (as elsewhere) all remained suspect.[52] Such purportedly pathogenic properties were in turn easily layered upon generic colonial disapproval of the native house, the state of sanitation in the colony, and the afterlife of classic nineteenth-century concerns about darkness, domestic disorder, and poor ventilation.[53] With the dismissal of the field rat as a source of plague, however, the question of the missing house rats from these plague dwellings assumed greater urgency.

Could it be, ventured van Loghem at this stage, that the rats inhabiting a plague house had already died inside?[54] This thought, exclaimed Deutmann, 'had struck no one before! But where might all these dead rats be?'[55] Launching into a 'systematic investigation of the houses where plague had occurred', van Loghem himself soon uncovered the remains of a rat beneath a slanting ceiling directly above the bed of a Chinese plague victim. 'On searching 5 surrounding houses on the same day, one nest with 2 plague rats was laid bare', which yielded several fleas carrying plague bacilli.[56] Elsewhere, de Vogel had the masoned ridge and tiled roof of a plague house removed, uncovering a rat nest 'containing the mummified corpse of a rat'. Upon splitting 'the bamboos of the native house', he discovered food remains 'that could have only been brought there by rats'.[57]

## The Plague House

The 'systematic' or 'methodological' search of the plague houses of Malang—a first reconnaissance, as it turned out—was possessed of a curious character. Like a diseased body dissected by the pathologist, the house was stripped: its hidden interiors revealed and pored over by a company of Dutch and Javanese physicians, bacteriologists, *doctor djawas*, and *mantri* inspectors. The quasi-anatomical nature of the investigation (conducted by and for physicians) is strikingly evident in a glass positive that appears to date back to this period (Fig. 8.1). The unknown photographer has homed in on the side of a roof from which the cover (tiles, in this case) has been removed. Segments of the bamboo rafters and laths have been cut

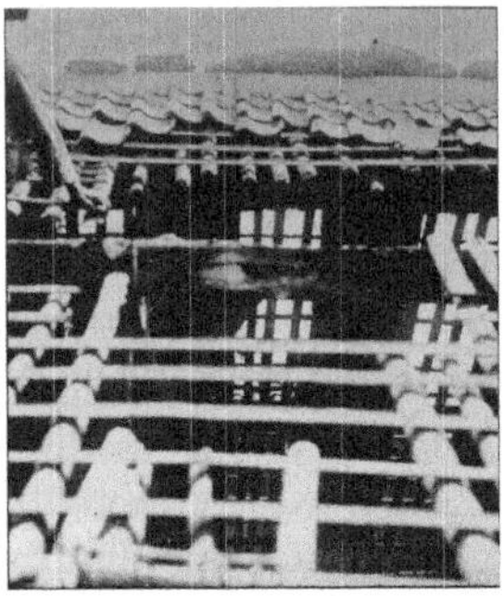

**Fig. 8.1** Plague house featuring a wooden beam containing a rat nest. (Source: University of Leiden Library, slide cabinet of Prof. P. C. Flu, drawer 22, slide 24)

away to provide a square frame centring a large, fissured wooden beam. To the left, a crouched figure appears to inspect the scene. The post-mortem of the plague house, so to speak, has yielded result. A description informs us that the beam is rotten and crowded with rat nests. Retrieved from a slide cabinet at the office of P. C. Flu, a professor of hygiene at the University of Leiden, the image was presumably used to instruct physicians bound for the Indies on the range of rat shelters within the Javanese house.[58]

Encouraged by their initial breakthroughs, de Vogel organised fifty *woningbrigades* (house brigades) across Malang to search plague houses as soon as they were identified.[59] Surgically, they removed the outer layers of *atap* and tiled roofs; peeled apart the double walls; broke open decorative door posts; and took a lancet to the bamboo poles of the frame and furniture. If their early findings had drawn attention to the roof as a shelter of the 'scansorial' rat, the brigades discovered 'that nests could be found in much closer vicinity to man'.[60] This newly identified pathology of the Javanese house warranted careful anatomic documentation, and was recorded in infographics, sketches, and perhaps most importantly—given their contested claim of providing unmediated access to truth—in photographs.[61] Indeed, de Vogel's government report on plague contained over forty photographs, only one of which had captured a plague patient against eight that captured nesting spaces of the rat about the house.[62] Separate reports by van Loghem, Deutmann, de Haan, and other health workers contained yet more images. The photographers are largely unknown. One photograph is specifically attributed to Deutmann. Several others were taken by Neville Keasberry, a Javanese photographer from Malang who specialised in producing stereoscopic images.[63] Most, however, appear to have been made either by the physicians themselves or by close assistants.

At the FEATM meeting in Hong Kong the following January, reported a French delegate, de Vogel provided a '*communication illustrée par des projections*' on plague in Java.[64] Some nineteen photographs demonstrated recent Dutch discoveries of the nesting spaces of the house rat, 'especially in the bamboo of which Javanese houses in this district are constructed.'[65] 'These are the rats dead in their nests on a bamboo ceiling just above the bed of a Chinaman who died of plague', he narrated, though the accompanying image in the conference proceedings has zoomed in on a single rat cadaver in its nest (Fig. 8.2). Embedded within a hollow beam cut open by the investigators, the creature's black, opened eye lends the scene

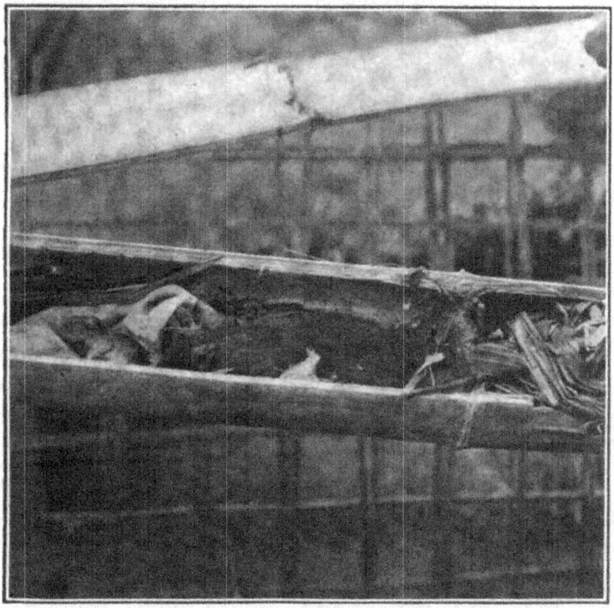

PLATE IX.

**Fig. 8.2** Mummified rat in the bamboo of a plague house in Malang. (Source: W. Th. de Vogel, 'The Connection Between Man and Rat in the Plague Epidemic in Malang, Java, in 1911', in Francis Clark (ed.), *Transactions of the Second Biennial Congress Held at Hongkong 1912* (Hong Kong: Noronha, 1913), Plate IX)

a grim power.[66] The spectacle of mummified rat cadavers in various stages of decomposition within pieces of bamboo soon became emblematic of the nature of the plague threat in Java.[67]

More disturbing even than finding rat nests scattered throughout the frame of the house itself was the rat's propensity to nest within the bamboos of the *balé-balé*: the iconic Javanese bedstead. 'One can hardly imagine a closer contact between man and rat', shuddered de Vogel.[68] Typically consisting of two large bamboo pieces (*watons*) connected by a raster of wooden or bamboo laths, the *balé-balé* was the principal item of furniture in a Javanese household as well as a stock object in Dutch literary and visual representations of the Indies. A photograph of an 'infected' specimen shown by de Vogel in Hong Kong depicts a *balé-balé* that had been dragged outside with two of its feet removed to angle the bench towards the camera.[69] A Javanese man has crouched beside the bed, the white hat balanced on his knee suggesting his position as a *mantri* with the house brigades. Both *watons* feature a horizontal incision along the top through which the trained eye can make out a score of nests hidden within them. On one rat retrieved from a *balé-balé* in Kaproe, van Loghem found twenty-two fleas containing 'typical' plague bacilli.[70] In Madioen, he encountered a 'fresh plague rat with 35 fleas' insolently lying dead upon the bed itself.[71]

The series of photographs, evincing how the Javanese house had come within the autoptic purview of Dutch colonial physicians and bacteriologists, imbued plague's historical, spatial association with the house with a distinct material quality. On one level, they merely provided examples of the rat's ability to live 'unobserved' in close proximity to humans. Simultaneously, they constituted unprecedented visual evidence of the nature of the 'plague house' that rendered its role in plague transmission visible to the naked eye. The house did not merely provide a space for 'spill-over' to occur or allowing transmission to take place, it facilitated it. Finally, on a metaphoric plane, the infestation of the hollow arteries of the house by plague rats mirrored the *Yersinia pestis* bacilli coursing through the bodies of the plague patient and—unlike them—could be seen without the complex mediations of the microscope. Shown in the context of his lecture on 'the connection between man and rat' in the plague outbreak in Malang, de Vogel deployed these images to argue that this connectivity rested in the very materiality of the house. In other words: he inserted a new bamboo link within the putative rat-flea-man transmission chain. How did the Javanese experience the violence done to their dwellings and possessions, meanwhile? How did they take to being evacuated from their villages? The government reports are characteristically silent on these matters, instead emphasising how residents understood 'the wisdom of our advice to leave the houses, where evidently the evil spirit had taken up its abode, and offered no resistance'.[72]

By August 1911, the aetiology of plague in Java had thus acquired a distinct spatial and material component. Its historic identity as a home or house disease was reiterated, but reconfigured to integrate the structure itself into plague's transmission scheme. Photographs unveiled the hollow, tubular bamboo as a crucial element in an emergent and distinctly Javanese plague ecology. Not only did it shelter rats, fleas, and bacilli near human occupants: bamboo helped to convey them. Unlike the elusive plague bacillus, the presence of dead rats and their nests was clearly visible to the naked eye. Dutch physician-photographers might fancy that the disease was not suggestively present 'at the edge of sight' but in full view.[73] Still, the powers of suggestion were not lost on them as they sought to reframe the Javanese plague threat by means of photography.

One such image—possibly taken by van Loghem—centres a horizontal beam protruding the side of a plague house in Malang (Fig. 8.3). A small opening is visible at the base of the pole, where the thin membrane separating bamboo segments has been gnawed through by rats. Such lesions, previously unnoticed, now marked the house as infested/infected: by rats if not by plague itself. The photograph suggests a more nuanced reading of the plague-dangerous features of the Javanese house. Graphic depictions of dismembered houses, carved up furniture, and split bamboo poles were joined by scenes with considerable suggestive potential. A close-up of the frayed, gnawed-on ends of woven bamboo mats serving as inner walls. A medium shot of a matted wall featuring a small hole, just above a *balé-balé*. A cavity at the base of a loam wall. An open space upon the ceiling or beneath the roof tiles.[74] Embedded within the voluminous Dutch plague reports, they intimate the presence of the rat within the house without showing the animal. They hint at passageways, and indeed: they suggest transmission.

A second photograph appearing alongside the former in a paper published by van Loghem in 1912 marks a similar vulnerability, this time inside the house (Fig. 8.4). The photographer has zoomed in on the convergence of two bamboo poles, a diagonal beam piercing a vertical support post. Our gaze is drawn towards an imperfection: the ragged edges above the original incision. As before, this newly ominous feature

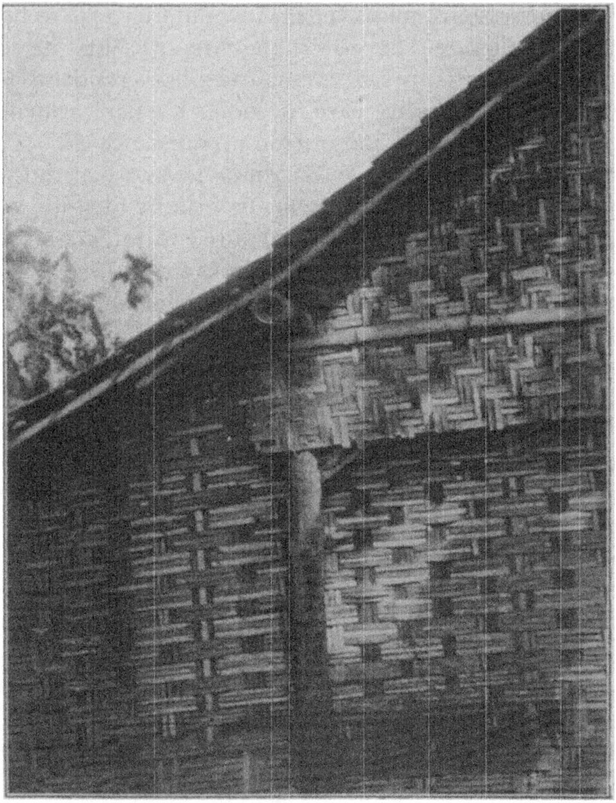

**Fig. 8.3** House with bamboos opened by the rat (outside). (Source: Wellcome Library, Closed Stores Journal, S4877, W. Th. de Vogel, 'Extract from the Report to the Government on the Plague Epidemic in the Subresidency of Malang (Isle of Java), November 1910 Till August 1911', *Publications of the Civil Medical Service in Netherlands India* 1a (Batavia: Javasche Boekhandel en Drukkerij, 1912), 30–111)

indicated how rats had forced entry into the hidden interior of the bamboo frame. In conjunction, however, these two images suggest a network. They have captured access and exit points in the expansive bamboo grid that dissected the Javanese house. These unmapped bamboo 'burrows' were grafted onto the traditional Javanese house and (to the Dutch) constituted a fatal threat.[75] Like the set of images of the rats embedded within

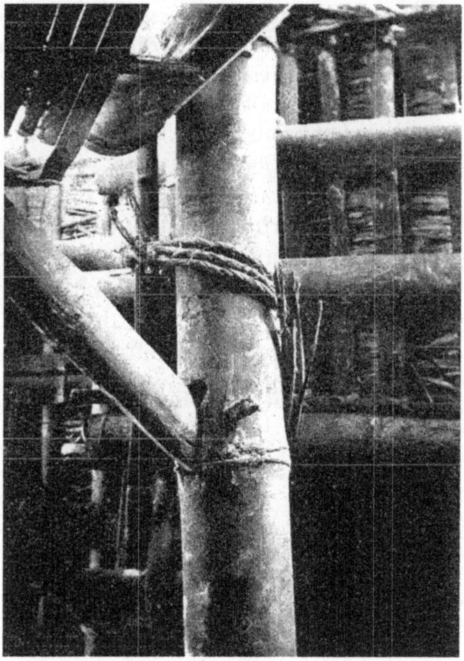

**Fig. 8.4** House with bamboos opened by the rat (inside). (Source: J. J. van Loghem, 'De Pest op Java', *Nederlands Tijdschrift voor Geneeskunde* 56 (1912), pp. 200–238)

the bamboo frame and furniture, described above, these two photographs became synonymous with plague in Java and saw reproduction in later publications. As late as 1936, the former would reappear in a plague manual edited by the prominent plague expert Wu Lien-teh.[76]

The questionable revelation that rats shared the bamboo dwellings of the Javanese in the context of plague elevated generic colonial anxieties for sanitation and the native house to new levels of specificity. Both in written and in visual sources, bamboo houses dominated other facets of plague and plague control in Java. For instance, a collection of nearly 200 'plague photographs' held at the *Museum Wereldculturen* contains some seventy distinct photographs of houses or villages (forty of which document different stages of 'improvement').[77] Indeed, the bamboo dwelling became an object of intense scrutiny. Providing an initial description of the 'typical' suspect one-storied building, van Loghem wrote:

> The outer frame ... usually consists of strong bamboos, less often of wood, and the vertical main poles (*soko goeroe*) bear horizontal beams (*blandar*).

> The inner frame also consists of vertical wooden or bamboo supports, which bear horizontal beams (*pengeret*) at right angles to the blandars. Vertical poles (*tiang*) rest on the blandar and pengeret which again support the ridge of the roof (*blandar woewoeng*). The inner frame is generally constructed of bamboos. … The inner room of the house very often has a horizontal ceiling (*pyan*) consisting of bamboo matting.[78]

A stylised blueprint accompanied this typology, illustrating what we might think of as the normal physiology of the bamboo dwelling (Fig. 8.5). But while the image illustrates how these various foreign building elements linked up, it does more than depict the base infrastructure of the house. Given the role now attributed to bamboo in spreading plague, and in conjunction with photographs such as those above, one might ask whether

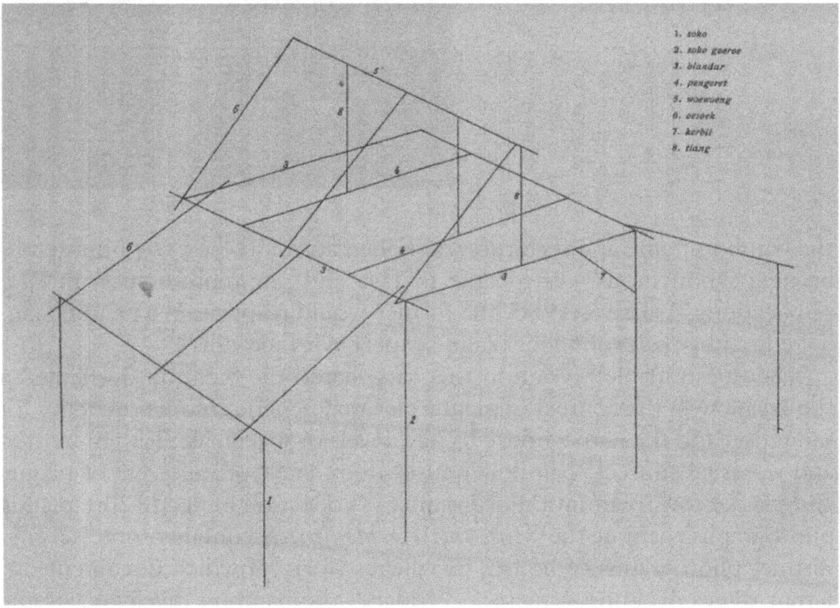


**Fig. 8.5** Schematic of a 'typical' bamboo dwelling in Malang. (Source: Wellcome Library, Closed Stores Journal, S4877, J. J. van Loghem, 'Some Epidemiological Facts Concerning the Plague in Java', *Publications of the Civil Medical Service in Netherlands India* 1b (Batavia: Javasche Boekhandel en Drukkerij, 1912), pp. 2–57)

what we are looking at is in fact a transmission diagram of sorts. The image resembles what the historian Lukas Engelmann might refer to as the epidemiologist's 'spatial diagram'—a 'hybrid' between a diagrammatic architectural sketch and a plague map that traced pathways of transmission and effectively sought to answer the question of 'how and why epidemics occurred only in some spaces'.[79] Specifically, the lines sketching out the 'strong bamboos' of the frame simultaneously worked to capture a vascular network that ran through the Javanese house and circulated rats, fleas, and their 'germs' to human occupants. The inclusion of the diagram within the government's plague reports for 1911 underscored the role seen to be played by the house in plague transmission, and added to a larger body of visual evidence that supported the highhanded intervention following this diagnosis.

The 'rat census' of plague houses in Malang had yielded valuable information on the preferred nesting spaces of the house rat.[80] Tabulated, this data hinted at significant differences in their distribution in individual districts that resulted from small variations in house design and the materials used in construction (Table 8.1). As the anthropologist Christos Lynteris suggested, such tables became 'entwined in an intricate exchange of opinions, arguments, and judgments' regarding the origin and transmission of plague.[81] In this case, tables not only helped to rearticulate the role of the house in plague transmission but suggested a possible intervention. The *pyan* yielding 50% of nests at Kotta Malang was a stylistic feature encountered in larger towns. The comparatively meagre 2.5% of nests hidden beneath the roofs of Karanglo flowed from the fact that 'the soil is more suitable for the production of tiles'.[82] By implication, the organic *atap* roofs prevailing elsewhere emerged as another hazardous material. These discrepancies, argued van Loghem, afforded a vital 'indication as to how plague may be forestalled'.[83] If rat extermination had proven both unfeasible and ineffective, this information suggested a 'powerful' alternative: to increase 'the distance between man and rat' by means of an alteration in the Javanese house.[84] 'The plague problem in a country where rats are once become infected', van Loghem concluded, 'thus resolves itself ... into a question of dwellings.'[85]

Citing beneficial results with interventions in the built environment 'in other countries', both de Vogel and van Loghem began to argue in favour of 'home improvement' as the most viable strategy against plague.[86] Identifying the horizontal *blandar* as the most popular nesting spaces of the rat (yielding a quarter of all nests), de Vogel had the ends of these

**Table 8.1** Places where rat nests were found in five plague infected districts as a percentage of a total of 2500

| | *Kotta Malang* | *Karanglo* | *Penangoengan* | *Ngantang* | *Toeren* |
|---|---|---|---|---|---|
| **Roof** | | | | | |
| *Atap* (thatch) | 0 | 2.5 | 13 | 19.3 | 2.2 |
| Ridge | 3.9 | 7 | 3.2 | 12 | 12.3 |
| *Oesoek* (diagonal bamboo) | 3.3 | 0 | 0 | 0 | 1.3 |
| *Pyan* (ceiling) | 50 | 1.5 | 2.3 | 0.5 | 11 |
| **Frame** | | | | | |
| *Soko* (vertical bamboo) | 1.7 | 1.6 | 0 | 1.5 | 0.3 |
| *Blandar* (horizontal bamboo) | 20 | 44 | 26.2 | 23 | 14 |
| *Pengeret* (inner bamboo beam) | 2.2 | 13.5 | 10.4 | 14.7 | 8 |
| Wood (rotten) | 0 | 0.5 | 4 | 1.4 | 1.3 |
| **Wall** | | | | | |
| *Gedek* (woven bamboo mat) and plinth | 9.4 | 7.4 | 14 | 10 | 12 |
| **Door** | 0 | 7.5 | 1.7 | 0.9 | 1.3 |
| **Floor** | 3.9 | 0.5 | 2.2 | 1.2 | 5 |
| ***Pogo* (storage space hanging from roof)** | 1.7 | 1.7 | 3.5 | 4.6 | 1 |
| ***Balé-Balé* (bedstead)** | 2.9 | 9 | 11.6 | 6.6 | 28 |
| **Stable** | 0.6 | 3.5 | 6.7 | 4.1 | 1.3 |

Source: J. J. van Loghem, 'De Pest op Java', *Nederlands Tijdschrift voor Geneeskunde* 56 (1912), pp. 200–238

Note: None of the numbers add up to 100%

'dangerous bamboos' sealed with cement or covered by sheet metal. A contest was held to design a rat-proof *balé-balé*.[87] The vacant spaces underneath the roof proved more challenging: replacing *atap* with tiles was an excellent start, but 'great financial difficulties were involved in introducing tiled roofs all at once in the whole district of Malang'.[88] Other parts of the house prone to infestation by the rat furthermore required not just a physical correction but also an alteration in the ways in which the house was both seen and inhabited: an attempt that, following David Arnold, we might think of as 'colonising the gaze'. The installation of 'movable' inner walls, for instance, provided access to the 'dead space' between the double matted walls. But without a 'desire'—or at least a 'habit'—to inspect such spaces among the occupants, this intervention was mute.[89] Almost immediately, home improvement was understood as a mere *precondition* to making the house rat proof and thereby impervious to plague.

Photographs documenting adaptations to 'plague houses' are plenty, but perhaps the most striking are those that at the same time hinted at this need to reform Javanese domestic practices. In one such photograph we see a *mantri* or *doctor djawa* standing inside an improved house.[90] The support post beside him is made of wood, the way the light falls through the roof suggests its cover is made of tiles rather than *atap*, but the *blandar* above the wall still appears to be made of bamboo. The anatomical undercurrents in the photographs that first implicated the house as a carrier of plague are again in evidence. Looking directly at the camera, the man discretely lifts up an inner wall newly mounted on hinges and invites us to inspect the previously obscured dead space behind it. Peeling away a layer of the outer surface of the house, he provides us with a glimpse of such spaces in a healthy state—cleared of rats or nests—redirecting our gaze to make us aware of the continued threat posed by such shelters. The image has a demonstrative faculty, casting blame on such spaces in propagating plague and reminding the viewer of the need to inspect such spaces regularly in order to maintain the 'health of the house' and—through it—the health of its occupants.[91] The scene bears a curious resemblance to the theme of the anatomical lesson (popular in Dutch early modern paintings) that depicted physicians instructing their students in the hidden workings of the human body.[92] Like them, colonial physicians sought to foster a new understanding of their subject—albeit the house rather than the body—that depended on inculcating a new, clinical mode of seeing. Sight, in turn, was intended to facilitate new behaviour.

Similar photographs likewise sought to redirect the gaze towards the various plague-dangerous features of the Javanese house and prompt an alteration in its habitation: foreshadowing more concerted plague 'propaganda' produced in subsequent years.[93] These images were used to instruct Indies' medical practitioners, *mantris*, and (within months) colonial and metropolitan populations. One image appearing alongside the former, for instance, shows a Javanese man holding down a newly installed hatch in the *pyan*.[94] Like the movable inner wall, this particular 'improvement' hardly rendered the ceiling immune to infestation by the rat, but made it accessible to regular inspection: keeping rats from nesting there at all, or allowing nests to be detected before they posed a threat. Both scenes thus demonstrated a physical correction of the house in combination with a performance of desirable behaviour—just as the entire set of home improvement photographs in Java would (in time) become a performance of the 'ethical' nature of Dutch colonial rule.[95] The educational purpose of such and other images produced in the wake of plague control is underscored by the fact that de Vogel and his associates quickly came to depend on their evidentiary aura to support their increasingly systematic intervention in the native house. In lectures before 'European planters ... high native officials, and influential rich Chinese', de Vogel employed lantern slides to 'plainly show how plague in the house rat is communicated to man, and how it can be prevented'.[96] Belatedly, it seems, such demonstrations were extended to the public of plague-stricken districts to legitimise home improvement and home inspection by colonial officials.[97]

## Colonising the Home

> It was already dusk underneath the heavy trees, the bananas lifted the cool, green paddles of their leaves, and under the stately canopy of the coco-palms sheltered the little bamboo houses, poetically Oriental, idyllic, with their *atap* roofs, their doors often already closed or, if open, framing a little black inward vista, with the vague outline of a *baleh-baleh*, on which squatted a dark figure.
>
> Louis Couperus, *De Stille Kracht* (1900), p. 155

About a decade before the plague outbreak in Malang, the celebrated Dutch author Louis Couperus had visited the *residentie* Pasoeroean. Using it as a model for the fictional district Laboewangi in his book *De Stille Kracht*, he provided an evocative image of the Javanese house in East Java. Little did Couperus suspect that the 'poetically Oriental' dwellings he conjured would become the object of such intense scrutiny a mere decade later—or did he perhaps recognise their subversive potential? Couperus pitched his European protagonists against an indefinable 'hidden' or 'silent' force in the Dutch East Indies: at once cultural, natural, and supernatural. The native bamboo house epitomised this resistance. Nestled beneath a verdant jungle canopy, it merged into Java's majestic nature and 'sheltered' its occupants from colonial oversight. Now, plague had subjected these 'idyllic' dwellings to unprecedented colonial surveillance and provoked an unparalleled intervention in the island's built environment: effectively severing this facet of the Javanese bamboo alliance.

Over the remainder of 1911, plague continued its erratic spread across East Java. Nevertheless, van Loghem returned to Amsterdam to join the newly founded Colonial Institute while de Vogel began a regular commute between Weltevreden and Malang.[98] Plague control was reduced to the evacuation of infected houses and villages, followed by home improvement. Leadership over this programme was ambiguously divided between various actors, and financially the scheme was a shambles.[99] Budgets for home improvement were vastly exceeded as they could simply not keep up with the steadily climbing number of houses slated for improvement.[100] The cost and logistics of tile production (and subsequently the supply of timber) proved to be the Achilles' heel of the operation.[101] The execution of home improvement, finally, was uneven.[102]

In November 1912, de Vogel returned to East Java to inspect ongoing control efforts. At Madioen, he identified potentially lethal oversights in the improvement scheme. The devil was in the detail, he stressed, accompanying his report with pen sketches to illustrate common mistakes. Holes cut into the horizontal *blandar* to allow them to be pierced by the vertical *soko* were too wide and not properly sealed off. Small pieces of bamboo placed diagonally between *soko* and *blandar* to improve stability—the *schoor*—was now made of 'split' or halved bamboo but installed convex (the hollow part upwards) thus providing a comfortable nesting space for the rat. Finally, the *soko* was often sealed only at the lower end but not at the top, while the installation of moveable inner walls and ceiling hatches presented myriad challenges that left the 'dead space' behind them uninspected. In Kediri and Toeloeng Agoeng, similar deficiencies were compounded by a lack of 'conviction' among the directing health officials in home improvement as an effective plague control strategy.[103]

Ultimately returning to Malang, de Vogel found that home improvement in the first affected kampongs had been nearly completed. The mistakes seen in neighbouring districts had been avoided, and the tide of plague had turned—temporarily, as it turned out. But at what cost? Plague and plague control, de Vogel observed not without pride, had effected a dramatic transformation of the landscape:

> Driving westwards along the road to Ngantang, I was struck by the changed appearance of the landscape that extends itself along the west slopes of the Ardjoeno up until Gabes near Patjet. Everywhere the red roofs of the homes in the various dessa's contrasted markedly to the green background, giving a somewhat European impression, and at the same time an impression of the wealth of population in the district, while in the past the atap-covered homes

> concealed the population density. The whole area was hit hard by plague last year. Now cases there are rare.[104]

Home improvement had not merely made the bamboo house rat proof: it had taken possession of the traditional Javanese dwelling and by extension its occupants and the land itself. Whereas the *atap*-covered houses had merged into the 'green background' and concealed the populace from colonial oversight (as Couperus had noted), the red-tiled roofs of 'improved' dwellings stood out clearly and hinted at the 'wealth of population' that had come into the purview of colonial governance. The Dutch colonisation of the Javanese home had at the same time transformed the Javanese landscape, leaving it to exude a quasi-European (and hence 'improved') atmosphere.

The contrast between the scenes described by Couperus and de Vogel was visually reproduced in Javanese plague photography. A photograph of uncertain date labelled 'the last plague house in Malang' captured a structure said to hide behind a railway embankment.[105] The house is drawn up of bamboo poles, bamboo laths, mats, and *atap*, and appears almost like an extension of the surrounding landscape. The structure is representative of the type of structure that caused the Dutch such great concerns and appears handpicked to demonstrate its potential dishevelment. Items including part of a matted wall are strewn around the structure, and a long bamboo pole of uncertain function (a drain? a support?) leads straight from the roof towards the grounds surrounding it: a veritable rat highway. A second dwelling in the left-hand corner almost escapes our notice; is obfuscated beneath the tropical foliage. Similarly, a group of two European physicians and a Javanese assistant are clearly positioned upon the embankment looking down upon the unfortunate dwelling, while two Javanese 'dark figures' are barely visible in front of the neighbouring house. Meanwhile, the 'quasi-European' landscape beheld by de Vogel from atop the Ardjoeno would not be dissimilar from a photograph of a newly founded village near Batoe in northwest Malang, the houses of which 'contrasted markedly' from the surrounding landscape by virtue of their tiled and uniformly shaped roofs.[106]

## Conclusion

In the months following the appearance of plague in Malang, Dutch health officials re-articulated longstanding associations between plague and the house. The initial confusion about the absence of plague rats in the affected districts and the subsequent discovery of dead and dying plague rats as well as rat nests scattered around the house implicated bamboo—Java's principal building material—as a new, material link in plague's aetiology in Java. The extensive documentation of the presence of the rat within the bamboo dwelling in diagrams, tables, but especially in photographs allowed the Dutch to reimagine the Javanese house not merely as unsanitary but as actively—*inherently*—pathogenic on account of its very materiality and design. Photographs of rats and nests embedded within the hollow bamboo frame or within the *balé-balé*, along with more suggestive images documenting passageways through the house utilised by the rat, situated the Javanese house at the centre of Dutch plague anxieties. They suggested a tubular network running through the house, conveying rats, fleas, and the plague bacilli they might carry to human occupants. Shown to officials, physicians, and colonial and colonised subjects in government reports, at hygiene fairs, scientific conferences, and meetings with affected populations, photographs were rallied as evidence of bamboo's lethal potential. Put differently, photographs helped insert a new bamboo link in plague's putative rat-flea-man transmission scheme in Java. At the same time, these photographs conspired with other scientific imagery to suggest a possible intervention. Home improvement was adopted as the principal strategy to contain plague in Java and subsequently systematised under the auspices of a dedicated Plague Service. Over the last decades of Dutch colonial rule, an estimated 1.6 million houses were either renovated or rebuilt. Millions more were subjected to periodic 'inspection', and countless Javanese were exposed to a concurrent hygiene campaign. The use of bamboo and *atap* in house construction was regulated, banned, and ultimately recuperated under the rising cost of wood and tiles—a cost largely placed upon the Javanese themselves in the form of advance loans. Distinctive Javanese roof designs such as the *joglo* and the *limasan* were dismissed as plague-dangerous and replaced with uniform saddle-shaped roofs in a further colonisation of the built environment. Already potentially subversive for its ability to elude oversight, in short, the newly identified role of the Javanese bamboo house in plague transmission suggested by Dutch plague photography justified an unprecedented degree of colonial scrutiny and intervention.

## Notes

1. W. Th. de Vogel, 'The Connection of Man and Rat in the Plague Epidemic in Malang, Java, in 1911'. In Francis Clark (ed.), *Transactions of the Second Biennial Congress Held at Hongkong 1912*, pp. 147–149 (Hong Kong: Noronha, 1913).
2. A. A. F. M. Deutmann, 'The Plague in Karanglo'. In *Publications of the Civil Medical Service in Netherlands India* vol. 1b, pp. 58–138 (Batavia: Javasche Boekhandel en Drukkerij, 1912), p. 115. Closed Stores Journal, S4877 (Wellcome Library).
3. De Vogel, 'The Connection between Man and Rat', p. 148.
4. Huu Ngoc and Lady Borton, *Cây Tre-Bamboo* (Hanoi: The Gioi Publishers, 2011), p. 13.
5. Susanne Lucas, *Bamboo* (London: Reaktion Books, 2013), pp. 30–100. See also: J. A. Loebèr, *Bamboe in Nederlandsch-Indië* (Amsterdam: De Bussy, 1909), pp. 5–24; Willem Wolters, 'Geographical Explanations for the Distribution of Irrigation Institutions: Cases from Southeast Asia'. In Peter Boomgaard (ed.), *A World of Water: Rain, Rivers and Seas in Southeast Asian Histories*, pp. 209–234 (Leiden: KITLV Press, 2007), pp. 223–224; Georges B. Cressey, *Asia's Lands and Peoples: A Geography of One-Third of the Earth and Two-Thirds of its People* (New York: McGraw-Hill, 1944), p. 536.
6. Alfred Russell Wallace, *The Malay Archipelago*, vol. 1 (London: Macmillan, 1869), pp. 120–121; Alfred Russell Wallace, 'On the Bamboo and Durian of Borneo'. *Hooker's Journal of Botany* 8 (1856): 225–230.
7. Timothy J. LeCain, *The Matter of History: How Things Create the Past* (Cambridge: Cambridge University Press, 2017), p. 134.
8. Pfijffer zu Neueck, *Skizzen von der Insel Java* (Schaffhouser: Franz Hurter, 1829), p. 28.
9. Bamboo utensils were ubiquitous in the Indies and served both everyday and ritual purposes. In Java, the use of bamboo knives by the *dukun* (indigenous healer) to cut the umbilical cord or perform circumcision was frequently commented on. Loebèr, *Bamboe*, pp. 43–46. See also: Liesbeth Hesselink, *Healers on the Colonial Market: Native Doctors and Midwives in the Dutch East Indies* (Leiden: KITLV Press, 2011), pp. 124–125.
10. *Koloniaal Verslag 1911*, p. 208. Index 2.02.21.01, no. 412 (Nationaal Archief).
11. Loebèr, *Bamboe*, p. 78; Lisa Drummond, 'Colonial Hanoi: Urban Space in Public Discourse'. In Laura A. Victoir and Victor Zatsepine (eds.), *Harbin to Hanoi: The Colonial Built Environment in Asia, 1840–1940*, pp. 207–229 (Hong Kong: Hong Kong University Press, 2014),

pp. 222–223; Robert Peckham, 'Matshed Laboratory: Colonies, Cultures, and Bacteriology'. In Robert Peckham and David Pomfret (eds.), *Imperial Contagions: Medicine, Hygiene, and Cultures of Planning in Asia*, pp. 133–134 (Hong Kong: Hong Kong University Press, 2013), pp. 123–147; Warwick Anderson, *Colonial Pathologies: American Tropical Medicine, Race, and Hygiene in the Philippines* (Durham NC: Duke University Press, 2006), p. 211; G. Verschuur, *Aux Colonies d'Asie dans l'Océan Indien* (Paris: Librairie Hachette, 1900), pp. 116–120.

12. Rice imports were speculated to have introduced plague rats and fleas from India or China to Java. In later years, local rice trade was thought to similarly disseminate plague within Java. The focus on the native plague patient or corpse in Hong Kong and Bombay was not evident to the same extent in Java. David Arnold, *Colonizing the Body: State Medicine and Epidemic Disease in Nineteenth-Century India* (Berkeley CA: University of California Press, 1993), pp. 200–239; Christos Lynteris, 'Suspicious Corpses: Body Dumping and Plague in Colonial Hong Kong'. In Christos Lynteris and Nicholas H. A. Evans (eds.), *Histories of Post-Mortem Contagion: Infectious Corpses and Contested Burials,* pp. 109–134 (Basingstoke: Palgrave Macmillan, 2017), pp. 119–122.
13. On plague's association with the home, see, for instance: Mary P. Sutphen, 'Not What, but Where: Bubonic Plague and the Reception of Germ Theories in Hong Kong and Calcutta, 1894–1897', *Journal of the History of Medicine* 52, no. 1 (1997): 81–113. On plague's urban identity, see: Myron Echenberg, *Plague Ports: The Global Urban Impact of Bubonic Plague, 1894–1901* (New York: New York University Press, 2010); Lukas Engelmann, John Henderson and Christos Lynteris (eds.), *Plague and the City* (London and New York: Routledge, 2018).
14. Prashant Kidambi, *The Making of a Modern Metropolis: Colonial Government and Public Culture in Bombay, 1889–1920* (Aldershot: Ashgate, 2007), pp. 64–69; Echenberg, *Plague Ports*, pp. 246–255; James C. Mohr, *Plague and Fire: Battling Black Death and the 1900 Burning of Honolulu's Chinatown* (Oxford University Press, 2005), pp. 69–98.
15. On plague's association with 'things', spaces, and soil, see, for instance: Robert Peckham, 'Hong Kong Junk: Plague and the Economy of Chinese Things', *Bulletin of the History of Medicine* 90, no. 1 (2016): 32–60; Christos Lynteris, 'A "Suitable Soil": Plague's Urban Breeding Grounds at the Dawn of the Third Pandemic', *Medical History* 62, no. 3 (2017): 343–357.
16. Terence Hull, 'Plague in Java'. In Norman Owen (ed.), *Death and Disease in Southeast Asia: Explorations in Social, Medical and Demographic History* pp. 210–234 (Oxford: Oxford University Press, 1987).

17. Christos Lynteris, Christos Lynteris, 'Tarbagan's Winter Lair: Framing Drivers of Plague Persistence in Inner Asia'. In Christos Lynteris (ed.) *Framing Animals as Epidemic Villains. Histories of Non-Human Disease Vectors*, pp. 65–90 (London: Palgrave Macmillan, 2019), p. 66. On epidemic photography, see: Christos Lynteris, 'The Prophetic Faculty of Epidemic Photography: Chinese Wet Markets and the Imagination of the Next Pandemic', *Visual Anthropology* 29, no. 2 (2016): 118–132.
18. On the ways in which photographs, diagrams, and inductive outbreak reports all contributed to a narrative plague model, see, for instance: Lukas Engelman, 'Making a Model Plague-Paper Technologies and Epidemiological Casuistry in the Early Twentieth Century', this volume.
19. A chapter on the demographic impact of plague of 1987 constitutes the most comprehensive history of plague in Java to date; Hull, 'Plague in Java', pp. 210–234.
20. Resident of Soerabaja, 'Report on the first occurrence, the recognition, and the control of plague in Soerabaja', 11 May 1911. Het Uitblijven der Wekelijksche Verslagen (Uitblijven), Manuscript Grote Bundel (MGB), no. 4674 (ANRI); Resident of Pasoeroean. 'Report on the occurrence of plague in the residency Pasoeroean', May 1911. Uitblijven, MGB, no. 4674 (ANRI); Suze de Vogel to Mother, 'Letter', 30 March 1911. Collection Willem Thomas de Vogel, D H 1568 (KITLV); 'Misschien toch pest', *Het Nieuws van den Dag voor Nederlandsch-Indië* (6 March 1911), p. 3.
21. J. de Haan, 'The Bacteriological Diagnosis of Plague in the District Malang'. In *Publications of the Civil Medical Service in Netherlands India* vol. 1a, pp. 2–29 (Batavia: Javasche Boekhandel en Drukkerij, 1912), pp. 2–4. Closed Stores Journal, S4877 (Wellcome Library).
22. The Dutch 'lived in anxious suspense' of plague following two cases at Deli in 1905. Earlier, in 1899, Dutch insurance companies expressed concern for its introduction following an outbreak at Penang. Plague was epidemic in Manila between 1899 and 1906. Still, preventative measures were scant and by 1911 the archipelago barely had two functioning quarantine stations; I. Snapper, 'Medical Contributions from the Netherlands Indies'. In Paul Kratoska (ed.), *South East Asia: Colonial History* vol. 3, pp. 129–152 (London: Routledge, 2001), pp. 141–142; J. J. van Loghem, 'Het Pestvraagstuk voor Nederlandsch-Indië', *Nederlands Tijdschrift voor Geneeskunde* 53 (1909): 44–51; Hans Pols, 'Quarantine in the Dutch East Indies'. In Alison Bashford (ed.), *Quarantine: Local and Global Histories*, pp. 85–102 (London: Palgrave Macmillan, 2016), pp. 90–91; Anderson, *Colonial Pathologies*, pp. 61–63; Soerabaja health insurance companies to Governor-General, 'Letter', 1899. Uitblijven, MGB, no. 4674 (ANRI).

23. De Haan, 'The Bacteriological Diagnosis of Plague', pp. 5–6; Algemeen Secretaris to Unknown, 'Telegram', 6 April 1911. Uitblijven, MGB, no. 4674 (ANRI).
24. Governor-General to Ministry of Colonies, 'Telegram no. 23', 2 April 1911. Geheim Archief Koloniën 1901–1962, index 2.10.36.50, no. 763 (Nationaal Archief).
25. David Arnold, 'Disease, Rumor and Panic in India's Plague and Influenza Epidemics, 1896–1919'. In Robert Peckham (ed.), *Empires of Panic: Epidemics and Colonial Anxieties*, pp. 111–129 (Hong Kong: Hong Kong University Press, 2015), pp. 111–112; Christos Lynteris, *Ethnographic Plague: Configuring Disease on the Chinese-Russian Frontier* (London: Palgrave Macmillan, 2016), pp. 121–122; Idenburg to de Waal Malefijt, 'Letter', 5 April 1911. Brieven van A. W. F. Idenburg 24 April 1907 t/m 25 Juni 1911, Box 5, Folder 38, Archief J. H. de Waal Malefijt Universiteitsbibliotheek Vrije Universiteit Amsterdam (VU).
26. J. de Bruin and G. Puchinger, *Briefwisseling Kuyper-Idenburg* (Franeker: T. Weber, 1985), pp. 238–242, note 2; W. Th. de Vogel, 'Extract from the Report to the Government on the Plague Epidemic in the Subresidency of Malang (Isle of Java), November 1910 till August 1911'. In *Publications of the Civil Medical Service in Netherlands India* vol. 1a, pp. 30–111 (Batavia: Javasche Boekhandel en Drukkerij, 1912), pp. 45–50. Closed Stores Journal, S4877 (Wellcome Library).
27. Lynteris, 'A "Suitable Soil"', pp. 343–357; Peckham, 'Hong Kong Junk', pp. 32–60; Lukas Engelmann and Christos Lynteris, *Sulphuric Utopias: A History of Maritime Fumigation* (Cambridge MA: MIT Press, 2020).
28. De Vogel, 'Extract', pp. 42–87; Governor-General to Ministry of Colonies, 'Telegram no. 334', 5 April 1911. Geheim Archief Koloniën 1901–1962, Index 2.10.36.50, no. 763 (Nationaal Archief); Resident of Soerabaja to Secretariat of the Governor-General, 'Letter', 14 May 1911. Uitblijven, MGB, no. 4674 (ANRI).
29. The civil medical service was only created in January 1911. As Idenburg put it in a letter to the Ministry of Colonies, 'no one had thought to expect the disease within the interior'; Idenburg to de Waal, 'Letter, 5 April 1911.
30. Governor-General to Ministry of Colonies, 'Telegram', 26 May 1911. Uitblijven, MGB, no. 4674 (ANRI).
31. De Vogel, 'Extract', pp. 32–33; Suze de Vogel to Mother, 'Letter', 4 May 1911. Collection Willem Thomas de Vogel, D H 1568 (KITLV).
32. Ministry of Colonies to Governor-General, 'Telegram no. 329', 3 April 1911. Geheim Archief Koloniën 1901–1962, index 2.10.36.50, no. 763 (Nationaal Archief).
33. Van Loghem, 'Het Pestvraagstuk voor Nederlandsch-Indië', pp. 44–51.

34. 'Genootschap ter bevordering van natuur-, genees-, en heelkunde te Amsterdam'. *Nederlands Tijdschrift voor Geneeskunde* 54 (1910): pp. 1798–1804.
35. This position differs markedly from the British handling of plague in Hong Kong and Bombay, where foreign scientists played prominent roles in plague science and control. The notion of 'viral sovereignty', floated by the Indonesian health minister during the H5N1 outbreak in the 2000s and, according to the anthropologist Celia Lowe, informed by 'postcolonial sensibilities', aptly captures Dutch reluctance to involve foreign scientists and foreign physicians in plague control; Celia Lowe, 'Viral Sovereignty: Security and Mistrust as Measures of Future Health in the Indonesian H5N1 Influenza Outbreak', *Medicine Anthropology Theory* 6, no. 3 (2019): 109–132.
36. Paul-Louis Simond, 'La propagation de la peste', *Annales de l'Institut Pasteur* 12, no. 10 (1898): 625–687.
37. N. H. Swellengrebel, 'Mededeeling omtrent Onderzoekingen over de Biologie van Ratten en Vlooien en over Andere Onderwerpen, die Betrekking Hebben op de Epidemiologie der Pest op Oost-Java'. In *Mededeelingen van den Burgerlijken Geneeskundigen Dienst* vol. 2:1, pp. 1–86 (Batavia: Landsdrukkerij, 1913).
38. Similar uncertainties about the role of the rat prevailed in plague outbreaks elsewhere, see, for instance: Anderson, *Colonial Pathologies*, pp. 61–63; Myron Echenberg, *Black Death, White Medicine: Bubonic Plague and the Politics of Public Health in Colonial Senegal, 1914–1945* (Oxford Heinemann, 2002), pp. 106–107; De Vogel, 'Extract', pp. 43–44.
39. J. J. van Loghem, 'Some Epidemiological Facts Concerning the Plague in Java'. In *Publications of the Civil Medical Service in Netherlands India* 1b, pp. 2–57 (Batavia: Javasche Boekhandel en Drukkerij, 1912), pp. 5–6. Closed Stores Journal, S4877 (Wellcome Library).
40. 'Genootschap ter bevordering van natuur-, genees-, en heelkunde te Amsterdam'. *Nederlands Tijdschrift voor Geneeskunde* 54 (1910): 1798–1804.
41. Well aware of the fraudulent practices this tactic had given rise to elsewhere in Asia. J. J. van Loghem, 'De Pest op Java'. *Nederlands Tijdschrift voor Geneeskunde* 56 (1912): 200–238; de Haan, 'The Bacteriological Diagnosis of Plague', pp. 6–24; Michael Vann, *The Great Hanoi Rat Hunt: Empire, Disease, and Modernity in French Colonial Vietnam* (Oxford: Oxford University Press, 2018).
42. Van Loghem, 'Some Epidemiological Facts', pp. 24–42.
43. De Vogel, 'Extract', p. 59; Van Loghem, 'Some Epidemiological Facts', p. 5.

44. Deutmann, 'The Plague in Karanglo', pp. 113–115.
45. Carl Mense ed., *Handbuch der Tropenkrankheiten* vol. 2, p. 403, as quoted in: De Vogel, 'The Connection between Man and Rat', pp. 147–148.
46. Nicholas A. Evans, 'Blaming the Rat? Accounting for Plague in Colonial Indian Medicine', *Medicine, Anthropology, Theory* 5, no. 3 (2018): 15–42, p. 16.
47. Van Loghem, 'Some Epidemiological Facts', p. 5.
48. Local Government Kediri to Governor-General, 'Letter', 11 May 1911. Uitblijven, MGB, no. 4674 (ANRI).
49. Van Loghem, 'Some Epidemiological Facts', p. 7; Deutmann, 'The Plague in Karanglo', pp. 114–115.
50. Plague's spread from dessa to dessa was frequently referred to as 'metastatic' over the first decade of plague control in Java. 'Verslagen van Vereenigingen', *Nederlands Tijdschrift voor Geneeskunde* (1912): 718–732, p. 720.
51. W. Th. de Vogel to Mother, 'Letter', 7 June 1911. Collection Willem Thomas de Vogel, D H 1568 (KITLV).
52. Peckham, 'Hong Kong Junk', p. 35; Ann H. Kelley and Aldumena Marí Sáez, 'Shadowlands and Dark Corners: An Anthropology of Light and Zoonosis', *Medicine Anthropology Theory* 5, no. 4 (2018): 21–49.
53. On this, see, for instance, the six-volume series *Kromoblanda* self-published between 1915 and 1923 by the philanthropist H. F. Tillema which proffered a critique of 'the question of living' in the colony.
54. This epiphany was widely attributed to van Loghem but remains undated.
55. Deutmann, 'The plague in Karanglo', p. 115.
56. Van Loghem, 'Some Epidemiological Facts', pp. 7–9.
57. Deutmann, 'The plague in Karanglo', p. 115.
58. Folder with image descriptions, no. 192 A. Paul Christiaan Flu Cabinet drawer 22, slide 24 (Leiden University Library).
59. Van Loghem, 'Some Epidemiological Facts', p. 14.
60. Ibid., p. 11.
61. On photography and its truth claims about 1900 see: Jennifer Tucker, *Nature Exposed: Photography as Eyewitness in Victorian Science* (Baltimore MD: The Johns Hopkins University Press, 2005), pp. 6–7; Susie Protschky, 'Camera Ethica: Photography, Modernity and the Governed in Late-Colonial Indonesia'. In Susie Protschky (ed.), *Photography, Modernity and the Governed in Late-Colonial Indonesia*, pp. 11–40 (Amsterdam: Amsterdam University Press, 2015), p. 19
62. De Vogel, 'Extract', Figure 1.
63. For instance, Figure 6 (burning of rats) in de Vogel's government report is also part of a stereoscopic set included in a photographic album by

Keasberry. Figure 1 (plague patient with buboes) in the same report is attributed to Deutmann. Neville Keasberry, "Bestrijden van ziektes door een zogaanaamde 'rattenoven' in Malang, Neville Keasberry, 1900–1935", RP-F-2001-17-113 (Rijksmuseum).

64. Rapport du médecin major de 2ème classe A. Gauducheau sur le Congrès de Hong Kong, p. 32. Ancien Fonds, Indochine, GGI, dossier 39146 (Archives Nationales d'Outre-Mer).
65. De Vogel, 'The Connection between Man and Rat', p. 148.
66. Ibid., Plate ix.
67. This particular photograph was reproduced in several media, including: Wu Lien-teh et al. (eds.), *Plague: A Manual for Medical and Public Health Workers* (Shanghai: National Quarantine Service, 1936), p. 437.
68. De Vogel, 'Extract', p. 78.
69. Van Loghem, 'Some Epidemiological Facts', Figure 2. See also: Deutmann, 'The plague in Karanglo', Figure H.
70. Van Loghem, 'Some Epidemiological Facts', p. 12.
71. Van Loghem, 'De Pest op Java', p. 211.
72. De Vogel, 'Extract', p. 62; Deutmann, 'The Plague in Karanglo', pp. 66–67; see also: G. A. Jansen Hendriks, *Een Voorbeeldige Kolonie: Nederlands-Indië in 50 Jaar Overheidsfilms, 1912–1962* (Amsterdam: unpublished PhD thesis at the University of Amsterdam, 2014), p. 52.
73. Shawn Michelle Smith, *At the Edge of Sight: Photography and the Unseen* (Durham NC: Duke University Press, 2013), p. 8. See also: Lynteris, 'A "Suitable Soil"', pp. 351–352.
74. For example, Van Loghem, 'Some Epidemiological Facts', Figure 4; De Vogel, 'Extract', Figure 22; Paul Flu cabinet drawer 22, slide 26–27.
75. This mapping of the rat shelters within the bamboo dwelling was contemporaneous with an investigation of the burrows of the tarbagan, another suspected host of plague, in Manchuria by the Sino-Russian plague commission. As Christos Lynteris has demonstrated, prominent plague physicians theorised that these burrows allowed not only the tarbagan but also the plague bacillus to survive the winter; Lynteris, 'Tarbagan's Winter Lair'.
76. Wu Lien-teh, *Plague*, p. 437.
77. Most of this collection is accessible online: http://collectie.wereldculturen.nl. On what constitutes a photograph of plague, see: Lukas Engelmann, 'What are Medical Photographs of Plague?', *Remedia* (August 14, 2017) https://remedianetwork.net/2017/01/31/what-are-medical-photographs-of-plague/#_ftn8 (accessed 14 August 2017).
78. Van Loghem, 'Some Epidemiological Facts', pp. 15–16.
79. Lukas Engelmann, 'Configurations of Plague: Spatial Diagrams in Early Epidemiology', *Social Analysis* 63, no. 4 (2019): 89–109.

80. De Vogel, 'The Connection between Man and Rat', p. 149.
81. Lynteris, 'Suspicious Corpses', pp. 119–122.
82. Van Loghem, 'Some Epidemiological Facts', p. 20.
83. Ibid., p. 20.
84. Ibid., p. 14.
85. Ibid., p. 14.
86. Deutmann, 'The Plague in Karanglo', p. 121. See also: Lukas Engelmann, John Henderson, and Christos Lynteris (eds.), *Plague and the City* (London and New York: Routledge, 2019); Kidambi, *The Making of a Modern Metropolis*, pp. 71–113.
87. Van Loghem, 'Some Epidemiological Facts', pp. 21–22; de Vogel, 'Extract', pp. 79–82.
88. Van Loghem, 'Some Epidemiological Facts', pp. 21–22.
89. This view was already expressed in the initial government plague reports published in 1912, but became more explicit in discussions of the Plague Service established in 1915. Van Loghem, 'Some Epidemiological Facts', p. 23; de Vogel, 'Extract', p. 79. W. J. van Gorkom, *Dienst der Pestbestrijding: Verslag over het Eerste Kwartaal 1915* (Batavia: Javasche Boekhandel en Drukkerij, 1915), pp. 73–74. 987 A 6, W. K. (Koninklijke Bibliotheek).
90. De Vogel, 'Extract', Figure 29.
91. Lynteris, 'The Prophetic Faculty of Epidemic Photography', pp. 118–132.
92. One might think, for instance, of Rembrandt's *Anatomical Lesson* (1632), depicting Nicholaes Tulp gingerly lifting up a bundle of muscles in the forearm of a corps to demonstrate the workings of the arm.
93. Maurits Bastiaan Meerwijk, *A History of Plague in Java*, C4 (in progress).
94. De Vogel, 'Extract', Figure 30.
95. Displayed at hygiene fairs, exhibits, and medical congresses in the Dutch East Indies, The Netherlands, and overseas as expressions of the productive nature of Dutch colonial rule, the entire set of Dutch plague photography overlapped or fed into still another genre of photography: the *camera ethica* discussed in: Protschky, 'Camera Ethica'.
96. Van Loghem, 'Some Epidemiological Facts', p, 23.
97. Newspaper coverage of the time suggests that the images shown during these lectures corresponded with those contained within the government reports from which the majority of photographs discussed here have been retrieved.
98. Vereeniging Koloniaal Instituut Amsterdam, *Tweede Jaarverslag, 1911–1912* (Amsterdam: De Bussy, 1913), p. 15.
99. W. Th. de Vogel to Mother, 'Letter', 31 August 1911. Collection Willem Thomas de Vogel, H 1568 (KITLV).

100. Resident of Pasoeroean, 'Memorandum', 11 August 1913. Terzijde Gelegde Agenda's (TZG), no. 6671 (ANRI).
101. 'Rapport omtrent de voorziening van dakpannen voor de woningverbetering bij de pestbestrijding in de residentie Soerakarta', continuation no. 10. MGS, no. 4958 (ANRI); Louis Otten to Governor-General, 'Letter', 13 February 1919. TZG, no. 7149 (ANRI); *Dienst der Pestbestrijding: Verslag over het Jaar 1919* (Batavia: Javasche Boekhandel en Drukkerij, 1920), pp. 50–53.
102. Resident of Pasoeroean, 'Memorandum'.
103. W. Th. de Vogel, 'Report', November 1912. TZG, no. 6671 (ANRI).
104. Ibid.
105. The University of Cambridge Repository-Visual Representations of the Third Plague Pandemic Photographic Database, Item: PhotoID_11649. https://doi.org/10.17863/CAM.32691
106. The University of Cambridge Repository-Visual Representations of the Third Plague Pandemic Photographic Database, Item: PhotoID_11632. https://doi.org/10.17863/CAM.32673

CHAPTER 9

# Making a Model Plague: Paper Technologies and Epidemiological Casuistry in the Early Twentieth Century

*Lukas Engelmann*

## Introduction

Writing a report on a plague outbreak at the beginning of the twentieth century seems to have required an extraordinarily inductive mind-set. Largely inconspicuous documents, structured by bureaucratic considerations, they were charged with the ill-defined task of providing a record of anything and everything possibly relevant to the outbreak. To provide an

Research leading to this chapter has been generously funded by the University of Edinburg Challenge Investment Fund 2018–2019. I would like to thank in particular my colleagues on the Plague.TXT project, Beatrice, Arlene Casey, Mike Bennett, Richard Tobin, Claire Grover and Iona Walker for their enthusiastic collaboration in reimagining the genre of plague outbreak reports.

L. Engelmann (✉)
University of Edinburgh, Edinburgh, UK
e-mail: lukas.engelmann@ed.ac.uk



C. Lynteris (ed.), *Plague Image and Imagination from Medieval to Modern Times*, Medicine and Biomedical Sciences in Modern History, https://doi.org/10.1007/978-3-030-72304-0_9

exhaustive account of an epidemic, authors needed to consider the value of many observations, integrate anecdotal vignettes and write themselves stories of onset, escalation and abatement. They had to utilise data from mortality registries, clinical case reports, as well as sanitary and meteorological authorities. They had to align their own observations with theories to tell a story of plague, a drama unfolding from the disease's arrival to the epidemic's climax and decline. Writing the report, after all, also sealed the epidemic to the archive and should provide a record of any lesson learned in order to contain the next one.

Although such reports accounted for plague outbreaks from places as disparate as Porto, San Francisco, Sydney, Hong Kong and Durban, outbreak reports of plague emerged as a peculiar global genre over the course of the third plague pandemic. As their authors collected and considered reports from previous epidemics and began to write with other epidemiologists, medical officers and public health commissioners in mind, reporting on plague grew into a distinctive practice of communication and collaboration. A record of events, data and observations tied together by a disease and a place, the reports began to shape ways of imagining the epidemic of plague as global object of research. Stories of outbreaks were transformed into narrative science, as descriptions and diagrammatic formalisations sought to identify general features of plague as epidemic. Authors turned explicitly to those aggregated aspects that pointed beyond individual cases of the disease or the efforts of medical classification. The reports also moved beyond general features of the places and locations in which the disease had struck. Instead, they contributed to the systematic consideration of the epidemic as a discrete object of knowledge.

Well into the nineteenth century, the term 'epidemic' was mostly used to describe the accumulated occurrence of cases of a specific disease.[1] A disease accrued epidemic quality when circumstances appeared to have led to a significant increase of cases within a confined place, such as a city.[2] Over the course of the nineteenth century, the term became embroiled within the ongoing controversies between sanitarians and contagionists.[3] An epidemic was either attributed to the environmental conditions, which led a disease to occur in a higher than usual frequency, or, with the bacteriological redefinition of communicable diseases, it was particular kinds of disease which had the capacity to become epidemic as they were attributed to a transmissible microbe.[4] However, the modern development of epidemiological reasoning aimed to reshape epidemics into an object of scientific knowledge in its own right, moving beyond the consideration of a mere quality of an insanitary location or an infectious disease. As

Crookshank, a British physician, argued in 1922, it was a particular curiosity in the English language that the noun of 'epidemy', appeared to be defunct. Hence, he argued, epidemiologist struggled to reshape the adjective descriptor 'epidemic' into an abstract and conceptual research object.[5] Such work of appropriating and developing a sense of *the epidemic* beyond it being either a quality of a specific location or a quality of a particular disease was, I argue, partially worked out in plague reports. This chapter introduces the outbreak report with a twofold agenda: first, I ask how plague has been configured in the outbreak report as an epidemic with discrete, characteristic and discernible patterns. Second, reconstructing this imagination of the plague epidemic as an object of scientific knowledge, I show how the outbreak report indicates the historical persistence of epidemiology as a narrative science and offers insights into an epidemiological casuistry, which persisted deep into the twentieth century.

The authors of the reports discussed in this chapter practised epidemiology as an empirical art, dedicated to inductive reasoning and correlative modes of thinking drawn from history, clinical medicine, anthropology, sociology and demography. To contribute to the making of an epidemic, reports summarised medical accounts of the disease, its natural history and its symptomatology. They covered significant aspects of the location of the outbreak, listing aspects ranging from weather trends over descriptions of the built environment to the social and cultural analysis of affected populations. Describing the chronology of events, the author would then seek to correlate aspects of the disease with coordinates of the local environment to set out general points of reference, which could explain why the epidemic occurred at this place and time. These elements were bound together to constitute the report's narrative, structured by a number of tacit conventions, common structures and rules.[6] With sections moving from questions of bacteriology to mortality rates, to quarantine measures, to outbreaks among rodents to summaries of the longer history of plague, the narrative correlated and combined disease, environment and population, thus allowing the epidemic to emerge as a configuration of these coordinates.[7] However, for this narrative to provide a formalised and ordered account of the epidemic, for it to become a scientific account, it required combination with instruments of abstraction and formalisation: tables, lists, graphs and maps.

The chapter will introduce the plague report as a paper technology, a tool and instrument with which epidemiological knowledge was noted, sorted, produced, stored and retrieved.[8] In this chapter, I will demonstrate how the report was utilised to develop, calibrate and stabilise models of plague, which were supposed to serve as a point of reference for other, forthcoming outbreaks. The chapter will emphasise that outbreak reports, comparable to clinical case reporting, relied on the combination of narrative and diagrammatic forms to extrapolate from the clinical case to the epidemiological case.[9] Lists of mortality rates, tables of containment activities, charts of climate data, graphs of incidence rates and maps of plague's spatial configuration accompanied the fine-grained narrative accounts in outbreak reports to constitute clusters, sub-epidemics and locations into units of reference and exemplary cases for the epidemic. Narratives offered comprehension and structured analytical insights from empirical observation. Lists, tables and diagrams proposed tacit formalisations of characteristic patterns, rhythms and repetitions. Maps and spatial diagrams finally configured plague as an epidemic, in which medical, bacteriological, environmental and demographic accounts coalesced within a place.[10] Equipped with narratives, diagrams and maps, outbreak reports thus offered rudimentary elements of a long-overlooked epidemiological casuistry.

The chapter will reconstruct the plague report as a pivotal instrument through which plague emerged as a scientific object of early twentieth-century epidemiological reasoning. To this end, a comprehensive report by the Australian Public Health Commissioner William (Bertie) Burnett Ham will serve as an example.[11] His account of a series of outbreaks from Queensland, Australia, emphasises the typical shape and structure of an outbreak report while also demonstrating practices of comparison and generalisation across many outbreaks. After a few general considerations on paper technologies and their historiography in science and clinical medicine, I will turn to Ham's report and begin with a general consideration of cases and casuistry therein. Second, I will consider his use of lists and tables to demonstrate his engagement with the social structure of the populations affected by the plague. Third, I will show how Ham used maps to infer a sense of what he calls infected localities. Finally, a closer look at his diagrammatic practices will allow further insight in how he generalised his broad correlations and considerations into a comprehensive model plague.

## An Epidemiological Paper Technology—the Plague Report

At the dawn of the twentieth century, plague had become a prominent object of scientific interest. While the responsible pathogen had been identified by Alexandre Yersin and Kitasato Shibasaburō in 1894, the following years saw a range of attempts in classifying the disease and its multiple variants of bubonic, pneumonic and septicaemic plague, as well as to separate the variable severity with which the disease developed.[12] However, beyond clinical classifications, questions also emerged regarding the driving forces, the identifiable causes of an outbreak of many cases within a constrained time and space, rather than of the individual case. A range of explanatory models, from sanitary conditions, over racial susceptibility to animal vectors were mobilised in the following decade, before a consensus around the disease's rodent host, and more specifically the rat as the principal culprit, was established.[13] But even equipped with the identified host, questions remained to explain, for example, why plague did emerge only in particular places, why outbreaks were more severe in some port cities and why some measures of containment worked where others had failed. Most of all, this was an endeavour to characterise the patterns of plague outbreaks over time and, as the Pasteurian Albert Calmette emphasised, an effort to systematically separate epidemic patterns from clinical patterns. While individual cases of plague usually exhibited a fast onset of fever, with a quick escalation, the onset of plague epidemics, Calmette had already observed in 1899 Porto, followed a different rhythm and was usually slow and continuous.[14]

The historian of medicine Charles Rosenberg has identified two conceptual frameworks through which epidemics have been explained up to the nineteenth century. The first, configuration, emphasised a systems view, in which epidemics were explained as 'a unique configuration of circumstances' of categorical equal significance.[15] Communal and social health was seen as a balanced and integrated relationship between humankind and environmental constituents, in which epidemics appeared not only as the consequence, but also as the origin of disturbance, crisis and catastrophe. Rosenberg's second framework, contamination, focused on a different view, which prioritised particular and identifiable causes for an epidemic event. Where configuration implies holistic concepts, the contamination perspective suggested a disordering element, a *causa vera*, suggestive of reductionist and mono-causal inferences. As Rosenberg emphasises, both of these themes have existed since antiquity in epidemiological reasoning, but it is particularly in the late nineteenth century, with the emergence of bacteriological science, when we usually see a proliferation of these themes into polemical dichotomies.

However, the plague report—and perhaps plague at large—did not neatly fit within this dichotomy. Despite the successful bacteriological identification of the pathogen, the epidemic did not lend itself to reductionist attribution of cause and effect. Understanding the puzzling configurations on the heel of the introduction of the contaminating pathogens was the fundamental purpose of plague reporting. The results thus present ways in which questions of contamination and configuration were integrated within a narrative account of the outbreak, interspersed with list, tables and graphs, seeking to include the entirety of relevant aspects.

To understand the narrative conventions that undergird plague reporting as a scientific practice, I turn to Mary Morgan and M. Norton Wise, who have proposed to think of three principle ways in which narrative has shaped the production of scientific knowledge. First of all, scientists have used narrative to bring together disparate aspects of their research, and to acknowledge objects from 'different sources and of different kinds'.[16] Through narratives, these aspects are brought together, situated in relation to each other to exhibit a correlation. Second, Morgan and Wise state that scientists have utilised narratives 'in making things known and understandable by revealing how, like a story, they "unfold in time."'[17] As tools to express the processual nature of their research objects, narratives lend themselves conveniently to develop a sense of chronology and development over time. Third, narratives serve most often in an explanatory nature, and this is particularly pertinent when scientific arguments include the integration of abstract models, graphs and images. Again, the history of epidemiology offers many instances, in which a 'similar model-narrative interrelationship appears in simulations of systems whose complexity defies an approach through a nicely unified mathematical model that is analysable in itself'.[18]

Much has been written on the role of lists and tables in the history of science and medicine, but here I seek to compare the outbreak report mainly to those paper technologies developed in the history of clinical medicine.[19] Hess and Mendelsohn have referred previously to the patient history as a paper technology, an instrument developed to infer generalisations from amassed individual observations in the historical development of clinical medicine.[20] Among other things, since the seventeenth century, indexing and sorting patient histories in tables and lists has been a key technique of prescriptive formatting to transfer a series of individual patient histories into a general observation. Importantly, 'prescriptive formatting on the administrative model standardised patient histories and (partly) divided them into elements; this made them comparable and combinable.'[21] And furthermore, as Hess and Mendelsohn stress, the 'non-narrative form of the table' then served to create a synopsis of a general narrative, moving from narratives of fevers in individual patients to a broad account of variations among individuals to infer, for example, about a fever rate within a year.[22]

A key element of Hess and Mendelsohn's argument referred to the tenacity of the clinical case as a persistent and significant element in the production of clinical knowledge. The consideration of the individual case as an empirical anchor for the broader exploration of disease classifications, as well as the careful construction of series of cases to study a disease more generally, have been important epistemological cornerstones of clinical knowledge production. I argue here that a similar inference can be made regarding the history of epidemiological reasoning. The case-based thinking of epidemiologists in the early twentieth century has been often overlooked and has remained out of sight as historians focused predominantly on medical statistics.[23] With John Forrester, it is helpful to recall the differences and perhaps opposition between the style of reasoning associated with statistics and quantification and a style fundamentally concerned with the analysis, association and organisation of cases. The 'rise of statistical thinking put in question the notion of the individual, through the very process of refining what it might mean to have knowledge of *a number of individuals*'; however, the notion of the individual has not disappeared in scientific practice per se.[24] As this might be obvious for sciences removed from the statistical craze of the early twentieth century—such as Forrester's subject, psychoanalysis—it is also true that the individual and exemplary continued to shape scientific discourse in epidemiology. As the reports on plague show, and as I will argue below, it is precisely in the combination and arrangement of diagrammatical and narrative practices, where we find 'thinking in cases' to have shaped the production of epidemiological knowledge significantly. After all, the report suggests, an epidemic is a cascade of cases, shaped from a collection of individual patient cases, generalised into the exemplary case of a localised outbreak as part of a series of outbreaks, which then constitute the epidemic.

## The Report on Plague in Queensland, 1900–1907

As example for this chapter serves a report about plague in Queensland, which covered eight outbreaks from 1900 to 1907. The report has been penned by Bertie Burnett Ham, a crucial figure to the development of Public Health Australia in the early twentieth century.[25] Ham was a fellow of the Australian Royal Sanitary Institute and acted as commissioner for public health in the state of Queensland. The report was presented to both houses of the Australian parliament and was issued and published by the Queensland Department of Public Health. According to Ham himself, the report offers a 'complete history of Plague in Queensland' and begins to fulfil the requirement of a 'more complete scientific investigation of Plague apart from laboratory inquiries'.[26] Such inquiries had been also championed in Australia by Ashburton Thompson, who authored over 200 reports on plague in New South Wales and Sydney.[27] Ham wrote his report in full awareness of a series of reports on previous outbreaks in Hong Kong, India, and the US, while he also refers to the comprehensive discussion about plague's epidemiology at two dedicated international conferences, with a third one planned for the following years.[28] Ham understood his report to belong to a global endeavour of collecting, organising and sharing observations and analytical insights about the epidemic.

The plague in Queensland remained a comparably small series of outbreaks. Soon after cases had emerged in Sydney in January 1900, the first cases of Queensland began in the ports of Rockhamton in April 1900. Just a few weeks later, cases appeared in Brisbane and the first outbreak spiralled to 136 cases within the state. The vast majority of cases there and the outbreaks to follow were confined to port towns and hardly any case was registered inland. Plague was widely considered by Ham as well as his contemporaries to have been imported by sea trade and then distributed by rats. However, questions remained regarding the types of plague they saw emerging and to the extent that different measures against plague in Queensland succeeded or failed.

The report itself is structured in four main parts, each of which contains dedicated subsections as well as appendices and notes. The first part offers an overview of the eight discrete outbreaks, which occurred in the state between January 1900 and June 1907. The second part was dedicated to the incidence of plague in different populations, separated by age, sex, race, and occupation. Here Ham also discussed the possible implication of seasonal and insanitary conditions favourable to the emergence of plague cases. The third section then focused on the clinical observation and the definition of plague, and contributed to the classification of the disease and its various clinical forms, as they had been recorded in Queensland. The fourth and by far longest section of the report presented the epidemiology of plague, discussing the infectivity of the pathogen, but mostly focused on the relationship between rodent epizootics, human and rat fleas so as to present and consolidate the emerging model of plague as a vector-borne disease.

Ham's report is an exemplary collection of observations, abstractions and analyses, which in form, shape and technique resembles over a hundred similar reports about plague from different locations and outbreaks of the third plague pandemic.[29] In the report, Ham utilised three different ways of shaping data into statements and arguments of epidemiological concern, all of which were carefully embedded in narratives of explanations, exploration and subtle conclusion. First, he used lists and tables, or what he called 'tabular statements', to shift case data and mortality circumstances into ordered representations. Here I will focus on how these tabular inferences engaged with the social order of plague, or in Ham's words, to understand the 'individual liability' to the disease within an epidemic. Second, Ham used maps to emphasise local clusters of outbreaks in Queensland and went on, like most epidemiologist, to mark disease incidence and occurrence on pre-existing ordinance maps.[30] These spatial diagrams, I argue, were used to shape the contours of what Ham called a disease of locality, or simply 'place infection', and demonstrate how the construction of plague as an epidemic utilised representation of locality. Finally, Ham used epidemic curves to infer a model of plague's waxing and waning as an annual interdependence of rat population, annual flea prevalence and resulting epizootics and epidemics. I will show how Ham used this epidemiological diagram to not only craft an analogy between different processes and systems, but also show how his explanation of diagrammatic circularity and annual repetition allowed for a model plague to emerge precisely through the graphic correlation of factors otherwise unconnected.

In short, the report demonstrated through the collected observation and analysis of eight years of plague in Queensland that all outbreaks shared repeating social characteristics, that they occurred within a spatio-temporal pattern, and that both of these characteristics might be best explained through a systematic consideration of a rat-flea model of plague. All three contributions to the classification of epidemic plague, I will show, were carried out through the exhausting collection of observations, the careful arrangement of data in tables and lists, the representation of abstract concepts in diagrams, and through the presentation and explanation of the same in narrative form. Before I turn to the three inferences of Ham's report, I first will introduce the way his narrative was built on the model of clinical case reporting, whereas Ham developed a way of thinking, which seamlessly switched from the individual case of a disease to the cases of outbreak and sub-epidemics.

## The Casuistry of Plague

'The case', Lauren Berlant argues, 'represents a problem-event that has animated some kind of judgement.'[31] The case sits at the heart of a form of expertise and explanation that offers insight into something bigger than the case itself and its exploration seeks to illuminate a characteristic commonality of the series, to which the case belongs. However, as Forrester reminds us, the historical emergence of case-based thinking in law, religion and medicine suggests a form of thinking and writing that remains inextricably attached to a specific individual: 'epistemically, the case will always be nailed down to the level of the individual.'[32] Regarding the epidemiological outbreak report, the question is then what position the level of the individual takes on and how Ham developed inferences about the epidemic from cases.

Traditional clinical case reports sit at the heart of Ham's account, however, they had been relegated to the report's third section. 'The Medical Aspects of Plague' begins with an overview of the disease's clinical features and summarises the consensus in diagnosing and classifying plague. The same section contains a long list of individual case notes, observations from cases of patients with plague between 1900 and 1907, listed and analysed to further refine the understanding of the disease in its variable appearances. Ham stressed in the beginning of the section that his list of case notes should not to be misunderstood as a contribution to the general literature on plague, but that his cases were carefully chosen to allow for further illustration and careful refining of the existing types and symptoms of the disease, as these appear within an epidemic.

As an example might serve here his expansion of the understanding of plague's impact on the human nervous system. Beyond the most iconic symptom, the bubo, fevers as well as general fatigue, Ham had collected a number of cases with significant demonstration of 'mania' and 'acute delirium'.[33] A rather general account of these symptoms was followed by abbreviated case notes, stating the number of cases, the year of the outbreak in which the cases occurred, gender, age and occupation, to then detail the clinical history of plague in these individuals. The same pattern was repeated for descriptions of the 'Circulatory Systems', the 'Skin' as well as for the 'Alimentary System', the 'Genito-Urinary System' and 'Pregnancy'.[34] While Ham considered the bubo to be a well-established and widely recognised sign of plague, he seized the opportunity of the sample of cases he gathered in outbreaks to expand knowledge about these supposed symptoms of plague, which had received only limited attention in literature.

In form and method, Ham's report follows the same conventions of clinical observations, as they had been in place for over two centuries. An administrative paper technology, concerned with keeping registers of cases, invested in tabular formatting and then sought to develop arguments not through the exemplification of special cases, but precisely through the series of grouped and clustered of cases.[35] Similar to the French clinician Pierre Louis, whose note-taking is the subject of Hess and Mendelsohn's work, Ham prefaced each group of observations with a small description of the principle characteristics to be learned from the group of cases, which were only published as abstracts. Ham justified the tabular nature of his case lists with reducing them to only containing the absolute necessary information to allow for comparison within the general purpose of the report.

It is rather unsurprising that the observation of an epidemic yielded to lists of cases of a disease. After all, an outbreak is characterised by the clustered occurrence of cases of the same disease. But what is the function of such clinical conventions within the report of an epidemic? One could well argue that the clinical case lists corroborate the identity of the disease within the outbreak and guarantee the unity of disease classification by integrating various appearances of plague. This yields to two interlocked epistemic effects. First, the report utilised the seriality of cases found in the outbreak to enhance clinical knowledge and diagnostic certainty about plague as a clinical entity. It lists the disease and its principal cause, and then groups common as well as uncommon symptoms, supported by highly abstracted case histories. Here, the epidemic in which these cases emerged provides an extraordinary frame of reference, allowing the author to explore uncertainties and grey areas in the appearance of cases without risking the unity of the disease class.

However, Ham's contributions moved beyond the refinement of the clinical picture of plague. Second, Ham's focus remained on the description and classification of an epidemic. The report emphasised therefore those aspects of the disease which were only recognisable in the accumulated occurrence of cases. One of these aspects, Ham emphasised, was the presence of a small series of mild and largely harmless cases of plague. In the absence of an outbreak, he argued, many of these would never have been recognised as plague. The condition was at the time called *pestis minor*, or in some cases *pestis ambulance*. As Christos Lynteris has argued, this 'benign but treacherous' form of the disease re-emerged in the 1890s as a category to describe those forms of plague which exhibited typical symptoms but left patients less debilitated and which usually allowed for a full recovery. However, patients of *pestis minor* were also considered to be silent carriers of the disease, seemingly enabling its unobserved spread and constituting a unique epidemiological threat to communities.[36]

Ham had already in 1902 pointed to the danger of misunderstanding *pestis minor* as a harmless form of plague, or indeed to mistake milder cases of plague with a different glandular condition. Once medical officers used the category of *pestis minor* to report from their observations, they were also more likely, so Ham feared, to let their guard down and to deviate from strict anti-plague measures. In this way, Ham renounced the category of *pestis minor* as anything but a variation in 'degree, not of kind'.[37] To underline this view, which had not changed since 1902, he reported from a series of cases, which had been observed by the local health officer in Mossman, Port Douglas, in the North of Queensland, where an outbreak in late 1907 had led to only two deaths and where all of the approx. sixty cases were reported to be *pestis minor*. However, both bacteriological analysis and experimental inoculation were mostly negative, suggesting that the majority of cases were not plague at all, but, so Ham reports, belonged to another, unknown disease: 'From information volunteered by people who have resided for many years in Port Douglas and Mossman districts, it appears that similar outbreaks of sickness among workers in the cane fields and bush are by no means uncommon.'[38] The symptoms, which were mistakenly understood to be mild cases of plague, were indeed signs of a common unclassified tropical fever.

All these clinical features and notes are listed within the third part of Ham's report. Ham, however, used these cases to develop a series of statistical observations, further reducing the case notes to principal and comparable features across all patients. It is in the first part of the report where the tabulated and enlisted data from clinical observation were fed into a different kind of case thinking. Under the title 'Statistics of Plague', Ham would introduce each of the eight outbreaks in Queensland between 1900 and 1907 as a discrete cluster, each constituting the case of an outbreak.[39] But before each individual outbreak and their repetitive characteristics were introduced on annual bases, Ham provided an introduction in the history of plague in Brisbane. This historical account framed the following annual descriptions as being structured by a temporal order in which the epidemic consisted of a chronological sequence of discrete outbreaks. With regards to the 'Origin and Mode of Spread', Ham first and foremost accounted for the geographical origin of plague. As the outbreak in Sydney in January 1900 was observed, Brisbane, due to the 'close and constant commercial communication' between the two cities, expected an imminent arrival of the epidemic. To pre-empt any outbreak, the extermination of rats commenced on February 26, however on March 5, the city bacteriologist nonetheless found plague bacilli in the corpse of a dead rat. Ham provides a detailed report on the exact circumstances under which the first human plague case was discovered, how the symptoms were found to be suggestive of plague, and how the observation of 'vomiting and giddiness', followed by a 'tender lump in the left groin' led the examining doctor to puncture the affected gland so 'smear preparations' were made to confirm the presence of plague in the laboratory.[40]

The general history of plague in Queensland then continues with the reconstruction of likely pathways of the disease. Ham reported anecdotally of waiters in hotels close to the port, as well as on rumours about sailors but then focuses on the overwhelming majority of cases, were patients had worked in 'shipping or the wharves, or lived in close proximity to the latter'.[41] He also quickly drew the conclusion, that the epidemic was in 'some way or other connected with locality'.[42] Over two thirds of all cases in Brisbane between 1900 and 1907 occurred within a one-mile radius from the General Post office, 'the principal thoroughfare of the capital'.[43] This thoroughfare was, however, not at all suspicious of facilitating human-to-human transmission, but the principle point of exploration to understand the city's hidden rat populations and their territorial occupation. To conclude his initial portrait of the 'mode of spread' of plague, Ham brings to the reader's mind a few 'incidental facts' about the city's sewage system (a 'highways along which rats migrated'), pointing out defective parts of the sewage system, which allowed for rats to enter into it freely from the wharf.[44]

This narrative account of how plague entered and dispersed across the city of Brisbane stands than in contrast to the techniques deployed to characterise each annual outbreak of plague from 1900 to 1907. Ham proceeded to portray each of the above through tabulated data, arranged to allow for the quick comparison and alignment of patterns, rhythms and distributions. However, this does not imply that narrative elements disappeared. The tables and lists were explained and each outbreak had its own synopsis, delivered through small texts, anecdotes and lively descriptions. Each discrete outbreak was introduced with a small preface, characterising one or two key aspects that Ham considered important, identifying the first and last case, overall mortality and seasonal qualities. The very first outbreak, in 1900, was introduced with the following statement: 'The characteristic features of the epidemic were its duration and the number of places in the State affected.'[45] The third outbreak, in 1902, 'ran a rapid and somewhat severe course'.[46] On the sixth outbreak in Brisbane, beginning in January 1905, Ham reported: 'As in former years, the epidemic in Brisbane was markedly periodic.'[47] We find moments of speculation and assumption, as Ham notes vaguely that 'it was felt that more efficient methods of destroying rats' were the aim of campaigns in 1907.[48] Furthermore, these prefaces sometimes integrate little vignettes, case histories of important cases, when Ham, for example, recalls the fate of a schoolboy living in a house adjacent to a fruit store, where numerous cases had been reported in 1907. Submitted to the hospital with a cervical gland and a lesion on the lip, the boy reported of 'a cat bringing dead rats from the incriminated store into the yard of his residence'.[49]

Each portrait of an outbreak runs across between one or three pages, but its focus lies predominantly on what Ham called 'tabulated statements' (Fig. 9.1). While not every outbreak was 'illustrated' with the same set of lists, each was accompanied by tabulated incidence rate in 'man and rats' over twelve months, and most often included a table with details on the 'situation of buboes', in which Ham separated different types of plague and distinguished their occurrence in male and female patients. For most outbreaks, mortality was also listed within age groups and we sometimes find a table listing cases by their geographic distribution. Towards the end of the decade, more and more tables are dedicated to epizootics among rats, separating cases by rat species or providing tabular statements on the destruction of mice and rats.[50]

The purpose of these tables is twofold, as the pages following on from the annual outbreaks show. Ham now summarised and compared data

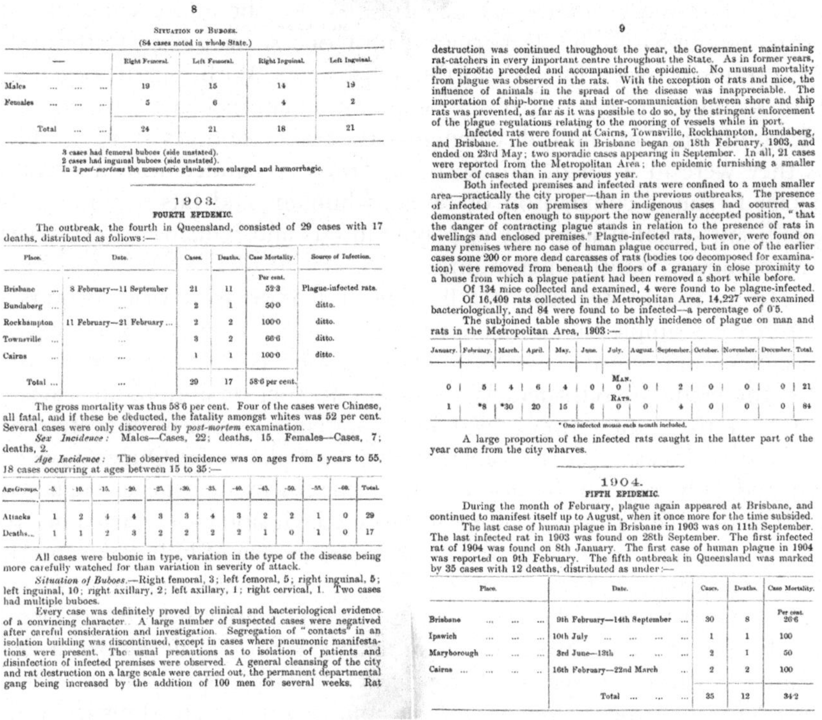

8

SITUATION OF BUBOES.
(84 cases noted in whole State.)

| — | Right Femoral. | Left Femoral. | Right Inguinal. | Left Inguinal. |
|---|---|---|---|---|
| Males ... ... ... | 19 | 15 | 14 | 19 |
| Females ... ... ... | 5 | 6 | 4 | 2 |
| Total ... ... | 24 | 21 | 18 | 21 |

3 cases had femoral buboes (side unstated).
2 cases had inguinal buboes (side unstated).
In 2 *post-mortems* the mesenteric glands were enlarged and hæmorrhagic.

1903.
FOURTH EPIDEMIC.

The outbreak, the fourth in Queensland, consisted of 29 cases with 17 deaths, distributed as follows:—

| Place. | Date. | Cases. | Deaths. | Case Mortality. | Source of Infection. |
|---|---|---|---|---|---|
| | | | | Per cent. | |
| Brisbane ... | 8 February—11 September | 21 | 11 | 52·3 | Plague-infected rats. |
| Bundaberg ... | ... | 2 | 1 | 50·0 | ditto. |
| Rockhampton | 11 February—21 February ... | 2 | 2 | 100·0 | ditto. |
| Townsville ... | ... | 3 | 2 | 66·6 | ditto. |
| Cairns ... | ... | 1 | 1 | 100·0 | ditto. |
| Total ... | ... | 29 | 17 | 58·6 per cent. | |

The gross mortality was thus 58·6 per cent. Four of the cases were Chinese, all fatal, and if these be deducted, the fatality amongst whites was 52 per cent. Several cases were only discovered by *post-mortem* examination.

*Sex Incidence:* Males—Cases, 22; deaths, 15. Females—Cases, 7; deaths, 2.

*Age Incidence:* The observed incidence was on ages from 5 years to 55, 18 cases occurring at ages between 15 to 35:—

| Age Groups. | -5. | -10. | -15. | -20. | -25. | -30. | -35. | -40. | -45. | -50. | -55. | -60. | Total. |
|---|---|---|---|---|---|---|---|---|---|---|---|---|---|
| Attacks | 1 | 2 | 4 | 4 | 3 | 3 | 4 | 3 | 2 | 2 | 1 | 0 | 29 |
| Deaths... | 1 | 1 | 2 | 3 | 2 | 2 | 2 | 2 | 1 | 0 | 1 | 0 | 17 |

All cases were bubonic in type, variation in the type of the disease being more carefully watched for than variation in severity of attack.

*Situation of Buboes.*—Right femoral, 3; left femoral, 5; right inguinal, 5; left inguinal, 10; right axillary, 2; left axillary, 1; right cervical, 1. Two cases had multiple buboes.

Every case was definitely proved by clinical and bacteriological evidence of a convincing character. A large number of suspected cases were negatived after careful consideration and investigation. Segregation of "contacts" in an isolation building was discontinued, except in cases where pneumonic manifestations were present. The usual precautions as to isolation of patients and disinfection of infected premises were observed. A general cleansing of the city and rat destruction on a large scale were carried out, the permanent departmental gang being increased by the addition of 100 men for several weeks. Rat

9

destruction was continued throughout the year, the Government maintaining rat-catchers in every important centre throughout the State. As in former years, the epizoötic preceded and accompanied the epidemic. No unusual mortality from plague was observed in the rats. With the exception of rats and mice, the influence of animals in the spread of the disease was inappreciable. The importation of ship-borne rats and inter-communication between shore and ship rats was prevented, as far as it was possible to do so, by the stringent enforcement of the plague regulations relating to the mooring of vessels while in port.

Infected rats were found at Cairns, Townsville, Rockhampton, Bundaberg, and Brisbane. The outbreak in Brisbane began on 18th February, 1903, and ended on 23rd May; two sporadic cases appearing in September. In all, 21 cases were reported from the Metropolitan Area; the epidemic furnishing a smaller number of cases than in any previous year.

Both infected premises and infected rats were confined to a much smaller area—practically the city proper—than in the previous outbreaks. The presence of infected rats on premises where indigenous cases had occurred was demonstrated often enough to support the now generally accepted position, "that the danger of contracting plague stands in relation to the presence of rats in dwellings and enclosed premises." Plague-infected rats, however, were found on many premises where no case of human plague occurred, but in one of the earlier cases some 200 or more dead carcasses of rats (bodies too decomposed for examination) were removed from beneath the floors of a granary in close proximity to a house from which a plague patient had been removed a short while before.

Of 134 mice collected and examined, 4 were found to be plague-infected.

Of 16,409 rats collected in the Metropolitan Area, 14,227 were examined bacteriologically, and 84 were found to be infected—a percentage of 0·5.

The subjoined table shows the monthly incidence of plague on man and rats in the Metropolitan Area, 1903:—

| January. | February. | March. | April. | May. | June. | July. | August. | September. | October. | November. | December. | Total. |
|---|---|---|---|---|---|---|---|---|---|---|---|---|
| | | | | | | MAN. | | | | | | |
| 0 | 5 | 4 | 6 | 4 | 0 | 0 | 0 | 2 | 0 | 0 | 0 | 21 |
| | | | | | | RATS. | | | | | | |
| 1 | *8 | *30 | 20 | 15 | 6 | 0 | 0 | 4 | 0 | 0 | 0 | 84 |

* One infected mouse each month included.

A large proportion of the infected rats caught in the latter part of the year came from the city wharves.

1904.
FIFTH EPIDEMIC.

During the month of February, plague again appeared at Brisbane, and continued to manifest itself up to August, when it once more for the time subsided.

The last case of human plague in Brisbane in 1903 was on 11th September. The last infected rat in 1903 was found on 28th September. The first infected rat of 1904 was found on 8th January. The first case of human plague in 1904 was reported on 9th February. The fifth outbreak in Queensland was marked by 35 cases with 12 deaths, distributed as under:—

| Place. | Date. | Cases. | Deaths. | Case Mortality. |
|---|---|---|---|---|
| | | | | Per cent. |
| Brisbane ... ... ... | 9th February—14th September ... | 30 | 8 | 26·6 |
| Ipswich ... ... ... | 10th July ... ... ... ... | 1 | 1 | 100 |
| Maryborough ... ... ... | 3rd June—13th .. ... ... | 2 | 1 | 50 |
| Cairns ... ... ... .. | 16th February—22nd March .. | 2 | 2 | 100 |
| | Total ... ... ... | 35 | 12 | 34·2 |

**Fig. 9.1** Ham's 'portrait' of the fourth epidemic in 1903, combining tables, lists and narrative

from all individual outbreaks to demonstrate statistical commonality. Tabular statements were now dedicated to common and generalisable patterns that have emerged across the eight outbreaks. The first two tables plot the geographical distribution, clustered by city and town. He could show that both case and mortality rate have been roughly stable within each location, while overall rates had declined after the first year's outbreak. The third table then compares the 'race incidence', showing with unmistakable clarity that plague was predominantly a disease of the white population, with incidence rate following roughly the percentages of ethnic diversity in Queensland at the time. We find the same summary tables for 'relative frequency of bubonic and non-bubonic plague', indicating a continued incidence of bubonic plague above 80 per cent in almost all years.[51] The age incidence has been plotted for the entire series of outbreaks to emphasise the clear accumulation of cases as well as deaths in the age groups between fifteen and thirty-five years. With regard to gender, Ham's tabular statement stresses the predominant incidence among men, with both cases and mortality being more than three times as high in men as they were in women. The section is then concluded with a series of detailed tabular comparison of cases in humans and rats, outlining the significant seasonal similarity of plague in rats and humans for every year with summary statistics per month.

The purpose of the section following on from year-by-year portraits of outbreaks is simple: all tables contribute to Ham's project of sorting local specificity and unusual aspects from those aspects that re-occur in every year the epidemic returned. Without any further definition, and lacking any narrative exploration, the tabular statements clearly indicate what Ham had discerned as key characteristics of plague in Queenstown: it affected with proportional accuracy the same towns almost every year. The pattern of 'race incidence' allows for little conclusion other than a disproportionate impact on white settlers. The frequency of bubonic, septicaemic, pneumonic and other types of plague was stable from year to year, just as the distribution between age groups and gender was almost identical throughout the outbreaks. In summary, Ham had developed a geographic pattern, a series of repeatable characteristics in the social distribution of the epidemic and he could demonstrate statistically the comparability between human and rat plague.

His report provides a great example for the transformation and transposition of individual, exemplary cases into data points, from which a new, second-order casuistry was developed to convey inferences about the epidemic. While the techniques deployed by Ham offer themselves for the common interpretation of reducing complex case files into well-defined and countable numbers to deploy medical statistics, it would ignore the persistence of narrative commitments to case-thinking in each of Ham's generalisations. While his collection of individual cases makes for a quite common arrangement of clinical characteristics, it is striking how he folded these aggregated cases into his descriptions of annual epidemics. These, due to the careful combination of general observations, tabulated case data and narrative explanations, appear themselves as cases, as 'problem-events', which require some kind of judgment. While Ham certainly moved from the study of variation among an individual (section three) towards the study of variations among a group of individuals (tabulated statements in section one), he sustained the notion of the annual epidemic as a discrete, discernible event, which could not be captured in statistical data alone. Once these events, or rather cases of the epidemic, were considered in their sequential appearance, and once their repeatable pattern was identified, Ham was able to pass on his judgement about plague epidemics more generally. This, I argue here, was how Ham used the report as a paper technology to translate medical casuistry into the field of epidemiology. In the following, this point shall be exemplified along the three principal inferences that Ham drew from his collected case data.

### *Individual Liability*

Equipped with the staggered casuistry from individual patients, to individual outbreaks and clusters, Ham then began the work of describing and defining his observations. The collected histories of patients and their tabulated representation in the first part became subject of an extensive discussion in the second part of the report, titled 'The Incidence of Plague'. A brief summary of geographical features was followed by an overview of the sanitary administration and the political circumstances, which led the author to gain his position as commissioner of public health.[52] The following descriptions, explanations, and interpretations were prefaced with a map of Queensland, showing the 'distribution of plague 1900–1907' with all affected towns underlined in red. With this picture defining the place of this localised epidemic, Ham offers then his interpretation of the data assembled and summarised in the previous part.

Ham had collected data on the 'individual liability' on the incidence of plague in regards to sex, age, race and occupation. First, on the issue of sex, he remarked that despite the consensus that plague affects both women and men in the same way, there had been a striking prevalence of cases and mortality among men, almost at a proportion of 4:1. With regards to age, Ham again noted a striking deviation from the medical consensus about age related liability to plague and identified a disproportionate amount of young people, mainly school children to have been affected. As he turned to the question of race, he pointed out that the incidence rate among the Chinese, which in medical literature was understood to be highly vulnerable to plague, paralleled the statistical distribution of ethnicities in Queensland at the time. Ham thus excluded any individual liability according to race and ethnicity to be significant. Finally, he turned to occupation, in which he identified the most important liability to understand the epidemic. Rather than stating his statistical finding as briefly, as with regards to the sex, age and race, he turned here to an extensive narrative explanation of his data.

Ham considered occupation to be a primary liability. Both the overwhelming gender disparity and the age distribution were, Ham argued, likely to have been influenced by the occupation of plague victims and subsequently due to the presence of rats. 'The superior liability of dealers in grain and produce, store employers, grocers, butchers, fishmongers, cooks, hotel servants, stable men' should be attributed to the presence of rodents within the vicinity of these businesses.[53] He remarked that over 36 per cent of all recorded cases between 1900 and 1907 were directly attributed to the food trades. He then went on to underline this characteristic by pointing to a cluster of cases which had emerged within the same food-related premise across different outbreaks.[54] All other occupations or groups of occupations (such as hospital staff or disinfection crews) were not significantly affected by the outbreaks, indicating to Ham a strong indication to understand the epidemic as a highly localised phenomena.

With these elaborations on the statistical incidence of plague, Ham could show the substantial difference between medical and epidemiological understandings of plague. He referenced medical opinions and consensus on plague's more or less indifferent infectivity with regards to sex, age and race, only to point to different incidence patterns that he had observed for the epidemic in Queensland. These incidence patterns, so he argued, were characteristic to the epidemic of plague and the distribution of cases along sex, age and race, would always deviate from a strict medical viewpoint, as they could only be explained through epidemiological considerations. These pointed then to the primary significance of occupation. To emphasise this inference from statistical data and to explain this epidemiological occurrence, Ham turned to narratives detailing the relationship between food trades and plague. He referred to instances of plague cases in warehouses in Rockhampton, in April 1908, and to a group of cases which emerged around an 'insanitary produce store' in Ipswich, May 1905. He brought in cases to illustrate and clarify his conviction that 'the frequent occurrence of plague-infected rats on premises or places of business [...] is highly significant'.[55] However, the cases he mobilised were not individual patients, but instead clusters of cases associated with a place. The convincing feature of these places was that the location itself clearly outranked any other mode of explanation, like sex, age or race. These clusters, or sub-epidemics, worked for Ham as exemplary outbreaks to refine the classification of the epidemic, similarly to how the exemplary case worked for clinicians to explore the classification of a disease. The clusters captured the significant features of general observations in such unique and convincing clarity, that their exploration as a 'text-book case' assumed an instructive and epistemic quality within the project of defining the epidemic as object of knowledge.

### *Place Infections*

To frame a closer exploration of these sub-epidemics and their places, Ham first considered some general seasonal characteristics of plague in Queensland. Every year, the epidemic curve reached its maximum in April and May, and was at its lowest point in the winter months of June and July. In most years, July to December was usually free of any cases, with human cases emerging back in 'moist or muggy season'.[56] This observation, so Ham argued, was valuable, as 'Queensland, with its even temperature and comparatively high humidity, is probably more constant than any other place within the same isothermal lines'.[57] While the seasonality of plague was well acknowledged in existing literature, a convincing explanation of the waxing and waning of plague over the year remained was lacking. Ham speculated if seasonal influences might impact directly on the 'bacterial flora', or if indeed one had to consider an 'adaptation [...] of the plague bacillus' both within the human host and within the 'particular environment of an infected locality'.[58]

It is here, in the close inspection of 'infected locality' where Ham emphasised the value of an epidemiological perspective.[59] While bacteriological research had offered insights as regards the plague bacillus, only the study of the bacterium in situ would allow understanding of what circumstances external to the bacterium itself were driving infections, which Ham considered questions of 'great epidemiological importance'.[60] To understand the factors within both the human host and the environment impacting the virulence of the bacteria, epidemiological research of infected locality was more important than any of the insights the bacteriological laboratory could offer:

> The bacteriologist rarely realises this limitation, from his point of view; or that the many factors that influence the life of specific bacteria in the outer worlds—e.g. strains or races which have different powers of resistance to external conditions, the neutralising effect of contaminating organisms, the

> tendency to adaptability, the variation in virulence, and the chances of escape from the infected organism—are always, as a matter of fact, epidemiologically operative.[61]

Ultimately, these questions are required to bear on the specific place, in which the epidemic emerged. To demonstrate how an epidemiological study of place was to be conducted, Ham considered 'localised or sub-epidemics'. One of these clusters emerged in Brisbane in January 1907 at Kingsholme, New Farm. Like other, comparable sub-epidemics, the outbreak was confined to one block, 'bounded by four streets' and yielded sixteen cases in total.[62] In the following part of the report, Ham wrote an astonishingly detailed and thorough description of the sub-epidemic. He clustered all cases in three sub-groups and offered brief abstracts on their age and occupation, as well as on their relation to the produce store at the centre of the block. The first six cases were all schoolboys, aged twelve to seventeen years, who were 'at home for the holidays'. Upon admission to the hospital with 'cerebral symptoms', headache and vomiting, they quickly developed swollen glands and laboratory confirmed plague in most of the children.[63] The second group of seven cases were all family members or children living within the same location. A quick search for rats proved futile. However, the third group were three members of the council's 'cleansing gang', deployed to remove 'putrefying and vegetable matter' from the premises.[64] Following on from these descriptions, Ham visualised the sub-epidemic with a detailed map (Fig. 9.2), in which the distribution of cases, the stages of their falling ill, and some indicators of the built environment were marked. As a plan of the infected area, the map encapsulated his narrative of an 'infected locality', and provided a spatial representation of this case of a sub-epidemic.

With maps, epidemiologist could create a two-dimensional abstraction of the complex relations of the outbreak. Maps like the one Ham used enabled the theoretical exploration of the relationship between the epidemic and its place. As Tom Koch has argued, it would be a mistake to understand such maps as representations of the outbreak, as pictures of research results. Rather, Koch emphasised in his history of the disease map that maps were most often used by epidemiologists to combine data and theories, to create a visual context in which theories could be tested. Similarly, Ham had drawn a device with which the spatial order of the infected area was combined with the chronological order of his casuistry. This two-dimensional devise drew together space and cases of plague to

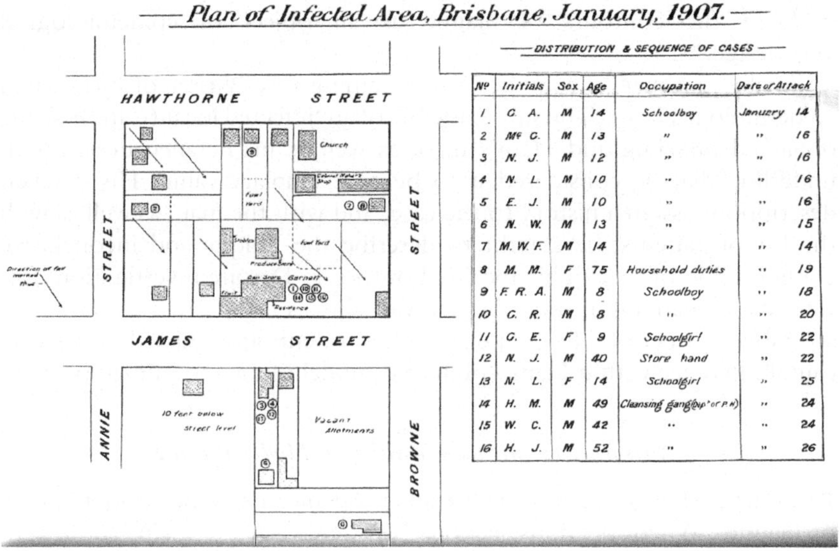


| Nº | Initials | Sex | Age | Occupation | Date of Attack |
|---|---|---|---|---|---|
| 1 | C. A. | M | 14 | Schoolboy | January 14 |
| 2 | Mc C. | M | 13 | " | " 16 |
| 3 | N. J. | M | 12 | " | " 16 |
| 4 | N. L. | M | 10 | " | " 16 |
| 5 | E. J. | M | 10 | " | " 16 |
| 6 | N. W. | M | 13 | " | " 15 |
| 7 | M. W. F. | M | 14 | " | " 14 |
| 8 | M. M. | F | 75 | Household duties | " 19 |
| 9 | F. R. A. | M | 8 | Schoolboy | " 18 |
| 10 | G. R. | M | 8 | " | " 20 |
| 11 | G. E. | F | 9 | Schoolgirl | " 22 |
| 12 | N. J. | M | 40 | Store hand | " 22 |
| 13 | N. L. | F | 14 | Schoolgirl | " 25 |
| 14 | H. M. | M | 49 | Cleansing gang ([illegible]) | " 24 |
| 15 | W. C. | M | 42 | " | " 24 |
| 16 | H. J. | M | 52 | " | " 26 |

**Fig. 9.2** Map of an 'infected locality' indicating the distribution and sequence of cases associated with the location

achieve a simplification of the messy and puzzling configuration with which this sub-epidemic had presented itself.[65]

Maps like these allowed the epidemiologist to draw together relevant factors to develop a theory while offering the empirical observation of the epidemic on the ground. In his map, Ham encapsulated fragments of the social order surrounding the food trade in the infected block. In the implied view from above, this social order merges with the spatial and geographical order of the city, integrating the built environment and its inhabitants into the contingent considerations of the dynamic of plague. Mapping the sub-epidemic framed the location as a point of inquiry. As Ham refused to consider human-to-human transmission, the density of cases in short succession in this block required an explanation in reference to the social and geographical qualities of the space. As a paper technology, the map derived its use and value precisely from the preparation of the infected space as an object of epidemiological research. Far from offering causal inferences and judgements about drivers and origins, it instead offered Ham a device for the open-ended combination of case data with spatial coordinates to let a network of relations emerge, which in turn outlined the contours of plague as an object of epidemiological knowledge.[66]

Finally, the map encourages the reframing of a cluster of cases as an epidemiological case. The sub-epidemic of an 'infected locality' in Brisbane remained puzzling and Ham's map, as well as his detailed description, underline that he considered it to be of exemplary value. His detailed descriptions assign a history to the case, and with the map, as well as with the list of patients, he sought to describe the unique and inconclusive configuration of this highly localised outbreak. Despite investigation, this case maintained an unresolvable individuality as it did not yield any considerable number of infected rats or other rodents and resisted a satisfying causal inference within Ham's preferred model of rat and human plague.

### *Rats and Humans—Making a Model Plague*

Part four of Ham's report is finally explicitly dedicated to the 'Epidemiology of Plague'. Where previous sections had described the main 'features of the disease', this section was now meant to illuminate the 'spread and communication of plague in Queensland' and contained the combination and analysis of previously established observations.[67] Ham began with a verbose description of what he considered important aspects of the disease's infectivity, further emphasising the strong correlation of location and the epidemic. The basic premise was stated clearly: 'Where several cases of plague occurred simultaneously, or within a short period of each other, the infectivity of the locality within which the attached resided, or worked, cannot be inferred from the fact alone that the locality, or even infected premises within such locality, harboured plague-infected persons.'[68] The main culprit in this puzzle was identified as the rat.

The connection between rats, their fleas and human plague was nothing new and, at the point of the report's publication, constituted a widely accepted explanatory model.[69] Although there was an abundance of circumstantial indicators that suggested a close relation between plague in rats and plague in humans, still many questions persisted on how epizootics and epidemics of plague were mediated. Ham abandoned any alternative theory, which 'was not conveyed by any methods common to other infectious disease'.[70] However, he conceded that, while the association of rat, fleas and human plague was long accepted, members of the Plague Commission in India had struggled to provide clear proofs of causal chains as they had not 'acquired the data necessary for the decision of the question'.[71]

Here, the data collected in Queensland were supposed to help solidifying that human epidemics were a consequence of epizootics in rats. To this end, Ham had set in place an ambitious rat-catching program, which offered data on plague in the rodents across the state not only in times of outbreaks but throughout the year. Further, he had instated a similarly ambitious program to measure the incidence of fleas across the state all year round. From these data, captured in daily bacteriological analysis of hundreds of rats and fleas, Ham argued that robust inferences could finally be made to the relation of epizootics and epidemics. To this end, he plotted the data in a diagram as an epidemic curve indicating both rodent and human incidence over the course of the year for each year from 1900 to 1907 in the metropolitan area of Brisbane. Week by week, the diagram visualised the equivalence between epizootic and epidemic, and for most years (1900, 1902, 1903, 1904, 1906) rats with plague were discovered roughly 1–2 weeks before the first human cases had developed.[72] The diagram visualised that year after year, cases in rats developed largely with similar dynamics to cases in human. Rise and fall of cases were the same here as there. The overall shape of the epidemic, as well as the epizootic curve, were largely comparable, indicating that both were governed by similar factors. Furthermore, Ham drew the conclusion from his data that the 'epizootic runs concurrently with and outlasts the epidemic' and that the 'epizootic invariably precedes the epidemic'.[73]

He further solidified his observation of the reoccurring interdependence of rat and human plague with a cycle-diagram, added to the appendix of the report (Fig. 9.3). With the innermost ring dedicated to 'rat prolificity [sic]', the second ring indicated the presence, between September and December, of what Ham called 'chronic rat plague'. This was data indicating the continuous presence of plague in isolated rats outside the epidemic curve. The next ring then plotted the incidence of flea population, growing in December and January, before acute rat plague and human plague, running from January to July, were indicated on the outer rings of the cycle-diagram. All together, the cycle-diagram provided a complete model that, according to Ham's observations, explained the waxing and waning of 'a plague-cycle in Brisbane'. While the rat population acted as a reservoir all year round, it was the seasonal growth of fleas (dependent on changing temperature and humidity) that preceded the development of acute rat plague, from which human cases followed and

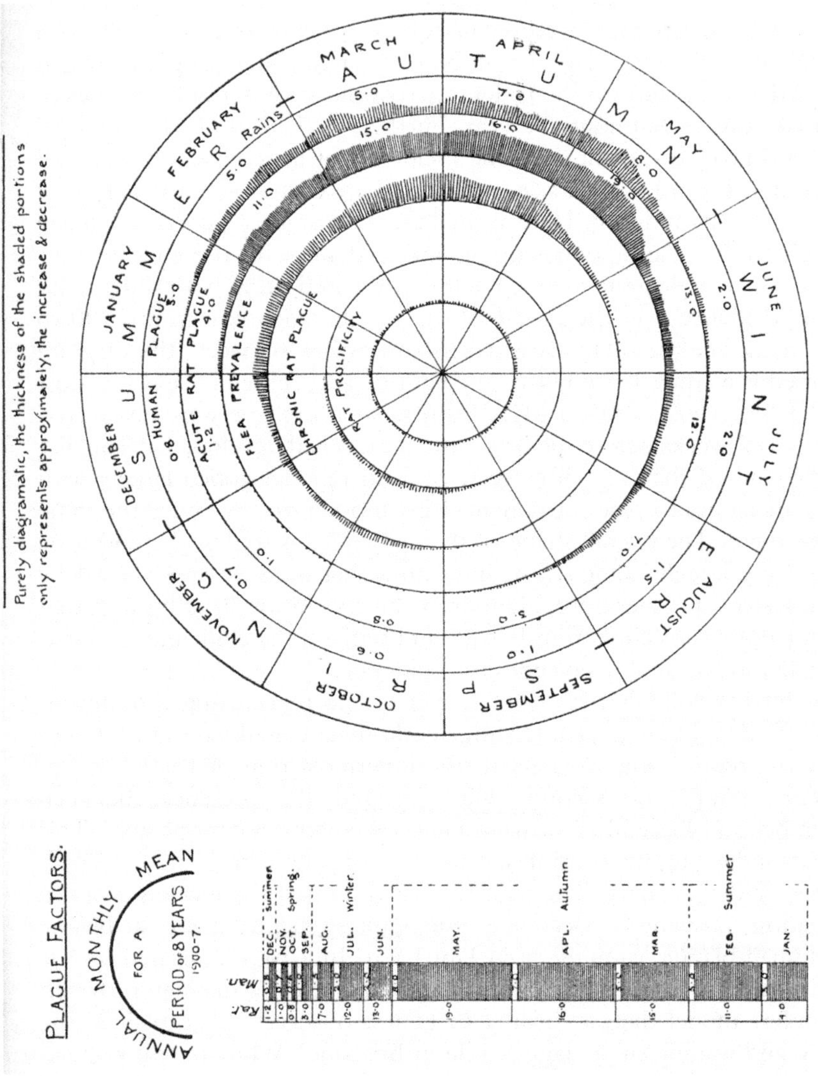


**Fig. 9.3** Diagram of a plague cycle in Brisbane, aligning interdependent variables as a predictable configuration of plague

which together with rodent cases fell rapidly as soon as the flea prevalence began to be reduced in June.

With his diagrammatic condensation, Ham expressed and illustrated the theoretical inference yielded by his study of eight consecutive plague outbreaks. Such diagrams were essential to the project for developing an object of epidemiological knowledge. With Daston and Galison, a diagram like Ham's, is first and foremost a 'working object'. Drawing it constitutes a kind of workbench to enable the parallel interrogation and manipulation of empirical observation and theoretical assumptions.[74] It allowed Ham to arrange his rich collection of observations, narratives, cases and data as a system, in which the relationship between variables was revealed. The aim of such diagrams was to give expression to the internal rationality of an epidemic and to open a pathway towards the development of the kind of mathematical expressions, which would allow the robust prediction of a plague epidemic.[75]

The diagram constitutes more or less the pinnacle of Ham's reporting on plague. With this graphic device, he summarised and condensed most aspects of his report in a neat diagrammatic argument. Reading from the outer rings inwards, he first considered occupational liability as an effect of the localised appearance of sub-epidemics. Second, these infected localities were in fact attributed to the presence of rodents, which had predominantly occupied food trades along the coast of Queensland. After demonstrating the statistical relation between epizootics and epidemics, Ham began to build a theoretical model of plague's epidemiology in the report's fourth part. The model, as most epidemiological models at the time, sought to order the vast array of interdependent variables to define and—in some cases—to predict the dynamics of the epidemic. To this end, Ham attributed the previously established observations as re-occurring effects of an annual plague cycle, which was ultimately driven by the seasonal variance in Queensland. The diagram thus evokes a model plague, which had been painstakingly developed from the cases of individual patients clustered in cases of localised sub-epidemics before these were aligned with rodent and insect ecologies. Like Ursula Klein's chemists, Ham used the diagram as a 'paper tool' to develop a representation of the invisible relations, dynamics and repetitions that he had distilled as significant drivers of plague.[76] The diagram projected and constituted the epidemic as an object of knowledge, which remained otherwise invisible. It took its capacity to construct and present the epidemic from both carefully arranged empirical observations and the robust theoretical interpretation of their relations.

In the diagram, the epidemic is neither just an accumulation of cases nor, however, a mere problem of insanitary environment or suspicious populations. Instead, the epidemic emerges through the diagram as a network of interdependent variables, a network of correlations spanning medical, environmental, social, and ecological domains. Such diagrammatic imagination of the epidemic as an object of knowledge suggests it to have been understood as a relational entity, assembled from a variety of domains and resources and anchored in the collection, interpretation and arrangement of cases. While the diagram's appearance might be suggestive of the prevalence of a statistical way of thinking in epidemiology in the twentieth century, I argue for it instead to be read as the representation of a refined epidemiological casuistry. The model's theoretical contribution to the epidemiology of plague was, just like general observation in clinical medicine, one that relied fundamentally on exemplary cases, convincing narratives, illuminating anecdotes, as well as on statistical data.

## Conclusion

In this chapter, I have revised perspectives on early twentieth-century epidemiology, which is often assumed to be a phase of quantification, focused on the utilisation of medical statistics. Contrary to this historical account, I have introduced the outbreak report as a paper technology, with which its authors did not only report to governments and authorities on the occurrence, development and abatement of a crisis, but in which they also contributed to the production of epidemiological knowledge. While dedicated to the production of statistical evidence, this epidemiological knowledge has retained a dedication to a way of reasoning familiar to nineteenth-century clinical medicine, fundamentally based in narrative conventions and dedicated to a thinking in cases.

I have shown that, in his exemplary report on plague in Queensland, Bertie Burnet Ham used narrative conventions and an epidemiological casuistry to enable the correlation of previously disconnected spheres and domains, to explain and interpret the data listed and tabulated throughout his report, and to emphasise the historical quality of epidemics as processes developing over time. I have shown that each of his arguments and contributions to a more generalised understanding of plague as epidemic required formalisations either in lists and tables, in maps and spatial or as topographic diagrams. Each of these were essential elements of the report in producing an imagination of plague as an epidemic independent of 'individual liability' other than occupation, prone to be organised in localised clusters or 'infected localities', and in principle driven by a series of interdependent variables that follow an annual rhythm. These perspectives and approaches, so Ham concluded with his report and diagram, are what make the epidemic an object of knowledge.

I have proposed to read the report as a paper technology to emphasise the outstanding significance of the case within the reasoning of Ham and other fellow epidemiologists of plague. I asked how at the beginning of the twentieth century, plague was imagined as an object of epidemiological knowledge. I have further asked how this production of epidemiological knowledge about plague might hold insights into the conventions of epidemiological reasoning at the time. This history points clearly to the significance of the much-overlooked role of case-based thinking in epidemiology. Within the report, the plague epidemic is not imagined as a statistical aggregation of cases of the disease, but as a discrete object of knowledge, which has been inferred from outbreaks and sub-epidemics, in turn described, analysed and arranged as cases of the epidemic.

Both cases of patients and cases of sub-epidemics bore multiple representations within the report as they were utilised in different parts of the text to refine clinical categories, to emphasise unique incidence patterns of the epidemic, and to construct a cyclical model of the waxing and waning of plague. While the report might appear as a detailed history of plague in Queensland, it is through the lens of the case within the report that Ham's casuistic techniques of listing, tabulation and diagrammatisation become significant to the imagination of plague as an epidemic in the modern sense. The open question is then: what exactly might be grasped within this framework as an epidemic case? If we were to return to Crookshank, who wrote a good fifteen years later on the first principles of epidemiology, one answer might be found in the abstract nature of the epidemic as an object of knowledge. If one were indeed to compare the concept of the epidemic with the concept of a disease, instead of considering the first as a mere quantitative exaggeration of the latter, one would have to consider the population to be the epidemic's patient. Given the constraints on time and space, in which such a delineation of a collective patient would have to be outlined, Crookshank suggested to think of this conceptual patient—and epidemiological case—perhaps as a 'time-population unit'.[77]

## Notes

1. Charles Rosenberg, *Explaining Epidemics and Other Studies in the History of Medicine* (Cambridge: Cambridge University Press, 1992); Olga Amsterdamska, 'Demarcating Epidemiology', *Science, Technology, & Human Values* 30, no. 1 (2005): 17–51.
2. Lukas Engelmann, John Henderson, and Christos Lynteris (eds.) *Plague and the City*, (London and New York: Routledge 2018).
3. Erwin H. Ackerknecht, 'Anticontagionism Between 1821 and 1867', *Bulletin of the History of Medicine* 22 (1948): 562–593.
4. However, hardly ever have these interpretations claimed a polemic opposition. Most were, as Worboys and others have pointed out, negotiated within a 'seed' and 'soil' metaphor and definitions of 'infectious diseases' assumed throughout the nineteenth century a broader framework then the mere presence of a micro-organism; Michael Worboys, *Spreading Germs: Disease Theories and Medical Practice in Britain, 1865–1900* (Cambridge: Cambridge University Press, 2000); Flurin Condrau and Michael Worboys, 'Second Opinions: Epidemics and Infections in Nineteenth-Century Britain', *Social History of Medicine* 20, no. 1 (April 2007): 147–158.
5. F. G. Crookshank, 'First Principles of Epidemiology'. In F. G. Crookshank (ed.), *Influenza: Essays by Several Authors*, pp. 11–30 (London: W. Heinemann 1922).
6. These common characteristics of plague reports as a global corpus are currently subject to collaborative research with Beatrice, Arlene Casey and Mike Bennet. See: Plague.TXT.
7. J. Andrew Mendelsohn, '"Like All that Lives": Biology, Medicine and Bacteria in the Age of Pasteur and Koch', *History and Philosophy of the Life Sciences* 24, no. 1 (2002): 3–36; Charles E. Rosenberg, 'What Is an Epidemic? AIDS in Historical Perspective'. In Stephen R. Graubard (ed.), *Living with AIDS*, pp. 1–17 (Cambridge, Mass.: MIT Press, 1989).
8. Anke te Heesen, 'The Notebook. A Paper-Technology'. In Bruno Latour and Peter Weibel (eds.), *Making Things Public*, pp. 582–589 (Cambridge, Mass.: MIT Press, 2005); Volker Hess and Andrew Mendelsohn, 'Sauvages' Paperwork: How Disease Classification Arose from Scholarly Note-Taking', *Early Science and Medicine* 19, no. 5 (2014): 471–503; Michael Lynch, 'Science in the Age of Mechanical Reproduction: Moral and Epistemic Relations between Diagrams and Photographs', *Biology and Philosophy* 6.2 (April 1991): 205–26.
9. Volker Hess and J. Andrew Mendelsohn, 'Case and Series: Medical Knowledge and Paper Technology, 1600–1900', *History of Science* 48, no. 161 (2010): 287–314.

10. Lukas Engelmann, '"A Source of Sickness": Photographic Mapping of the Plague in Honolulu in 1900'. In Lukas Engelmann, John Henderson, and Christos Lynteris (eds.), *Plague and the City*, pp. 149–168 (London and New York: Routledge, 2018).
11. Bertie Burnett Ham, *Report on plague in Queensland, 1900–1907 (26th February 1900 to 30th June 1907)* (Brisbane, Queensland, 1907).
12. Alexandre Yersin, 'La peste bubonique à Hong Kong', *Annales de'l Institut Pasteur* (Paris) 8 (1894): 662–667; Christos Lynteris, 'Pestis Minor: The History of a Contested Plague Pathology', *Bulletin of the History of Medicine* 93, no. 1 (April 2019): 55–81.
13. William John Ritchie Simpson, *A Treatise on Plague; dealing with the historical, epidemiological, clinical, therapeutic and preventive aspects of the disease* (Cambridge: Cambridge University Press, 1905); Christos Lynteris, *Ethnographic Plague: Configuring Disease on the Chinese-Russian Frontier* (Basingstoke: Palgrave Macmillan, 2016).
14. Albert Calmette, 'The Plague at Oporto', *North American Review*, 171, no. 524 (July 1900): 104–111; Albert Calmette and A.T. Salimbeni, 'La peste bubonique (Etude de l'épidémie en Oporto en 1899)', *Annales de'l Institut Pasteur*, 13 (1899): 865–936. Within the same period, epidemiology was reconstituted as a field of scientific knowledge production, increasingly aiming for the development of a discrete scientific method, shaping and defining the boundaries of a new and unique medical science. See, for example, Amsterdamska, 'Demarcating Epidemiology', p. 31.
15. Rosenberg, *Explaining Epidemics*, p. 295.
16. Mary S. Morgan and M. Norton Wise, 'Narrative Science and Narrative Knowing. Introduction to Special Issue on Narrative Science', *Studies in History and Philosophy of Science* Part A, 62 (2017): 1–5, p. 2.
17. Ibid., p. 2.
18. Ibid., p. 3.
19. See, for example, Bruno J. Strasser, 'Collecting Nature: Practices, Styles, and Narratives', *Osiris* 27, no. 1 (January 2012): 303–340; Lisa Cartwright, *Screening the Body: Tracing Medicine's Visual Culture*, (Minneapolis: University of Minnesota Press, 1995). C. Coopman, J. Vertesi, Michael Lynch, and Steve Woolgar (eds.), *Representation in Scientific Practice Revisited.* (Cambridge, Mass: MIT Press, 2014).
20. Hess and Mendelsohn, 'Case and Series'; Hess and Mendelsohn, 'Sauvages' Paperwork'.
21. Hess and Mendelsohn, 'Case and Series'.
22. Ibid.
23. Ian Hacking, *The Taming of Chance* (Cambridge: Cambridge University Press: 1990); J. Rosser Matthews, *Quantification and the Quest for Medical*

*Certainty* (Princeton, NJ: Princeton University Press, 1995); Magnello and Hardy, *The Road to Medical Statistics.*

24. John Forrester, 'If p, then what? Thinking in cases', *History of the Human Sciences* 9, no. 3 (1996): 1–25, p. 3.
25. M. John Thearle, 'Dr B Burnett Ham', *Medical Journal of Australia* 161, no. 1 (1994): 55–57.
26. Burnett Ham, *Plague in Queensland*, p. 8.
27. New South Wales. Department of Public Health, New South Wales. Department of Public Health, J. Ashburton Thompson, and London School of Hygiene and Tropical Medicine, *Report on the outbreak of plague at Sydney [1900–1907]/by J. Ashburton Thompson, Chief Medical Officer of the Government and President of the Board of Health*, (1900).
28. Arthur Stanley, George Ford Petrie, and Richard P. Strong, *Report of the International Plague Conference Held at Mukden, April, 1911* (Manila: n.p, 1912).
29. For a preliminary list of reports, see Lukas Engelmann, 'Mapping Early Epidemiology: Concepts of Causality in Reports of the Third Plague Pandemic 1894–1950'. In E. Thomas Ewing and Katherine Randall (eds.), *Viral Networks: Connecting Digital Humanities and Medical History*, pp. 89–118 (December 2018).
30. Tom Koch, *Disease Maps. Epidemics on the Ground*, (Chicago IL: Chicago University Press, 2011); Engelmann, 'A Source of Sickness'; Lucas M. Mueller, 'Cancer in the Tropics: Geographical Pathology and the Formation of Cancer Epidemiology', *BioSocieties* 14 (May 2019): 512–528.
31. Lauren Berlant, 'On the Case', *Critical Inquiry* 33, no. 4 (2007): 663–672, p. 663.
32. John Forrester, 'On Kuhn's Case: Psychoanalysis and the Paradigm', *Critical Inquiry* 33, no. 4 (2007): 782–819, p. 810.
33. Burnett Ham, *Plague in Queensland*, p. 63.
34. Ibid., p. 64 ff.
35. Hess and Mendelsohn, 'Case and Series'.
36. Lynteris, 'Pestis Minor'.
37. Burnett Ham, *Plague in Queensland*, p. 65.
38. Ibid., p. 68.
39. Berlant suggests thinking of these abstract entities as metacases. Berlant, 'On the Case'.
40. Ibid., p. 1.
41. Ibid., p. 3.
42. Ibid., p. 3.
43. Ibid., p. 3.
44. Ibid., p. 4.
45. Ibid., p. 4.

46. Ibid., p. 6.
47. Ibid., p. 12.
48. Ibid., p. 12.
49. Ibid., p. 18.
50. Ibid., pp. 10, 14.
51. Ibid., p. 23.
52. Ibid., p. 33.
53. Ibid., *Plague in Queensland*, p. 39.
54. Ibid., *Plague in Queensland*, p. 39.
55. Ibid., *Plague in Queensland*, p. 39.
56. Ibid., *Plague in Queensland*, p. 43.
57. Ibid., *Plague in Queensland*, p. 43.
58. Ibid., *Plague in Queensland*, p. 43.
59. The term was used by Ham. For discussion in relation to the use of the term in British India see: Prashant Kidambi, '"An Infection of Locality": Plague, Pythogenesis and the Poor in Bombay, c. 1896–1905', *Urban History* 31, no. 2 (2004): 249–67.
60. Ibid., *Plague in Queensland*, p. 47.
61. Ibid., *Plague in Queensland*, p. 47.
62. Ibid., *Plague in Queensland*, p. 50.
63. Ibid., *Plague in Queensland*, p. 51.
64. Ibid., *Plague in Queensland*, p. 53.
65. Koch, *Disease Maps*, p. 13.
66. On maps as spatial diagrams in epidemiology see Lukas Engelmann, 'Configurations of Plague: Spatial Diagrams in Early Epidemiology', *Social Analysis* 63, no. 4 (2019): 89–109 and Lukas Engelmann, *Mapping AIDS: Visual Histories of an Enduring Epidemic* (Cambridge: Cambridge University Press, 2018).
67. Burnett Ham, *Plague in Queensland*, p. 119.
68. Ibid., p. 119.
69. Michael Kosoy and Roman Kosoy, 'Complexity and Biosemiotics in Evolutionary Ecology of Zoonotic Infectious Agents', *Evolutionary Applications* 11, no. 4 (June 2017): 1–10; Christos Lynteris, 'Zoonotic Diagrams: Mastering and Unsettling Human-Animal Relations', *Journal of the Royal Anthropological Institute* (NS) 23, no. 3 (2017): 463–485.
70. Burnett Ham, *Plague in Queensland*, p. 120.
71. Ibid., p. 126.
72. Ibid., diagram following p. 126. Similar techniques were used, for example, by the Hong Kong Principal Medical Officer, William Hunter, who also plotted epizootics and epidemics of plague within the same epidemic curve to visualise the similarity of temporal patterns between plague in rats

and humans. William Hunter, *A Research into Epidemic and Epizootic Plague* (Noronha & Co: Hong Kong, 1904), p. 86ff.
73. Burnett Ham, *Plague in Queensland*, p. 127.
74. Lorraine Daston and Peter Galison, 'The Image of Objectivity', *Representations* 40 (1992): 81–128.
75. Such diagrams are comparable to the efforts invested in modelling in economics at the time. See, for example, Morgan, 'Models, Stories and the Economic World', p. 25.
76. Ursula Klein, *Experiments, Models, Paper Tools: Cultures of Organic Chemistry in the Nineteenth Century* (Berkeley CA: Stanford University Press, 2003).
77. Crookshank, 'First principles of epidemiology'.

CHAPTER 10

# Ethnographic Images of the Plague: Outbreak and the Landscape of Memory in Madagascar

*Genese Marie Sodikoff*
*and Z. R. Dieudonné Rasolonomenjanahary*

The bubonic plague appeared in Madagascar in 1898, two years after France colonised the island. By the early 1930s, the colonial state had begun to photograph its achievements in controlling seasonal plague outbreaks. Like other epidemic photographs of the third plague pandemic, the colonial images provide evidence of progress. They depict newly built hospitals and lazarettos to isolate the sick, men fumigating houses, vaccination campaigns, scientists examining rodents in the laboratory, and so on. At the same time, the photographs of Madagascar avert the gaze from images of hospitalised patients or spaces thought likely to cause zoonotic spillover events, such as filthy alleyways where rats roam or markets

G. M. Sodikoff (✉)
Rutgers University, Newark, NJ, USA
e-mail: sodikoff@rutgers.edu

Z. R. D. Rasolonomenjanahary
Moramanga, Madagascar



C. Lynteris (ed.), *Plague Image and Imagination from Medieval to Modern Times*, Medicine and Biomedical Sciences in Modern History, https://doi.org/10.1007/978-3-030-72304-0_10

crowded with live wild animals.[1] As Christos Lynteris argues, such epidemic photography bridges the gap between images of microbes and images of landscapes in which zoonotic spillover events occur. It enables the viewer to infer the routes of the pathogen and means of transmission, or the conditions of epidemic possibility, even while the pathogen and its movement remain invisible.[2]

The focus of the archival photographs from Madagascar, however, is mainly on solutions to outbreaks, not causes. The images tell a tale of France's civilising mission and project the triumphalism of science and modernisation.[3] Moreover, they do not deceive. Pasteurian scientists did manage to arrest the plague's expansion over the island for decades.[4] Yet the record also shields from view dimensions of the plague that were politically sensitive for the colonial state. Imagery of consent and compliance by Malagasy people, exemplified by subjects waiting calmly in long queues to be injected with the plague vaccine (Fig. 10.1), contradicts what historians know of Malagasy responses to plague control measures of the early twentieth century and resistance to medical authorities.[5] The archive lacks

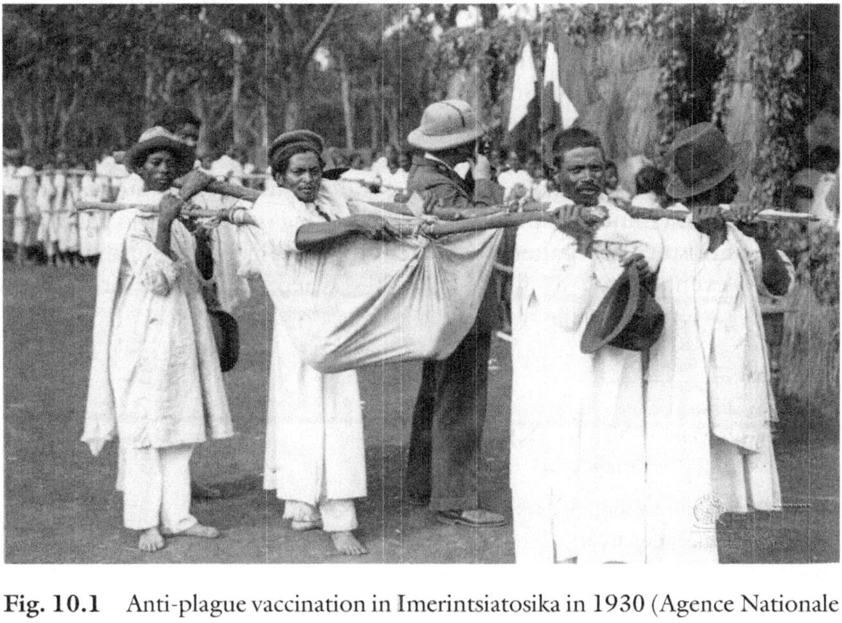

**Fig. 10.1** Anti-plague vaccination in Imerintsiatosika in 1930 (Agence Nationale d'Information Taratra Anta, Antananarivo, Album AS 5, No. 26)

images of newly built plague cemeteries of the era and other representations of death, with a couple of notable exceptions, such as a coffin being transported out of a village by masked men (Fig 10.2). Death was a political problem, not so much the fact of death but the way it was epidemiologically managed.

We are interested in this other invisible dimension of plague outbreaks, the experiences and perspectives of colonial subjects who suffered a high death toll and whose fear of infection was often compounded by draconian plague control measures. In Madagascar, rural people resisted the state's policies concerning the treatment and burial of the plague dead because these policies disturbed the relationship between deceased ancestors and living descendants.

Colonial-era mortuary rules remain in force and continue to complicate plague control efforts. Malagasy rural resistance to these rules, the despair, fear, anger—in short, the emotional register and embodiment of epidemics—has not abated. During our collaborative ethnographic study of the plague in Madagascar, we have sought to collect photographs that depict

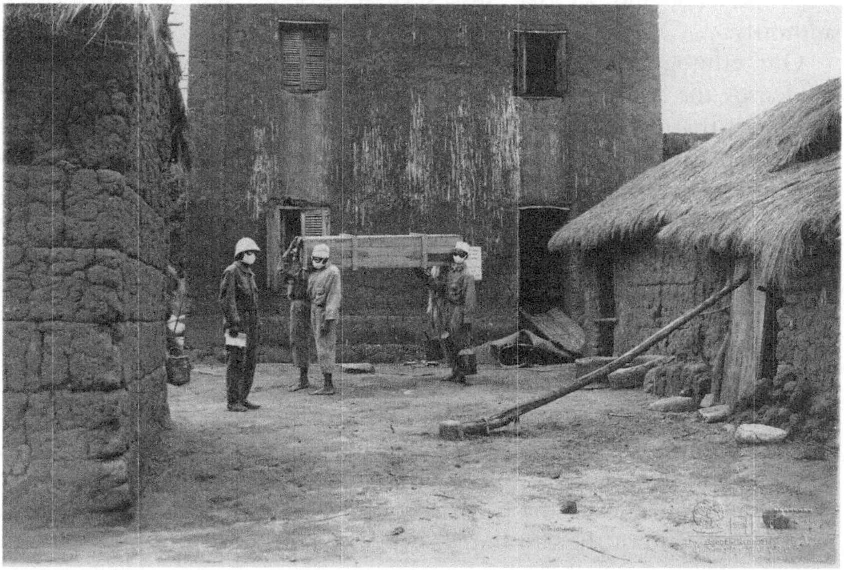

**Fig. 10.2** Inhumation for truck transport of a plague victim in 1930 (Agence Nationale d'Information Taratra Anta, Antananarivo. Album AS 5 No. 9)

survivors' outbreak imaginary. Our photographs aim to make visible the experiences that lie outside the frames of reference of colonial plague photography. The ethnographic photos, however, do not speak for themselves; they require contextualisation. As Deborah Poole writes, 'the "hard" visual evidence of ethnographic photography or film is intimately, even inextricably, bound up with the "soft" testimonial voice (or "subjectivity") of the ethnography.'[6]

We cannot say decisively how our identities and roles influenced what plague survivors revealed to us and when, but we do know that with each successive encounter over three years, our interlocutors shared new details. The recollections that guided our photography offer evidence of the plague's impact on one family, bringing to light by inference the toll of the outbreak on its members. Our photographs dwell on loss and the anxieties caused by profane burials, as this was the dominant theme of survivors' recollections. If masked faces constitute a dominant visual trope of the plague for states and scientists, then the plague pit, a hastily dug grave for one or multiple plague victims, represents the dominant trope for plague survivors. The image of the plague pit sums up all that is wrong with mortuary policies, and all that stokes rural people's wariness of medical authority.

Our ethnographic photographs of meaningful sites and objects are guided by the accounts of survivors of a localised outbreak in the central highlands that occurred in 2015, and that we investigated one, two, and three years later. This retrospective and emic photographic record seeks to provide a counterweight to the presentist, authoritative imagery of plague control and plague science. For plague survivors, a term that refers to those who have either survived the disease or have lost loved ones to it, the landscape becomes embedded with painful memories and a sense of foreboding about the future. This sense of dread stems less from the possibility of another plague outbreak than from the possibility of vengeance by deceased ancestors, who resent their exile to a cold and lonely grave.

For Malagasy subjects, to be buried in the familial tomb is of paramount importance, and the thought of being buried in the dirt, alone, leaves people aghast. They fear for themselves, and they mourn for their dead who must endure the hardship of posthumous exile. Plague pits in Madagascar are not the vast burial sites of the medieval and early modern plague epidemics, which, Lukas Engelmann argues, serve as a bio-historical archive, containing 'indices and traces that allow for various interpretations of and relations to the past'.[7] In Madagascar, the pits contain anywhere from one cadaver to several related plague victims. They represent 'spatial anchors' of outbreak narratives, to use Keith Basso's term for place-making. Excavating these recent plague pits would enlighten us to the circumstances of the burials if we had no first-hand witnesses. Pits conjure past outbreaks for survivors and create a sense of pervasive danger. They are 'out of place', disrupting the moral code of descendants caring for their dead ancestors and in turn being protected by them.[8]

Through documentary photographs of survivors' narratives of the intense days of one outbreak, we began to see the landscape with new eyes. The 2015 outbreak of bubonic and pneumonic plague occurred in a several hamlets of Moramanga District, where the predominant population consists of subsistence farmers who identify ethnically as Bezanozano and Betsimisaraka. Most residents live in one-room, mud-walled, thatch- or tin-roofed houses with few possessions. Access to the nearest medical clinic requires walking at least several miles, and the District hospital in Moramanga town is approximately ten miles from the 2015 outbreak zone. The authors, Genese Sodikoff, a white North American anthropologist, and Dieudonné Rasolonomenjanahary, a Malagasy ethnographer and now-retired branch manager of the Pasteur Institute in Moramanga town, interviewed survivors of this outbreak who had lost eight relatives in rapid succession. We concentrate here on the perspectives of one couple, Jean-Paul and Baotine (pseudonyms) who undertook herculean efforts to get the sick to the hospital and handled most of the necessary expenses. As this couple and other family members answered our questions about events of the outbreak, we photographed them, as well as various sites and objects, to supplement our fieldnotes. As a whole, our photographs offer a partial, inadequate, and hindsight perspective of an epidemic from the rural margins. The stories they illustrate echo the earliest encounters of rural Malagasy encounters with state health agents as recorded in archival documents.

## Official and Unofficial Representations of the Plague

Existing photographs of early twentieth-century Madagascar anti-plague efforts depict an organised and efficient campaign. As Christos Lynteris and Ruth J. Prince explain, medical photography had been deployed to configure the third plague pandemic, originating in Hong Kong in 1894, as 'scientifically intelligible and actionable'.[9] Those early images project advances in plague science: medical workers in white caps, gowns, and facemasks (all signs of 'hygienic modernity'), scientists in laboratories dissecting pestiferous animals, newly built isolation buildings for plague patients (*lazerets*), Malagasy subjects standing in long queues for mandatory vaccinations, teams trekking over grasslands carrying gear to disinfect contaminated homes, and masked workers disinfecting buildings and burning contaminated objects.[10] The images would have reassured viewers in Paris. To the extent urban Malagasy people also saw such photographs in the capital's newspapers, they may have also felt reassured given the panic caused by plague outbreaks.

Although written accounts survive, there is a lack of visual evidence of acts of rural resistance that complicated public health efforts. Dr Georges Girard, Director of the Institute Pasteur of Madagascar (1920–1944) and Director of Plague Services of the Pasteur Institute of Paris (1941–1954), writes,

> One knows only too well ... that the natives have never willingly bent to measures of prophylaxis, which they do not understand and which they consider vexatious. Avoiding sanitary practice, escaping vaccinations, having the freedom to carry on with traditional burials that are incompatible with hygiene, all is done to render inoperative the regulation concerning pestilential infections, the plague first and foremost.[11]

Official photographs of Malagasy people offer rare glimpses of the infected, and one can only presume certain individuals are plague patients. One image, for example, depicts a Malagasy man and woman circa 1930 sitting outside an isolation building wearing facemasks. The image shown above of a vaccination queue foregrounds a man hoisted up in a carrier. Missing is imagery of the ill in their homes, of the deceased, of funerals and confrontations between villagers and pith-helmeted officials. Indeed, cameras were elite apparatuses in the early twentieth century, inaccessible to the vast majority of Malagasy people. This limited the possibility of the emergence of 'vernacular photography', referring to the ordinary photographs bought or taken by everyday people.[12] The same holds true in a majority of rural districts today.

French officials would have also been disinclined to depict imagery of hopelessness or rebellion.[13] Moreover, rural resistance to the state was usually clandestine. Many simply did not inform authorities of suspected plague cases in order to avoid having patients or the deceased removed from villages.[14] The genre of medical photography that emerged in the 1930s in Madagascar approached the subject of the plague through the 'civilising mission' narrative.

Although the colonial state undertook plague control measures as early as 1898, when the pathogen was introduced at the eastern port town of Tamatave, a more intensive hygiene regime consisting of burning infected houses, rat eradication, vaccination, the regulation of mortuary practices, and the screening of dead bodies for plague developed after the disease flared up in the populous capital, Antananarivo, in 1921. The pathogen had gradually migrated upland via the railway.[15] It made a dramatic appearance there between June 24 and July 22, 1921, when forty-six members of two extended families who had attended a wedding together rapidly died of pneumonic plague.[16]

After 1925, not only would the bodily fluids of suspected plague cases be screened for evidence of the disease after death, but anyone who died of a chronic disease would also be screened owing to the state's distrust of the medical information provided by families. For confirmed cases of the plague, sanitary agents or prisoners would take away the bodies and place them in special cemeteries. Historian Faranirina Esoavelomandroso (now Rajaonah) writes that the vision of these foreigners with otherworldly facemasks stealing the dead was 'a nightmare' for people of Antananarivo.[17] This interpretation puts a very different cast on the image, intended by administrators to project hygienic rigor and scientific advancement (Fig. 10.2).

The ethnographic photographs, taken nearly a century later, remain in dialogue with the colonial past insofar as the Ministry of Health continues to implement old policies concerning the treatment and internment of plague victims. Although many contemporary plague victims in Madagascar express distrust of medical authorities and hospitals, often believing that medical staffers intentionally infect patients, the majority of rural people with whom we spoke willingly consumed prophylactic antibiotics during plague outbreaks when and if they had the opportunity.

The 2015 outbreak in Moramanga District resulted in fourteen cases and ten deaths, and eight of the deceased were related.[18] Although small in scope, the outbreak devastated one extended family. These photographs have helped to orient us to the landscape as it is envisioned by survivors. Travelling overland on foot and by car with the family members of plague victims enabled us to interpret the terrain. The photographs materialise survivors' accounts, such as when we were shown, for example, the footpath over which Zoky Lahy, in the throes of death, was hoisted onto a wide plank and carried miles by his brothers-in-law to the hospital.

In some instances, we requested permission to photograph during conversations with plague survivors. In other instances, people requested photographs as they posed with their families, as these are rare possessions for the rural poor. In one example, a woman we visited in her home in August 2016 asks us to take a picture of her holding up a photograph of her family, which shows the son she lost to the plague outbreak (Fig. 10.3). The photograph represents her agency in memorialising her son and recording her lingering grief.[19] It also offers her a material souvenir of a moment in her life history, a year after the plague outbreak. For us, the ethnographers, the photograph records, however inadequately, a suppressed narrative of the plague, the face of grief that remains invisible in the official archive. In addition, it offers visual evidence of health

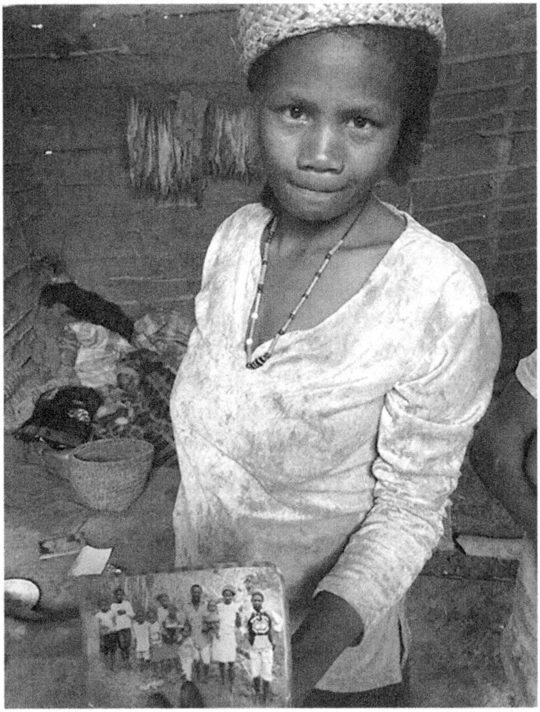

**Fig. 10.3** Woman displaying a photograph of the son who died in 2015 plague outbreak in Moramanga District, Madagascar (photograph by Genese Sodikoff, 2016)

inequities in Madagascar: here, the material conditions of rural poverty that exacerbate plague mortalities.

Resistance to public health recommendations today revolves around regulations for funerary practices. Although scientists of the Pasteur Institute of Madagascar and the Ministry of Health have not been able to locate the original legal code in the books, the state has mandated that plague victims must be buried in separate graves rather than familial tombs, and that they must not be ritually exhumed for 'secondary burial', called *famadihana* (known in English as 'the turning of the bones' ritual) for a period of at least seven years. In usual circumstances, the *famadihana* is performed for deceased adults at least one year after death, once the corpse has dried out. Different localities have customary *famadihana* cycles anywhere from three to eleven years. The *famadihana* is considered the means by which deceased adults become firmly established in the realm of ancestors.

Among the island's diverse ethnic groups, the interment of deceased relatives in familial tombs, which in the central highlands resemble small stone and concrete houses, and regular acts of tribute to the ancestors ensure the continuity of tradition and social identity. Tribute to ancestors averts potential misfortune arising from their feelings of neglect. If mortuary rituals are disrupted, therefore, even for benevolent reasons such as protection against the plague, living descendants risk inciting ancestors' vengeance.

State prohibitions against interment in the tomb and *famadihana* for a period of at least seven years have made death by plague especially traumatic for rural residents who still practice *famadihana* (it is rarely practiced by urban dwellers anymore).[20] The state's over-cautious waiting period has intended to contain the spread of plague in bodies and the environment. Plague control measures concerning mortuary practices were guided by untested hypotheses about the survival of *Yersinia pestis*, the plague bacterium, in human remains, soils, and tombs, and on objects.[21]

Colonial scientists' fears about the long-term survival of bacteria in tombs justified the placement of the plague dead in separate graves to ensure remains would not be exhumed for the *famadihana* ritual, at which time ancestors are removed for celebratory rites and then re-interred. But scientific studies to determine the actual the length of time live bacteria can survive in a corpse have stalled for decades due to a lack of resources.[22]

During our interviews in Moramanga District in 2016, 2017, and 2018 with survivors of the 2015 outbreak, we learned that residents constructed an alternative aetiology of the plague outbreak that involved a curse inflicted by a man who was jilted by a woman in that family. The fact of the curse was proven by a diviner's discovery of seven personal objects—pieces of hair tied up with strips of cloth—of the deceased victims in a stream that flowed near the home of Jean-Paul and Baotine (Fig. 10.4).

Among the eight relatives of Jean-Paul and Baotine who died between August 22 and 26, 2015 was their eldest son, a young teenager. Since residents understood that the disease did not stem solely from germs spread by rats and fleas, as public health agents had informed them, they felt that efficacious treatment required a combination of antibiotics, hospitalisation, and consultation with a diviner, who provided a tincture of a medicinal plant.

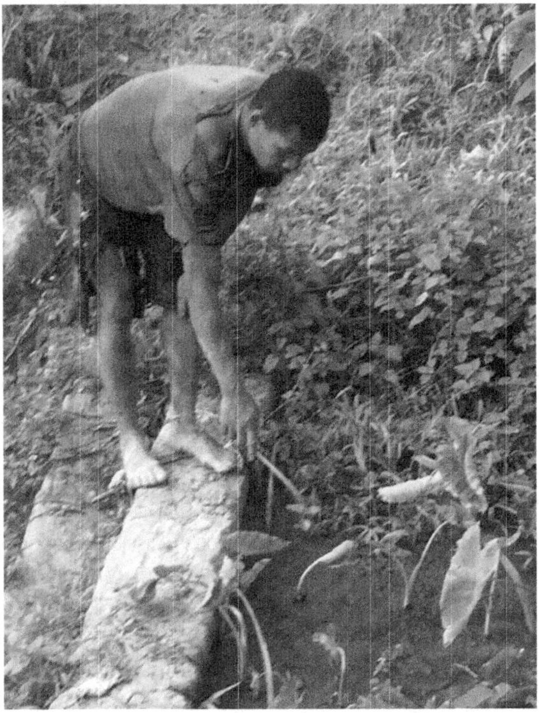

**Fig. 10.4** Jean-Paul indicating where cursed objects were found in the stream by the diviner shortly after the deaths of seven family members (photograph by Genese Sodikoff, 2016)

For the hospitalised patients who ultimately died there, too far gone to recover with streptomycin, their burial in a communal pit represents a mortifying and grievous affront to the tradition of sacred rites for ancestors. Jean-Paul and Baotine recounted that they and their son were visited at night by the dead, especially Jean-Paul's mother, who is buried in more distant plague pit, and his sister, buried with three of her male relatives in a pit at Tangaina. The ancestors complained of feeling cold; they admonished their relatives for leaving them in the pit to suffer. The plague pit had become a haunt, a site of unresolved emotion and disconsolate ancestors.

## The Plague Pit as Key Image

Although the plague pit is a landscape feature that memorialises past trauma, it also binds the ancestors to the present in a frightening way, not as invisible, benevolent beings who co-exist with the living, as they are usually imagined, but as miserable, demanding ghosts. This relationship of the plague pit to the future is configured by the risk of ancestral retaliation. We focus especially on the plague pit as a feature of a post-outbreak landscape that archives the event for survivors while foreshadowing future misfortune if the family does not redress the wrong of their burial outside of the familial tomb.

The photographs of the plague pit in which four relatives are buried have compelled us to look closely at the surrounding landscape and to take notice of what is missing or present. These details, we learned, matter. The environment of the plague pit and the objects left there construct a narrative of political critique, a condemnation of a ministerial policy that would exile ancestors to a cold limbo.

### *The Plague Pit, 2016*

To understand this nondescript image of scrubby ground and ripped corners of blue plastic blue tarp, we needed to hear survivors' stories about what lies beneath (Fig. 10.5). Each body in the plague pit tells a tale of that person's last days, but Jean-Paul and Baotine said the most about one, Zoky Lahy, Jean-Paul's sister's husband. He was the eldest male placed in the grave. His skeleton lies there in a plastic body bag that was too short, so his feet had to stick out. Baotine is agitated when she tells us this.

Jean-Paul recounted the day he carried Zoky Lahy to the hospital at Moramanga with the help of Zoky Lahy's brothers. They made a carrier

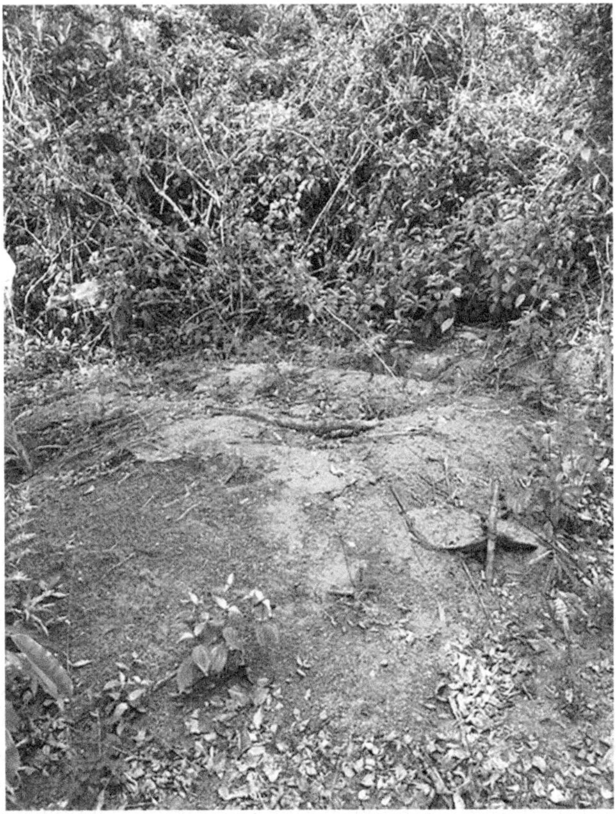

**Fig. 10.5** The plague pit in which four relatives who died of plague were buried in 2015 (photograph by Genese Sodikoff, 2016)

with a wood plank. They lifted Zoky Lahy on the plank and hoisted it on their shoulders, alternating sides, one man in front, the other in back:

> Me and by brother-in-law switched off carrying Zoky Lahy. We had to carry him because he couldn't walk anymore; and he almost died on the way. While we were carrying him on the path, he fell off the palanquin (*filanjana*). That's when he started vomiting a lot of blood. Actually, there were three of us shouldering Zoky Lahy, and the three of us are healthy to this day. The blood that spilled on the ground in Beravina wasn't swept up, but we let it dry on its own.

> We re-lifted Zoky Lahy onto the palanquin and continued on our path to the hospital. At that moment, Zoky Lahy asked us to go faster, attempting to survive long enough to get treatment. (Sodikoff, fieldnotes, August 1, 2016)

They arrived at the hospital, carrying him that long way, maybe ten miles in all. They did not say, but we believe that if they had tried to hitchhike, they would have probably been refused once drivers knew the illness of the patient. His symptoms may have been obvious, and in the midst of plague outbreaks, people are fearful.

Ultimately, Zoky Lahy died in the hospital, fifteen minutes after their son died. Jean-Paul and Baotine can recall precise details of those frantic few days. They told us the birthdays of each patient who died in the hospital (except for Pascal, their son), and the hour at which each of their four relatives succumbed in the hospital. We only noticed later in fieldnotes the gaps that surrounded the story of Pascal. We did not know his exact age, or what he said to his parents while ill. We knew that Jean-Paul carried Pascal on the footpath towards the hospital. We do not know if Baotine walked beside him with their other son, but she was present in the hospital. We asked about symptoms, and Jean-Paul described them. Overall, the parents were not forthcoming about Pascal's ordeal, a silence that speaks volumes about grief. Baotine recited the times of death for the four hospitalised patients, and when recalling these objective facts, seared in her memory, she was able to utter her son's name:

> Pika on Tuesday, August 25th at 9 am. Neny Mena at 12:15 am [Wednesday]. Pascal at 12:30 pm, and Zoky Lahy at 12:45 pm.

Since they died there and not in the village, these four—two men, a male youth, and a woman—were handled by the hospital. Jean-Paul explained how the burial of the four victims was profane:

> Before burial, the body [*razana*, 'ancestor'] is supposed to be washed and wrapped in a shroud. But for people who died of the plague, these steps were not followed. They were not treated like they ought to have been treated at a funeral. There were no burial shrouds but instead they were put in body bags that did not completely cover the whole body. Zoky Lahy's feet were sticking out. At the time, there were no rites (*fomba fomba*) performed or speeches made. But immediately a hole was dug and the bodies (*faty*, 'corpses') put inside: Lepika, Neny Mena, Pascal, and Zoky Lahy. That's to

say, that the four were put all together in the same grave at the top of a hill in Tangaina. (Sodikoff, fieldnotes, August 2, 2016)

The selection of words 'ancestor' and 'corpse' suggests how Jean-Paul processed the treatment of their relatives. What ought to be done for an ancestor was not. The relatives were treated like 'dead bodies'. It is the duty of the living to wrap a loved one warmly in a white burial shroud and to lay down offerings of coffee, sugar, rum or tobacco beside the head of the deceased. Ancestors are sensitive to perceived slights. People constantly acknowledge their ancestors' presence and pay tribute during agricultural tasks and village festivities with offerings of cooked rice and chicken and drops of rum. To neglect these tributes leads to a sense of despondency and fear of misfortune. Episodes of actual misfortune are always interpreted through the lens of filial neglect: Did the son or daughter forget to leave a meal for the ancestors before sowing the field? Did they hoard their money or spend it in drink rather than sacrifice a cow for the ancestor when it was high time?

In the case of the burial of the four plague victims, numerous infractions troubled the family. The mingling of sexes is deemed incestuous, and the lack of burial shrouds and offerings cruel, as was the disinfection of the corpses with bleach and the insertion into body bags. These material facts are inaccessible to us, but these images trouble Jean-Paul and Baotine when they look upon the profane grave.

After the hospital staff disinfected the bodies, the couple accompanied the medical staffers in a vehicle. They drove to a known potter's field that harbours the remains of 'strangers' to the area; that is, Malagasy residents who died far away from the land of their ancestors (*tanindrazana*). Their families intend to one day repatriate the remains home, but this usually requires years of saving. This potter's field lies off the side of a paved road on a steep hill.

On our first visit with Jean-Paul and Baotine in August 2016, Jean-Paul hacked away with his machete at the overgrowth obstructing the footpath to the plague pit. We arrived at a clearing and beheld a mound of dirt with pieces of blue plastic tarp sticking out. The tarp had covered the grave's surface, and now mud covered that after heavy storms. Baotine expressed dismay at its condition. That tarp was the only thing protecting the bodies from the elements.

Jean-Paul explained that a year earlier, he had assisted the hospital orderlies in digging a hole of about three meters deep. The depth of the grave became a point of interest for scientists at the Pasteur Institute when they heard this detail. The state recommends a depth of twelve meters for plague burials to ensure that rats cannot burrow towards the corpses and potentially absorb live bacteria. The hospital staffers did not follow this policy, and who would? Without the proper tools and manpower, digging to that depth, if humanly possible, would have taken days. The four bodies were laid beside each other, heads pointed northwards. The hospital staffers did not permit the couple to make speeches to the ancestors (*kabary*) or to leave offerings. They were hurried, eager to be away from pestilent corpses and this place.

Jean-Paul and Baotine had returned to the site a couple of weeks later with the blue plastic sheet to protect the bodies from the elements. What was visible from the surface a year later—the mound, the overgrowth, and the torn plastic sheet—added insult to what they knew was underground: the incestuous mixing of sexes, the body bags in lieu of winding sheets, including the bag with Zoky Lahy's feet exposed, and the absence of wood planks beneath or on top of the bodies to protect them from the cold earth. In a future excavation, these extraordinary signs would tell a grim story about the circumstances of the burial while prompting interpretations of what the present and absent objects, and their arrangement in space, mean for the living.[23]

As we were leaving, Baotine walked behind Genese Sodikoff to catch her if she slipped on the path. Baotine spoke softly to herself, addressing the deceased and murmuring something about *paraky* (tobacco). Baotine explained later that she was making a promise to the ancestors that next time she would bring the things each of them liked.

The couple explained that the hospital authorities made it clear they were obliged to wait seven years before transferring the bodies to their tomb. Yet they did not see how they would be able to amass the necessary funds to transfer all four bodies in 2022. The undertaking would require a large amount of labour and cash to organise the transportation of the bodies. They would also need to sacrifice at least one cow in order to open the tomb, as custom demands, and to provide a feast of rice and meat for the extended family. Jean-Paul complained that his own siblings and father habitually did not contribute their fair share. Nevertheless, the haunting by Jean-Paul's sister and mother made them eager transfer their remains sooner than the required waiting period.

### *The Plague Pit, 2017*

The following year, in August 2017, we visited Jean-Paul and Baotine and asked if they would like to leave offerings at the plague pit as they had discussed the year before. They enthusiastically agreed, and we arranged to pick them up at the roadside village nearest to their hamlet and head to Tangaina. We, the authors, would provide the offerings of rum and chewing tobacco, the items they requested, as well as a new tarp with which to protect the grave's surface. Jean-Paul and Baotine sent word to Jean-Paul's father, siblings, and their families so that all would convene there the following day.

The doctor from the hospital accompanied us to the village where we planned to pick up Jean-Paul and Baotine in our rented vehicle. The doctor came with us as a precaution because foreigners who show a morbid interest in Malagasy gravesites attract a lot of unwanted and sometimes aggressive attention. Rumours have long circulated that middlemen and foreigners steal human bones in Madagascar for medicines and beauty products. The doctor prepared to reassure anyone who asked about us. Eventually, the family was reunited at the gravesite. Jean-Paul and his brothers cut vegetation to clear weeds and to open up the space. The men and women then came together, and Jean-Paul's father began speaking to the ancestors, thanking them for their blessings and explaining the predicament they have suffered (Fig. 10.6). Jean-Paul and Baotine also made speeches. They then snapped off several larges leaves from a tree, twisted them into cones, and poured swallows of rum into each 'cup'. They opened the pouch of chewing tobacco and sprinkled a bit on a leaf on the ground.

The men then began digging a small trench around the grave with their spades. The women lifted the new clear tarp over the wide mound of dirt. It billowed and floated down over the grave's surface, and the group weighted down the borders with stones that were tucked inside the shallow trench. In less than an hour, the funeral was finished.

We walked back to the road. The family was in good spirits. Njaka, a brother of Jean-Paul, whom we had heard about but not yet met, was eager to tell us his own experience of the plague. He had been infected and was hospitalised for two weeks. Unconscious, he did not realise his family members were dying around him. Luckily, he survived, but it took months before he regained his strength. He thanked us, the ethnographers, for facilitating the belated funeral, but also gently chided us for not

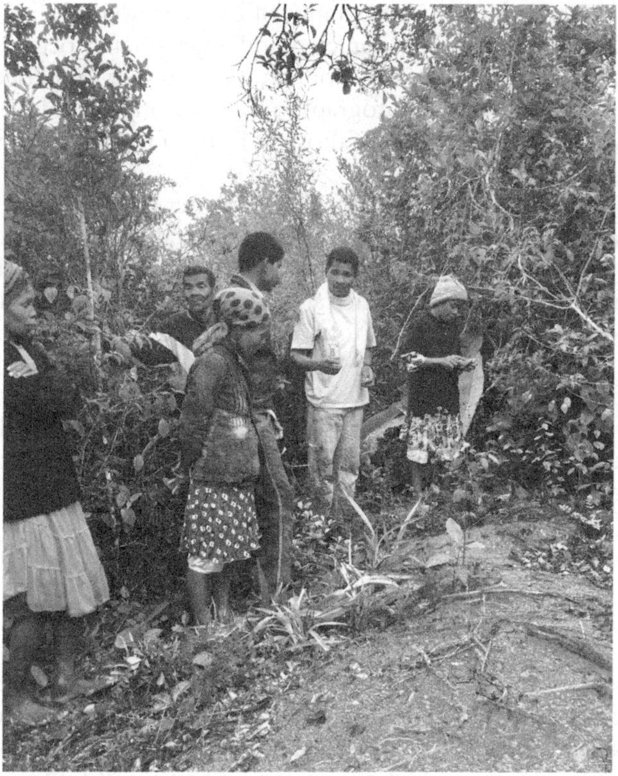

**Fig. 10.6** Family members perform a proper funeral for the deceased at the plague pit (Photograph by Genese Sodikoff, 2017)

notifying their father first. Jean-Paul listened silently, stone-faced. He felt that his and Baotine's efforts were never recognised or appreciated. He complained later that his siblings treat him badly, believing he puts on airs.

The funeral rites performed on that day in August 2017, which we captured in photographs, gave the family some relief, though they knew it was only a temporary fix. However inadequate and partial, the photographs represent more than an ethnographic record. They also capture the fleeting, present moment of that collective sense of relief from the standpoint of a family dealing with the moral consequences of state mortuary policies. The photographs thus capture 'evidence of affect', to use Elizabeth Edwards' term for the partial and fleeting record of historical subjective experience that photographs offer the viewer. She argues that if 'photographs are not merely depicted and appropriated occurrences and scenes, but an inscription of the moment which is that "experience someone lived through", then photographs mark not only the photographer's standpoint but a point of view of those in front of the camera, even if that moment is asymmetrical'.[24]

For us, the ethnographers, the before and after photos of the plague pit offer visual evidence of a profane burial, ancestral caretaking, and efforts by the living to offer solace to the dead. Material objects in the photographs, including the new plastic tarp, a glass bottle of clear rum, and a half-spilled-out plastic pouch of tobacco on a bright green leaf, offer testament that the landscape was transformed by a belated funeral in a field of neglected souls. The earlier absence of offerings gave us insight into how the survivors perceived the site before this intervention. All of those missing objects were glaring absences, and the absences brought into presence those fraught four days of the 2015 plague outbreak. The gestures of Jean-Paul and Baotine that bade us to 'look here' and 'look there' opened a window onto this other archive of the plague, this landscape of memories.

## Conclusion

In the image of the plague pit, past and present collide. The temporality of the image is recursive, a haunting, leaving survivors unable to unhinge their thoughts from the site and all it signifies until the matter is resolved. Due to the high value placed on maintaining ancestral traditions in rural districts of Madagascar, particularly funerary rites, the resentments of residents towards the state for being put in this predicament with their ancestors mirrors the affective fallout of every plague outbreak since such regulations were enacted in the early twentieth century.

Georges Bataille writes with reference to war that the state's systems regularly work to prune back the possibilities of catastrophe, often at the expense of the individual's safety and survival. 'Each civilized unit (thus civilisation)', he argues, 'proclaims the primacy of its undertakings—by which it means to secure the future—over all considerations of feelings.'[25] In colonial Madagascar, the state prioritised the public health over what the French considered unhygienic practices associated with the cult of ancestors, thereby subordinating sentiment to science for the sake of curtailing a catastrophic death toll. The colonial photographic record projects this mission and temporalises the plague as an arrow pointing towards its eradication.

As the colonial medical system sought to quell the plague, among its most contentious activities were the rounding up Malagasy subjects for vaccinations (thought by many to cause death) and the removal of the dead from villages to be buried alone in distant graves. While the magnitude of anti-plague efforts was deemed necessary for the colony's well-being, the insensitivity of the measures to the well-being of the dead would be the lens through which rural Malagasy people saw their own futures, their own profane burials. Many believe that the exclusion of plague dead from the familial tomb also poses an existential threat. Zoe Crossland's words about past generations still ring true: 'the fear of drawing reproach (*tsiny*) from the ancestors and other spirits meant that the future could not be overtly imagined without putting oneself at risk.'[26]

For rural highland communities, the problem of plague-centred mortuary rules is not merely the dread of the coldness and loneliness of the plague pit. It is also contemplation of a protracted period of suspense, trapped in the liminal spacetime between the first and second burials, unable to fully enter the realm of ancestors. The image of this uncertain posthumous future weighs on the minds of plague survivors.

## Notes

1. Christos Lynteris, 'The Prophetic Faculty of Epidemic Photography: Chinese Wet Markets and the Imagination of the Next Pandemic', Visual Anthropology 29, no. 2 (2016): 118–132.
2. Christos Lynteris and Ruth J. Prince, 'Anthropology and Medical Photography: Ethnographic, Critical and Comparative Perspectives', *Visual Anthropology* 29, no. 2 (2016): 101–117, p. 106.
3. Elizabeth Edwards, 'Photographs and the Sound of History', *Visual Anthropology Review* 21, no. 1–2 (2015): 27–46; Annabelle Wienand, 'Santu Mofokeng: Alternative Ways of Seeing (1996–2013)', *Safundi: The Journal of South Africa and American* Studies 15, no. 2–3 (2014): 307–328.
4. By 1935, scientists in Antananarivo had developed an effective vaccine against the plague, called the E.V. vaccine (from a family name, Evesque, but known familiarly by Malagasy as 'Enfant Vazaha', or 'white people's child'): P. Coulanges, 'Cinquantenaire du vaccin antipesteux EV. (Girard et Robica)', *Archives Institut Pasteur Madagascar* 50, no. 1(1982): 169–184.

5. Branwyn Poleykett, 'Ethnohistory and the Dead: Cultures of Colonial Epidemiology', *Medical Anthropology* 37, no. 6 (2018): 472–485, https://doi.org/https://doi.org/10.1080/01459740.2018.1453507; Faranirina Esoavelomandroso, 'Résistance à la médecine en situation coloniale: la peste à Madagascar', *Annales* 36, no. 2 (1981): 168–190.
6. Deborah Poole, 'An Excess of Description: Ethnography, Race, and Visual Technologies', *Annual Review of Anthropology* 34 (2005): 159–179, p. 168.
7. Lukas Engelmann, 'The Burial Pit as Bio-historical Archive'. In Christos Lynteris and Nicholas H.A. Evans (eds.) *Histories of Post Mortem Contagion: Infectious Corpses and Contested Burials*, pp. 189–211, (London: Palgrave Macmillan, 2018), p. 190.
8. Keith Basso argues that place-making, including naming features of the landscape, is a means of inventing history and constructing personal and social identities: Keith H. Basso, *Wisdom Sites in Places: Landscape and Language Among the Eastern Apache* (Albuquerque, NM: University of New Mexico Press, 1996).
9. Lynteris and Prince, 'Anthropology and Medical Photography', p. 102.
10. Christos Lynteris, 'Plague Masks: The Visual Emergence of Anti-Epidemic Personal Protection Equipment', *Medical Anthropology* 37, no. 6 (2018): 442–457, p. 443.
11. G. Girard, 'Dépistage Post Mortem de la Peste par Ponctions d'Organes', *Bulletin World Health Organization* 5 (1952): 109–116, p. 110.
12. Geoffrey Batchen, *Each Wild Idea: Writing, Photography, History* (Cambridge, MA: The MIT Press, 2001); Pamila Gupta and Tamsyn Adams, '(Vernacular) Photography from Africa: Collections, preservation, dialogue', *Critical Arts* 32, no. 1 (2018): 1–12.
13. In other French colonial cities, officials actively suppressed public information about plague outbreaks: Branwyn Poleykett, 'Public Culture and the Spectacle of Epidemic Disease in Rabat and Casablanca'. In Lukas Engelmann, John Henderson and Christos Lynteris (eds), *Plague and the City*, pp. 159–172 (London and New York: Routledge, 2019).
14. A substantial archive of late nineteenth- and early twentieth-century photographs of village life in Madagascar was compiled by foreign missionaries and French colonial administrators interested in Malagasy culture, particularly Raymond Decary. They generally depict regional differences in dress, hairstyles, bodily adornment, buildings, and other elements of material culture, but they do not construct Malagasy life in political terms. On early missionary photography in Africa, see Christraud M. Geary, 'Missionary Photography: Private and Public Readings', *African Arts* 24, no. 4 (1991): 48–100; for a study of the photography of missionary Reverend William Ellis in Madagascar, and the context of religious rivalry between the French and British at the time, see Simon Peers, *The Working of Miracles. William*

*Ellis—Photography in Madagascar 1853–1865* (London: The British Council Visual Arts Publications, 1995). On the ethnographic photography of colonial administrator, Raymond Decary, who served for twenty-seven years in Madagascar beginning in 1916, see Martine Balard and Edmond Maestri, 'Raymond Decary (1891–1973) ou Madagascar mis en collections', *Outre-mers* 88, no. 332–333, (second semester 2001): 207–229.

15. Suzanne C. Chanteau, Pascal Boiser, Elisabeth Carniel, Jean Bernard Duchemin, Jean Marc Duplantier, Steve M. Goodman, Pascal Hanschumacher, Isabelle Jeanne, Stephane Laventure, Philippe Mauclère, René Miglianai, Dieudonné Rabeson, Lila Rahalison, Noelson Rasolofonirina, Lala Rasifasoamanana, Bruno Rasoamanana, Maherisoa Ratsitorahina, Jocelyn Ratovonjato, Marie Laure Rosso, Jean Roux, and Adama Tall, *Atlas de la peste à Madagascar* (Paris: Institut de Recherche pour le Développement, Institut Pasteur, Agence Universitaire de la Francophonie, 2006).
16. Brygoo, E.-R. 'Epidemiologie de la peste à Madagascar', *Archives de l'Institut Pasteur de Madagascar* 35 (1966): 1–219.
17. Esoavelomandroso, 'Résistance à la médecine en situation coloniale', p. 177.
18. Genese Marie Sodikoff, 'Zoonotic Semiotics: Plague Narratives and Vanishing Signs in Madagascar', *Medical Anthropology Quarterly* 33, no. 1 (2019): 42–59; B. Ramasindrazana, V. Andrianaivoarimanana, J.M. Rakotodramanga, D.N. Birsell, M. Ratsitorahina, and M. Rajerison, 'Penumonic Plague Transmission, Moramanga, Madagascar, 2015', *Emerging Infectious Diseases* 23 (2017): 521–524.
19. Richard Vokes discusses how AIDS patients in Uganda in the 1990s created photo albums of themselves in their last days as a means of conveying agency and affecting their social worlds: Richard Vokes, 'On Ancestral Self-Fashioning: Photography in the Time of AIDS', *Visual Anthropology* 21, no. 4 (2008): 345–363.
20. Maurice Bloch, *Placing the Dead: Tombs, Ancestral Villages, and Kinship Organization in Madagascar* (Long Grove, IL: Waveland Press Inc., 1993); David Graeber, 'Dancing with Corpses Reconsidered: An Interpretation of *Famadihana* (in Arivonimamo, Madagascar)', *American Ethnologist* 22 (1995): 258–278; Pier Larson, 'Austronesian Mortuary Ritual in History: Transformations of Secondary Burial (*Famadihana*) in Highland Madagascar', *Ethnohistory* 48, no. 1–2 (2001): 123–155.
21. Christos Lynteris and Nicholas H.A. Evans 2018. 'Introduction: The Challenge of the Epidemic Corpse'. In Christos Lynteris and Nicholas H.A. Evans (eds), *Histories of Post-Mortem Contagion: Infectious Corpses and Contested Burials*, pp. 1–25. (London: Palgrave Macmillan, 2018).

22. Today, scientists at the Pasteur Institute of Madagascar are preparing to begin an experiment that will test whether and how long plague bacteria can survive underground in rat corpses in order to understand the pathogen's dormant risk in the environment (Dr Minoarisoa Rajerison, Head of the Plague Unit, Institute Pasteur de Madagascar, personal communication, August 2019).
23. In her semeiotic analysis of Malagasy tombs, graves, landscapes, and ruins, Zoë Crossland discusses the concept of 'presence' in the work of archaeology, entailing the interpretation and interplay of material signs at different temporal registers. Zoë Crossland, *Ancestral Encounters in Highland Madagascar: Material Signs and Traces of the Dead* (Cambridge: Cambridge University Press, 2014), p. 179.
24. Elizabeth Edwards, 'Anthropology and Photography: A Long History of Knowledge and Affect', *Photographies* 8, no. 3 (2015): 235–252.
25. Georges Bataille, 'Concerning the Accounts Given by the Residents of Hiroshima'. In Cathy Caruth (ed.) *Trauma: Explorations in Memory*, pp. 221–235 (Baltimore: The Johns Hopkins University Press, 1995), p. 229.
26. Zoë Crossland, *Ancestral Encounters in Highland Madagascar*, 179.

# Index[1]



[1] Note: Page numbers followed by 'n' refer to notes.


C. Lynteris (ed.), *Plague Image and Imagination from Medieval to Modern Times*, Medicine and Biomedical Sciences in Modern History, https://doi.org/10.1007/978-3-030-72304-0

## Q



## R

The manufacturer's authorised representative in the EU is Springer
Nature Customer Service Centre GmbH, Europaplatz 3, 69115 Heidelberg,
Germany. If you have any concerns regarding our products, please
contact ProductSafety@springernature.com

Printed and bound by CPI Group (UK) Ltd, Croydon, CR0 4YY
30/09/2026
02225451-0013